Thermoreversible gelation of polymers and biopolymers

to my parents

to Chantal, Aurélie and Delphine

Thermoreversible gelation of polymers and biopolymers

Jean-Michel Guenet
Laboratoire d'Ultrasons et de Dynamique des Fluides Complexes
CNRS URA 851-Université Louis Pasteur
STRASBOURG
FRANCE

ACADEMIC PRESS
Harcourt Brace Jovanovich, Publishers
London San Diego New York Boston
Sydney Toronto Tokyo

ACADEMIC PRESS LIMITED
24–28 Oval Road
London NW1 7DX

United States Edition published by
ACADEMIC PRESS INC.
San Diego, CA 92101

This book is printed on acid-free paper

A catalogue record for this book is available from the British Library
ISBN 0–12–305380–3

Typeset by Mathematical Composition Setters Ltd, Salisbury, Wiltshire
and printed in Great Britain by T. J. Press (Padstow) Ltd, Padstow, Cornwall

CONTENTS

PREFACE

Thermoreversible gelation, a growing area of polymer science, encompasses many different systems from synthetic polymers to biopolymers. With the exception of gelatin gels, it may be regarded as a relatively new branch of macromolecular science since most of the relevant investigations have been carried out over the last 25 years. The abundant literature on the subject, scattered in a number of different journals, consists only of reviews that generally cover one particular aspect. To date, no single work has attempted to gather together the current knowledge on both polymers and biopolymers thermoreversible gels – the aim of this book is to take up this challenge.

This book is chiefly concerned with experimental results since virtually no theories have been specifically developed so far for these gels. It is organized into three main chapters. In each chapter polymers and biopolymers are treated separately, although this is rather an arbitrary choice and stems from the usual distinction made by scientists (this distinction appears also in the book's title). Although it is true that both types of polymers involve different approaches or methods, one will, however, discover throughout this book that the gelation phenomena often bear more resemblance to one another than one would have imagined. To take but one example, the formation of polysaccharides–solvent (water) complexes in gels has its counterpart in some polymeric gels.

The first chapter is dedicated to gel formation and thermal behaviour. If they have been established experimentally, the temperature-concentration phase diagrams are supplied. In many cases these diagrams are of invaluable help for understanding the gelation mechanism and give a good indication of the actual gel molecular structure. The second chapter, on morphology and molecular structures, deals primarily with the gel state, but also provides information, whenever available, on the pregel state that is either on aggregates formed below the critical gel concentration or on structure(s) existing above the gelation temperature. An understanding of the gel and/or the pre-gel structure at different scales and levels not only provides us with the opportunity to discuss in more detail the gelation mechanism deduced from the phase diagram, but also throws some light on the origin of the mechanical and rheological properties which are presented in the third, and final chapter.

As already stated, thermoreversible gelation is still a comparatively new branch of polymer science, and therefore a multidisciplinary appproach should be adopted when studying a case about which little is known. It would seem particularly unsuitable to confine physical gelation, as it is all too often done, to polymer crystallization and to its related concepts or to the concepts developed for chemical gels. The multidisciplinary approach also implies being prepared to tackle the problem with many different techniques. Accordingly, for the sake of completeness, I have included some basic theories or calculations related to what, for me, represent the most important techniques, especially when these have been and are still used for investigating several gel systems. As it would have been both inappropriate and awkward to insert them in the middle of the presentation of results and data, they have therefore been collected in four appendixes at the end of the book which will hopefully provide useful information on phase diagrams and related questions, diffraction by helices, scattering by semi-rigid molecules, and elasticity of rigid networks. A researcher specialized in one of these four domains may find these appendixes a trifle too basic and schematic, however, their primary aim is to give scientists and research students some idea of the fundamental concepts and parameters of a field about which they have little or no knowledge and give them the opportunity of understanding, and appreciating the importance of what has been achieved elsewhere. Hopefully, this structure of three main chapters and four appendixes will make the book appealing to the widest possible audience.

Finally, I would like to express my deepest gratitude to all my colleagues that have been kind enough to provide information, suggestions, advice and, above all, their knowledge on many topics. My thanks go particularly to Dr. Cyril Rochas, who guided me throughout the literature devoted to the gelation of biopolymers, and whose comments on the book were more than helpful to me. Dr. Madeleine Djabourov is also gratefully acknowledged for advice concerning gelatin gels. I benefitted much from discussions with Dr. Jean-Pierre Munch and Dr. Michel Delsanti, particularly those concerning the definition of a gel. Discussions with Pr. E.D.T. Atkins on the helical structure of polysaccharides, with Dr. Annette Thierry on thermal analysis, with Dr. Bernard Lotz on helix diffraction, with Dr. Jean Doublier on amylose gels and with Dr. Gregory B. McKenna on mechanical properties were also most helpful. My thanks also go to Drs. Monique Axelos, Sauveur Candau, Nazir Fazel, Marc K. Pein, Pierre Terech and Jean-Claude Wittman for discussions on special topics. When, ten years ago, I introduced the field of thermoreversible gelation in Strasbourg I received enthusiastic support from Drs. S. J. Candau, B. Lot, C. Wippler and J.C. Wittman. I owe them a debt of gratitude. I am also greatly indebted to Professor Andrew Keller with whom I started working in this domain during a post-doctoral stay at Bristol University. Last, but not least, I want to express all my gratitude to my wife, Chantal, and to my twin-daughters, Aurélie and Delphine. For many months, they have accepted without complaint, life with a rather absent-minded computer-bound husband and father. My wife, who is a molecular biologist, has also been kind enough as to provide me with references that would not have been otherwise available.

Introduction

'La dernière chose qu'on trouve en faisant un ouvrage est de savoir celle qu'il faut mettre la première.'
Blaise Pascal in Les Pensées.

If one were asked how to retain a liquid, it would probably occur to one that any plain container would do and that such a question is definitely silly. However, there does exist another way in which large amounts of liquid can be kept 'solid', namely by using a gel. A gel possesses the unique property of incorporating and retaining a proportion of liquid molecules outweighing by far the proportion of the basic component. In some cases a gel can contain up to 99% solvent.

The widely accepted **topological definition** of a gel is therefore a three-dimensional network constituted of basic elements connected in some way and swollen by a solvent. As a rule, only systems wherein the solvent represents the major component are considered as gels.

Among the large number of systems that are liable to form three-dimensional networks (e.g. steroids, vanadium oxide, polymers), gels prepared from either synthetic polymers or biopolymers have received and are still receiving considerable attention. These gels, referred to as polymeric gels, break down into two main categories depending upon the process whereby the elements of the network are connected: **chemical gels** and **physical gels**.

In **chemical gels** the connection usually occurs through covalent bonds. Covalent bonds are very strong links so that a cross-link usually consists of one multifunctional molecule. As a result, the junction zone possesses a size similar to the monomer unit and may be regarded as point-like as far as macroscopic properties are concerned. Also, as covalent bonds are not only localized on the junction point but also constitute the polymer chains joining the junction points, these gels are 'heat-irreversible': heating this type of chemical gel to the point where covalent bonds break up. This ultimately entails irreversible degradation which impedes the re-formation of a similar system.

There do exist gels that can be termed reversible chemical gels. These occur through ion complexation. One well-known example is the borax–galactomannan system (see for instance Pezron *et al.*, 1988).

In **physical gels** bonding between chains occurs through van der Waals interaction. This has two major consequences: (1) to be stable, van der Waals bonds require cooperativity which implies that the junction domains are not point-like but extend into space and (2) the energy involved in van der Waals interaction is of the order of kT so that these gels are 'heat-reversible'. The latter property is often called **thermoreversibility** which leads to these systems being called either **physical gels** or **thermoreversible gels**. This book is devoted to this category of gels.

The **topological definition** of gels has given birth to other definitions which involve macroscopic properties. The easiest properties to investigate remain the mechanical properties. Scientists felt that the topological definition implied that a gel ought to behave mechanically as a solid provided that the solvent is not expelled when stress is applied. This yields the **mechanical definition**: a gel should display solid-like behaviour. This definition can be translated into more scholarly terms: a gel should be characterized by an elastic modulus at zero frequency. This in turn implies that a gel ought to be self-supporting, i.e. left on its own the initial shape ought not to change in time. And here, scientists got into trouble with physical gels. The problem was summarized in the late 1920s by Dorothy Jordan Lloyd (1926) who wrote: 'The colloïdal condition, the gel, is one which is easier to recognize than to define.' This sentence is now well known and describes reality so closely that it is widely used in many introductions to both papers and lectures. It has become, to some extent, the first principle of the science dealing with thermoreversible gelation. This is so because the mechanical definition seldom applies in the strict sense to most of the physical gels that are regarded as such. For instance, some gels deform almost permanently after being submitted to mechanical stress, which more resembles the behaviour of a liquid. And yet, from their macroscopic aspect, nobody would question their gel status and would immediately dismiss the idea of a very viscous solution.

Scientists always feel uncomfortable when no clear-cut, accurate definition is available to describe a system. However, in this case the idea of a nice sound definition will have to be abandoned, as there probably exists a continuum from the (almost) true gel state to the true liquid state. Where to put the borderline may be exceedingly tricky. As a result, every author will probably give his own definition of a physical gel, and thus some would disregard entities for which others would fight tooth and nail to have considered as gels.

Obviously, the systems described in this book must rely upon a definition that is not thought and, above all, not claimed to be universal. It, however, allows one to include most of the systems that are currently regarded as gels. This definition turns out, in fact, to consist of a set of different criteria, some of which were already considered by Hermans in his review (Hermans, 1949).

(1) The topological definition is considered valid. This excludes systems that consist of spherulites for instance, since the notion of network is not fulfilled.

According to the definition given in the dictionary a network is: a large system of lines, tubes, wires, etc., that cross one another or are connected with one another (*Longman Dictionary of Contemporary English*, 1987).

---

(2) Gel formation or gel melting should proceed via **first-order transition** arising from the creation of a minimum of molecular order out of the initial, disordered solution. This excludes phase-separated glasses which are still amorphous despite the fact that they may show a network morphology (see for instance Arnauts and Berghmans, 1987).

This criterion also implies that there exists a well-defined temperature, called gelation temperature or gelation threshold, below which the polymer–solvent mixture no longer behaves as a viscoelastic solution. At this temperature, many properties show a discontinuity. In particular, the elastic behaviour is seen at lower and lower frequency unlike for a viscoelastic solution of equivalent concentration for which this property is only observed at high frequencies.

(3) A gel immersed in an excess of preparation solvent should either be unaffected or swell but in no case dissolve or disaggregate. This behaviour rests upon experimental observations.

(4) Finally, from a practical viewpoint, a gel is a system that can be taken out from the vessel wherein it has been prepared without losing its shape or integrity. This property differentiates a gel from a viscous solution or from a Bingham fluid.

While criteria 1 and 2 should be unreservedly obeyed, criteria 3 and 4 may be less strictly applied. For instance, 'weak' gels may obey the first two criteria but would probably not fulfil the last two and would be unjustly ruled out.

The discovery of physical gelation probably dates back to the human discovery of fire, as gelatin gels can simply be produced by boiling bones, skin or any animal tissue in water and by cooling down the surrounding solution. In many places, such as the British Isles and Japan, physical gels constitute an important part of people's everyday food. When and why these peoples began to use them is difficult to say. It may find explanation in the fact that these peoples all live near seashores where seaweeds containing 'gelifying' biopolymers such as agarose or carrageenans are abundant.

Scientifically speaking, physical gels have only recently aroused the interest of scientists although some work was already achieved over a century ago; the very first investigations of these systems were carried out by von Nägeli (1858). He systematically studied natural material such as starch and postulated the existence of a discontinuous granular structure. Later, Hardy (1899, 1900) realized that gelation of biopolymers may originate in the occurrence of a liquid–liquid phase separation, a concept periodically rediscovered every twenty years or so. While gelatin gels received comparatively continuous attention, other physical gels such as those formed with polysaccharides have only been systematically studied for the past twenty years.

Meanwhile, new gels have been discovered: those prepared from synthetic polymers in organic solvents. Apparently, the first system of this kind ever to be studied was poly(vinyl chloride) (PVC) gels. The very existence of these gels has shaken the belief that physical gelation is a unique property of biopolymers in water, the latter being a special medium on account of the occurrence of hydrogen bonding. The making of physical gels with systems in which hydrogen bonds are conspicuously absent has opened up new fields of research and has also triggered the quest for some universality. Thermoreversible gels from synthetic polymers have been investigated only recently. Originally, most of them were obtained as by-products of experiments intended to investigate crystallization processes and were, accordingly, considered highly undesirable. Fortunately, curiosity won over displeasure.

Industrial applications of biopolymer gels are essentially found in the food industry where they are used as thickeners. They accordingly play an important role in the world economy. Conversely, gels prepared from synthetic polymers have only found limited applications so far, probably because of their relative 'youth' compared with biopolymer gels. One of these applications is spectacular: the making of high-modulus polyethylene fibres by spinning a gel. There is no doubt that in the near future these gels will lead to more and more industrial developments, especially when materials with special properties are sought.

Three main areas have been explored over the past twenty years: the elucidation of the gelation mechanism(s), the molecular structure together with the gel morphology and the rheological behaviour. Therefore this book is structured in the same way: three main chapters the first one dealing with *gel formation: thermal behaviour and phase diagrams*, the second one with *gel molecular structure and morphology* and the last one with *mechanical properties*.

1 Gel formation: thermal behaviour and phase diagrams

'Qu'importe le flacon, pourvu qu'on ait l'ivresse!'
Alfred de Musset

1.1 INTRODUCTION

Whereas in some systems such as aqueous gelatin gels there are just two possible states, i.e. the gel state or the liquid state, and accordingly one transition temperature, in most of the polymeric systems gel formation competes with the natural tendency of some polymers to form chain-folded crystals. This situation occurs, for instance, in isotactic polystyrene solutions as reported by Lemstra and Challa (1975) and Girolamo *et al.* (1976). A polymeric solution, depending on the thermal treatment, may give either a transparent gel, as in biological macromolecules, or a turbid crystal suspension (provided, of course, that refractive indices of the solvent and the polymer do not match). In a sense, despite the fact that polymers have a simpler chemical structure than biopolymers, the gelation process is more complex. Most of the time, while for biopolymers the gelation phenomenon can be studied under conditions that can be termed 'at equilibrium' (see for instance Godard *et al.*, 1978), this is not the case for polymers that possess the capability of crystallizing in the form of chain-folded lamellae. In these systems gelation usually occurs at temperatures below the crystallization temperature which implies that the solution must be quenched as rapidly as possible (the rapidity depending upon the crystallization rate) so as to by-pass the crystallization phenomenon. Interestingly enough, however, there are some non-crystallizable polymers (that is, they do not produce chain-folded crystals) that also display gelation behaviour. These are usually atactic such as atactic polystyrene (aPS) or PVC. These polymers may bridge the apparent gap between synthetic and natural macromolecules.

In this chapter the gel formation conditions are detailed as a function of various parameters such as the solvent type, the polymer concentration or even in some cases the mechanical perturbations applied to the solution. The thermal behaviour of the gel, that is the behaviour on heating, will be examined in parallel. Temperature–concentration phase diagrams, when available, will also be added to the discussion. It is worth emphasizing that these phase diagrams, whenever they can be determined, are invaluable in throwing some light on the gelation mechanism and, to some extent, in giving an idea of the short-range molecular structure. For instance, a phase diagram can indicate whether the gel molecular organization consists of a polymer–solvent compound, in which case the polymer and the solvent ought to be 'organized' together at the molecular level, or of a solid solution for which the polymer is organized without the solvent. (A short description on phase diagram properties may be found in Appendix 1.)

Besides, gel formation is not only followed and investigated by means of calorimetry. Results gained from other techniques that also demonstrate changes of state (e.g. mobility as in nuclear magnetic resonance (NMR), helix content as in polarimetry), will also be reported whenever appropriate. These techniques are more or less directly related to the molecular structure. However, in this chapter description of the results will be restricted to formation conditions, i.e. temperature, concentration, sample type and solvent type. Detailed analyses of the molecular structure will be postponed to Chapter 2.

Very often there is an abundant literature on pregels, that is the aggregates formed below the critical gel concentration, C_{gel}. Sometimes, as is the case for poly(methyl methacrylate) (PMMA) systems, pregels are even far more documented than are the gels. Since aggregation in dilute solutions has evidently a strong bearing on the gelation phenomenon it will be deservedly discussed.

Synthetic polymers will be discussed first followed by some biopolymers. Whenever appropriate, the analogies or the differences between either type of macromolecule will be emphasized.

1.2 SYNTHETIC POLYMERS

The reader of a scientific book expects to find some classification or, at least, some indication about the domain under consideration. This may appear, at the moment, to be difficult with physical gels from synthetic polymers since most of the experimental reports convey the impression that each system is, at first sight, one of its kind. In fact, these seemingly unrelated gelling systems may be easily classified into two major types:

(1) system I, forming gels by **solvent induction**, e.g. polystyrenes;

(2) system II, forming gels by crystallization (**crystallization induced gelation**), e.g. polyethylene or PVC.

As hinted at by these definitions, this classification relies upon the gelation mechanism. In system I, crystallization is usually not the phenomenon taking place in the early stage of the gelation mechanism, whereas it is for system II. Also, in system I the solvent type is an essential parameter as the solvent is often reported to participate in the physical junctions, thus forming a polymer–solvent compound. On the contrary, in system II, at least for the early stage of gelation, the physical junctions are said to consist of the usual crystalline form as observed in the bulk crystalline state of these polymers.

As a result, unlike system I, gel formation for system II is little sensitive to solvent type. It ought to be clearly stated here that the statement 'little sensitive to the solvent' is only valid at equivalent solvent quality towards the polymer, that is, at constant Flory's interaction parameter, χ_1. It is otherwise obvious that the solvent quality will affect many properties such as the melting behaviour. This solvent quality effect has been studied by Flory who derived a theoretical relation that relates the melting point of a polymeric–solvent system forming only a solid solution to the polymer concentration and to the Flory's interaction parameter χ_1 (Flory, 1953):

$$\frac{1}{T_{m_p}} - \frac{1}{T^0_{m_p}} = \frac{\boldsymbol{R} V_p}{\Delta H_{0_p} V_s} (v_s - \chi_1 v_s^2) \tag{1.1}$$

where T_{m_p} is the actual melting temperature of the solid solution formed at a solvent volume fraction v_s and $T^0_{m_p}$ is the melting temperature of the pure polymer (in Flory's book (Flory, 1953) this temperature is the melting temperature of the most perfect crystals). $\boldsymbol{R}$ is the gas constant and ΔH_{0_p} is the polymer melting enthalpy per mole in the pure crystalline state and V_p the monomer molar volume.

If the system is dilute ($v_s \simeq 1$) but the condition

$$v_s - \ln(1 - v_s) \ll x v_s (1 - \chi_1 v_s)$$

is fulfilled (x is the ratio of the molar volume of the polymer to the molar volume of the solvent) then relation 1.1 can be approximated to (Harrison *et al.*, 1972):

$$\frac{1}{T_{m_p}} - \frac{1}{T^0_{m_p}} = \frac{\boldsymbol{R} V_p}{\Delta H_{0_p} V_s} (1 - \chi_1) \tag{1.2}$$

While χ_1 may vary with polymer volume fraction, v_p, and with temperature, which accordingly implies a more complicated behaviour, Flory's relation definitely indicates that the better the solvent, the larger the melting point depression with increasing solvent content (for further details see Appendix 1).

It turns out that polymers possessing bulky side groups fall into system I while those without bulky side groups fall into system II category. How this classification is related to the presence or absence of a bulky side group will be more clearly understood from a knowledge of the gel molecular structures which are described in Chapter 2.

1.2.1 Solvent-induced gels

(a) Polystyrenes

Polystyrene comes in three different forms: atactic, isotactic and syndiotactic:

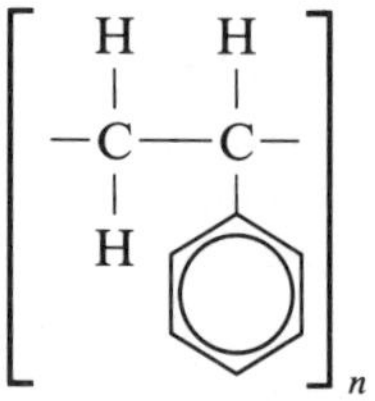

$$\left[\begin{array}{cc} H & H \\ | & | \\ -C- & -C- \\ | & | \\ H & \end{array}\right]_n$$

Atactic polystyrene does not display any crystallinity in the bulk state. **Isotactic polystyrene** is a semi-crystalline polymer characterized by slow crystallization rates and low crystallinity (30–50%). Its melting point, which depends on the crystallization conditions, lies somewhere between 160 and 250°C (Keith *et al.*, 1970; Overbergh *et al.*, 1976). **Syndiotactic polystyrene** is a newly synthesized material possessing a higher melting point (270°C) and also higher crystallinity (60–70%) (see for instance Immirzi *et al.*, 1988).

The glass transition temperature is not very sensitive to tacticity ($T_g \simeq 95^\circ$C).

So far only atactic polystyrene and isotactic polystyrene thermoreversible gels have been intensively studied.

(i) Isotactic polystyrene

As already underlined in the introduction, polystyrenes have been in the forefront of the interest in the gelation phenomenon of synthetic polymers. Not only does isotactic polystyrene (iPS) form gels but also the atactic version. The recently synthesized syndiotactic form can also produce gels. For the latter only a few results have been obtained so far so this will not be discussed.

The discovery of the gelatin properties of iPS was made during the study of crystallization behaviour in solutions. Lemstra and Challa (1975) reported unexpected behaviour as the temperature of the solution was decreased. With polymers such as polyethylene the crystallization rate increases with decreasing temperature while for polystyrene a maximum is observed near 30–40°C. In concentrated solutions, this effect gives rise to the formation of a gel instead of the usual suspension of chain-folded crystals.

Girolamo *et al.* (1976) studied the thermal behaviour of these gels prepared in a mixture of *cis*- and *trans*-decalin. They found that the gel has a much lower melting point than the crystal suspension. The latter melts at around 120°C in a 5% (w/w) solution while the former melts at around 60°C. A large hysteresis is needed to form the gel. As a matter of fact, annealing a few degrees below the gel melting temperature leads to the formation of the crystal suspension. Girolamo *et al.* (1976) reported that as a rule the solution must be cooled down to well below 20°C. This temperature happens to be close to the θ temperature for atactic polymer solution. As a result Wellinghoff *et al.* (1979) suggested that gelation might arise from a liquid–liquid phase separation frozen in at its early stage by crystallization. This concept became popular probably because the liquid–liquid phase separation was thought to proceed via spinodal decomposition, a mechanism which, according to Cahn and Hilliard (1959, 1965), gives off a network structure.

However, from results obtained by means of differential calorimetry, Guenet and McKenna (1988) have questioned this statement. The gel formation process produces a noticeable exotherm. By extrapolating the value of the onset of the exotherm gained at finite cooling rates to zero cooling rate one can determine the 'equilibrium' gelation temperature. Guenet and McKenna have obtained for 15% concentrated gels in the three following solvents:

$$T_{gel} = 17.2 \pm 0.3°C \text{ in } cis\text{-decalin}$$
$$T_{gel} = 22 \pm 0.3°C \text{ in } trans\text{-decalin}$$
$$T_{gel} = 20 \pm 0.3°C \text{ in 1-chlorodecane}$$

These experiments have also been carried out with iPS samples of differing molecular weights but no significant discrepancy has been observed. Clearly, the gelation temperature is molecular weight independent which is in disagreement with the liquid–liquid phase separation hypothesis. As a matter of fact, as the critical demixing temperature, T_c, is dependent on chain length, T_{gel} should have varied with T_c had this mechanism determined the gelation behaviour. There is, however, little doubt that in either decalin liquid–liquid phase separation interferes with the gelation process when the solution is quenched to low temperature. As a matter of fact quenching at 0°C and below implies that gelation occurs, for a given range of concentrations, within the miscibility gap since the θ temperature amounts to 12°C in *cis*-decalin and 18°C in *trans*-decalin (Berry, 1966). Yet, as will be detailed in Chapter 2, the gel molecular structure does not result from any effect of liquid–liquid phase separation (through either spinodal decomposition or nucleation and growth).

The study has been extended by Guenet and McKenna (1988) to different polymer concentrations for *cis*-decalin and *trans*-decalin. The resulting temperature–concentration phase diagrams of gel formation are given in Figure 1.1. As can be seen the phase diagram obtained for *cis*-decalin is

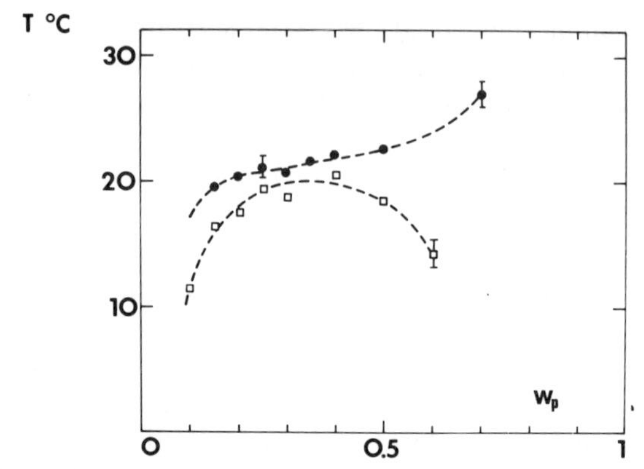

Figure 1.1 Formation phase diagram: onset of gel formation exotherm as a function of polymer concentration. □, iPS–*cis*-decalin gels; •, iPS–*trans*-decalin gels. Results obtained from DSC experiments carried out at 5 °C/min (with permission from Guenet and McKenna, 1988).

markedly different from the one for *trans*-decalin. This would be surprising for classical crystallization but it is quite explicable if a polymer–solvent compound is involved in the formation of these gels. In view of the shape of the phase diagrams, one expects to be dealing with a congruently melting compound in *cis*-decalin and with a compound characterized by a singular point in *trans*-decalin (see for instance Reisman, 1970 and Appendix 1). The study of the gel formation enthalpy as a function of polymer concentration gives further support to the concept of a polymer–solvent compound (Figure 1.2). In both systems a maximum is seen which is located at $C_{pol} = 30\%$ (w/w) in *cis*-decalin and at $C_{pol} = 40\%$ (w/w) in *trans*-decalin.

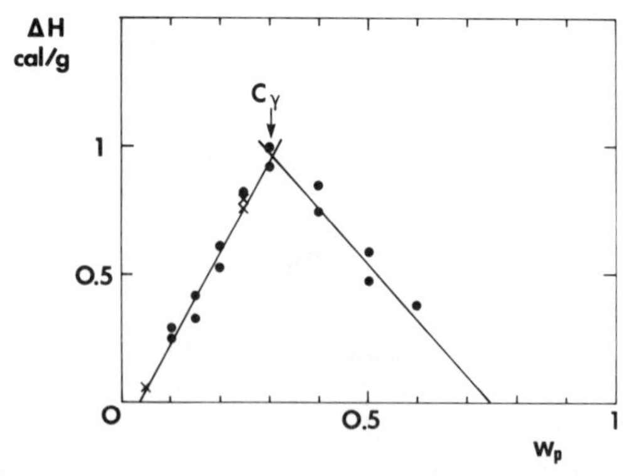

Figure 1.2 Gel formation enthalpy (per g of gel) in iPS–*cis*-decalin systems as a function of polymer concentration: •, fraction FK2 ($M_w = 7 \times 10^4$, $M_w/M_n \simeq 2.1$); x, fraction FK1 ($M_w = 3.2 \times 10^5$, $M_w/M_n \simeq 2.8$) (with permission from Guenet and McKenna, 1988).

On either side of this maximum the variation is linear as expected. This maximum corresponds to the stoichiometric composition of the compound. It specifically means that there are 1.75 *cis*-decalin molecules per monomer unit and 1.15 *trans*-decalin molecules per monomer unit. Guenet and McKenna (1988) have carefully checked that this result does not arise from kinetic effects due to high viscosity by showing the absence of a significant outcome when varying the cooling rate.

It is worth commenting on the gelation point further. Recent experiments by Klein *et al.* (1990b) have allowed determination of the gelation threshold by isothermal calorimetry for the system iPS–*cis*-decalin (Figure 1.3). These results are in excellent agreement (to within 1°C) with those obtained by Guenet and McKenna (1988) by means of scanning calorimetry. This suggests that this threshold may be regarded as defining a kind of equilibrium state. As a matter of fact, once the system is below a given temperature, crystallization in the usual form of chain-folded crystals cannot take place at all. Nevertheless it is not a truly equilibrium state, in the thermodynamic sense, since it cannot be obtained by slowly cooling the solution from high temperature.

Interestingly, when iPS–*trans*-decalin solutions are annealed just above the gelation threshold, the usual chain-folded crystals with a melting temperature of about 120°C are formed. Conversely, in iPS–*cis*-decalin solutions while spherolitic structures are formed, the melting point remains virtually unchanged, that is 50–55°C. This result will be discussed in the next chapter which deals with the morphology and molecular structure.

A systematic study as a function of concentration has also been carried out by Guenet and McKenna (1988). They found that the melting enthalpy is dependent on the rate of heating and also the molecular weight. This

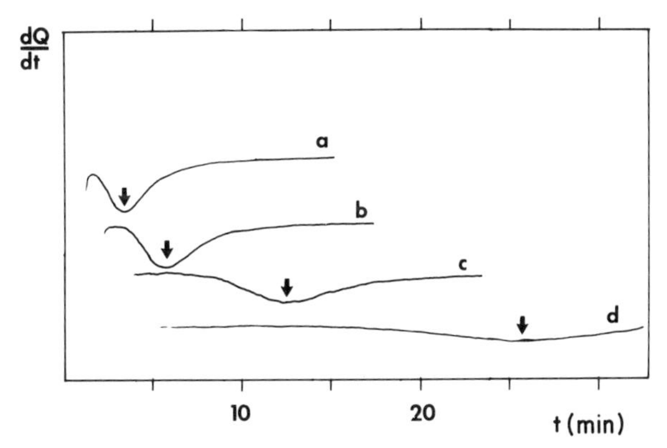

Figure 1.3 DSC isothermal exotherms obtained at different temperatures for an iPS–*cis*-decalin solution (C_{pol} = 30% w/w), annealing at: a = 18°C, b = 19°C, c = 20°C, d = 21°C (with permission from Klein *et al.*, 1990b).

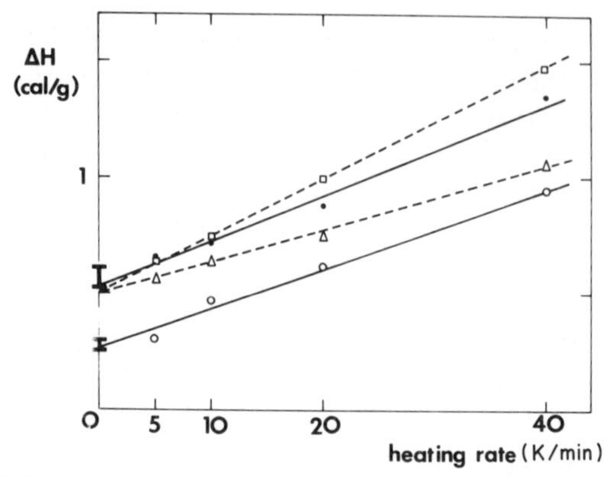

Figure 1.4 Gel melting enthalpy (per g of gel) as a function of the heating rate for iPS–*cis*-decalin gels. Bars drawn on the ordinate stand for the values found for the gel formation enthalpies for $C_{pol} = 10\%$ and $C_{pol} = 20\%$, respectively; □, FK2 with $C_{pol} = 40\%$; △, FK1 with $C_{pol} = 40\%$; •, FK2 with $C_{pol} = 20\%$ and ○, FK2 with $C_{pol} = 10\%$ (with permission from Guenet and McKenna, 1988).

unexpected result seems to be related to the viscosity of the solution obtained from the melting gel. In fact, the extra heat of melting is of lower magnitude with higher molecular weight samples at a given polymer concentration (Figure 1.4). Guenet and McKenna (1988) suggest that this extra heat might arise from the collapsing of the sample when the gel turns into a solution, thus giving rise to a kind of constraint release.* It then seems reasonable to assume that the constraint release should be enhanced with lower viscosity solutions for which the collapsing takes place in shorter times than for higher viscosity solutions.

Still, as shown in Figure 1.4, extrapolation to zero heating rate gives the same values as those determined on cooling, regardless of the sample's

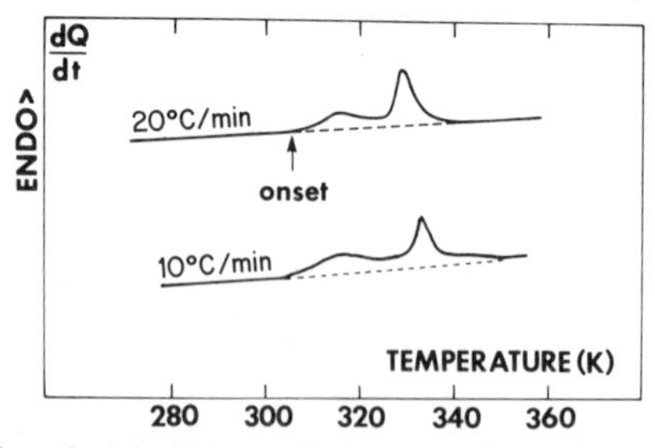

Figure 1.5 Typical endotherms recorded on melting an iPS–*cis*-decalin gel that had been formed by a quench at 0°C for 30 min. $C_{pol} = 20\%$; heating rates as indicated (with permission from Guenet and McKenna, 1988).

* This is a purely mechanical effect which is often seen when a perturbation is applied to the DSC detection head.

molecular weight. Typical melting endotherms obtained with *cis*-decalin gels are drawn in Figure 1.5. As can be seen, there are two maxima. The first maximum located at low temperature does not vary with concentration whereas the second one does. The first maximum originates in a monotectic transition due to gel formation taking place within the miscibility gap ($T_{prep} = 0°C$ in *cis*-decalin). (The monotectic transition is explained in detail in Appendix 1.)

It is important to call the reader's attention to the mention by Guenet and McKenna (1988) of the absence of a well-defined melting endotherm for gel samples that have been molten and re-formed **within the DSC pan** by a rapid quench to 0°C (rapid gelation conditions). Conversely, a gel that has been molten and re-formed in the DSC pan by annealing at 14°C (slow gelation conditions) gives a well-defined melting endotherm whose enthalpy is close to the formation enthalpy (Guenet, unpublished). One may probably suspect that gels formed at 0°C are far from being perfect, and hence the melting endotherm spread over a large temperature range preventing it from being observed.

Phase diagrams obtained in either decalin after extrapolation of the melting temperature to zero heating rate are given in Figure 1.6. The shape of each diagram is quite similar to the one obtained on cooling and, therefore, is consistent with the existence of polymer–solvent compounds. As with the formation enthalpies, melting enthalpies exhibit a maximum as a function of polymer concentration. Values found for the stoichiometric concentrations are 28% (w/w) for *cis*-decalin and 38% (w/w) for *trans*-decalin which are in close agreement with those determined on cooling (inset of Figure 1.6). It should be noted that the values of the stoichiometric composition account for the fact that iPS–*cis*-decalin gels melt at temperatures lower than iPS–*trans*-decalin gels: the higher the solvation, the lower the melting point. Further, these data do not obey Flory' relation (1.1) which states that no marked discrepancy between either gel melting points should be observed with preparation solvents of nearly identical χ_1 parameter.

Here, it is important to emphasize that Guenet and McKenna have defined the upper melting limit in their phase diagrams by taking the end of the gel melting endotherm. Obviously, this may not correspond to the visible gel melting temperature which occurs when macroscopic connectivity is lost. At this temperature the melting of all the physical junctions has not necessarily taken place. These considerations, however, do not change the conclusions as to the nature of the system but can be of importance when it comes to calculating the liquidus lines.

Studies carried out by Guenet *et al.* (1985) with various solvents support the existence of polymer–solvent compounds. These authors have shown that in some solvents the gel **melting point is metastable** as would be expected with incongruently melting compounds. They have shown

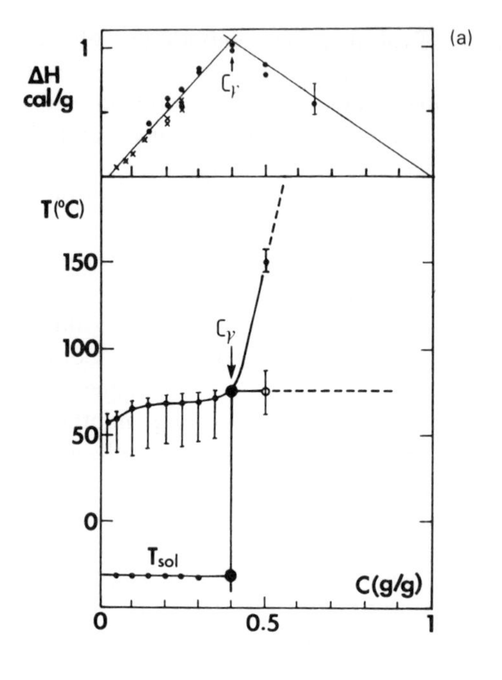
(a)
ΔH cal/g
1
C_γ
T(°C)
150
100
50
0
T_{sol}
C(g/g)
0
0.5
1

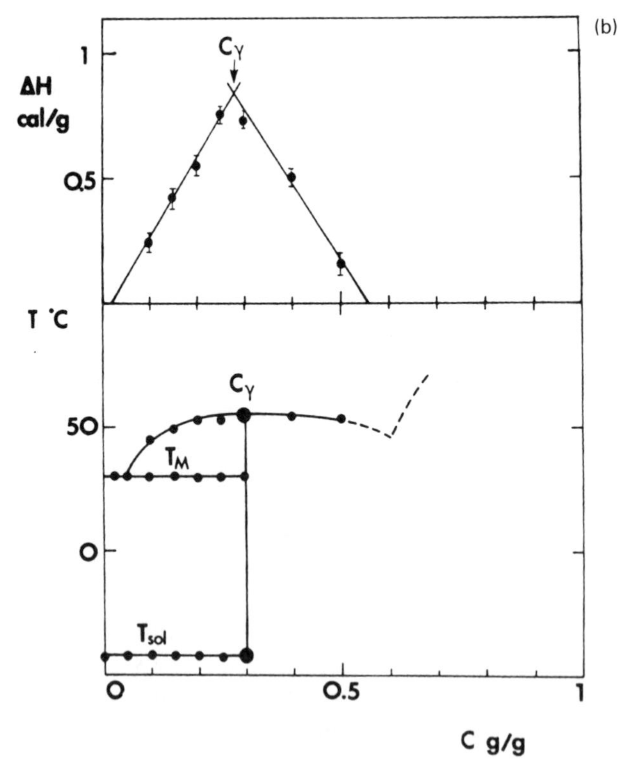
(b)
ΔH cal/g
1
0.5
C_γ
T °C
50
T_M
0
T_{sol}
0
0.5
1
C g/g

that there is a spontaneous transformation of the gel nascent structure into the 3_1 helical form (see Chapter 2.2.3.1.a) **without gel melting** in 1-chlorodecane, 1-chlorododecane, diethyl malonate or diethyl oxalate. This is illustrated by two sets of experiments.

(1) At high heating rates the low-melting peak associated with gel melting is predominant over the high-melting peak corresponding to the fusion of the 3_1 form. Conversely, by decreasing the heating rate the low-melting peak virtually vanishes while the high-melting peak is now the largest (see Figure 1.7). This thermal behaviour shows that the 3_1 form grows at the expense of the nascent gel form.

(2) In Figure 1.8 is shown the effect caused by annealing the gel sample just below the low-melting peak. While this peak is visible on a nascent gel, it has disappeared after subsequent annealing.

Guenet *et al.* (1985) also mention that the ease with which the transformation occurs is solvent dependent. This does not happen in either decalin, is quite slow in 1-chlorodecane, very rapid in 1-chlorododecane and diethyl malonate and exceedingly rapid in diethyl oxalate. It should also be noted that this transformation can be observed visually since a clear or translucent gel becomes opaque afterwards.

The transformation according to these authors can also be observed macroscopically. While the gel in the nascent form is soluble in good solvents such as toluene or THF (tetrahydrofuran), it becomes insoluble at room temperature after transformation, as are polystyrene crystals in the 3_1 form.

Finally, it is of interest to mention gel-swelling experiments performed by Klein *et al.* (1990*a*). These experiments consist of immersing a piece of gel of known weight, P_0, into an excess of preparation solvent. The solvent absorption is followed as a function of time which allows determination of an equilibrium weight, P_∞, once the swelling behaviour attains a plateau regime. This determines an equilibrium concentration, C_{eq}, which is related

Figure 1.6 Temperature–concentration phase diagrams obtained on heating for iPS–*trans*-decalin gels (a) and iPS–*cis*-decalin gels (b). For the former system the phase diagram has been established from results obtained at 5 °C/min by considering the maximum of the endotherm (the width of the endotherm is given by the bars), while for the latter temperatures extrapolated to zero heating rates have been used (T_M represents the onset of the low-melting endotherm and is a monotectic transition, while T_F stands for the end of the high-melting endotherm). T_{sol} represents the solvent melting temperature (measured after the gel has been formed), C_γ the stoichiometric concentration. The gel melting enthalpies (per gram of gel) are given on top of each figure as a function of polymer concentration by means of a Tammann's-type diagram (for *trans*-decalin, x = fraction FK1 and • = fraction FK2 (see caption of Figure 1.2 for details) (with permission from Guenet and McKenna, 1988).

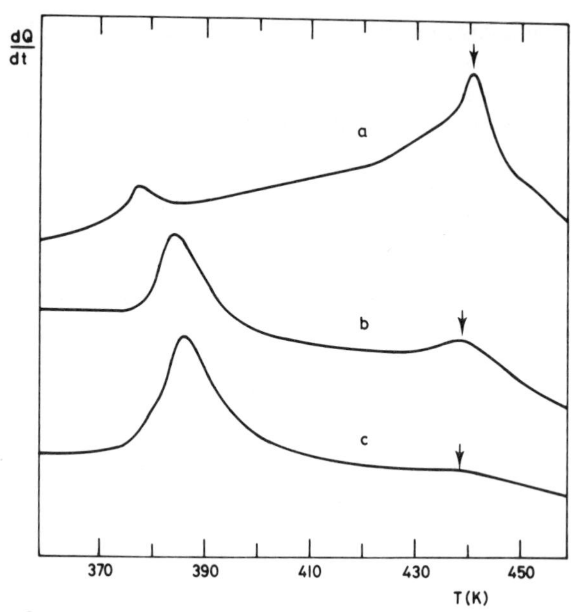

Figure 1.7 DSC thermograms obtained at different heating rates for an iPS–1-chlorodecane gel ($C_{pol} = 5\%$) prepared through a quench at $-5°C$. a = 20°C/min, b = 40°C/min, c = 80°C/min (with permission from Guenet *et al.*, 1985).

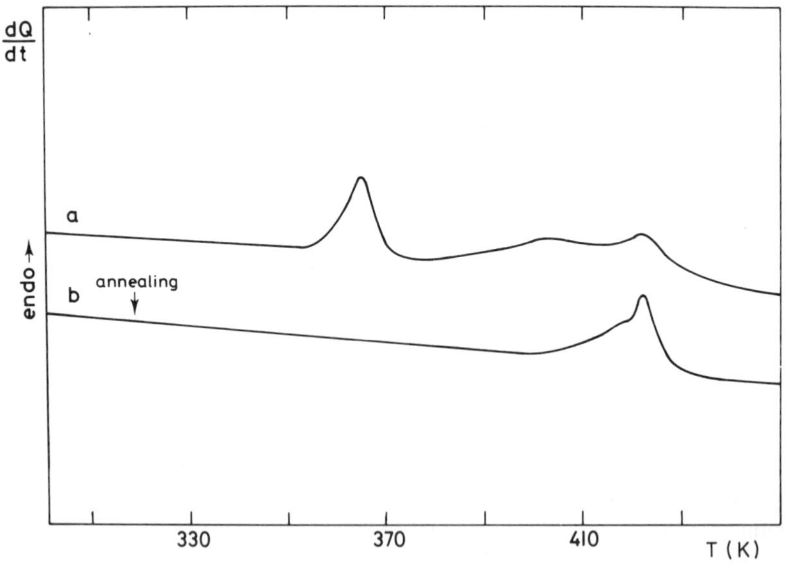

Figure 1.8 DSC thermograms of a gel prepared in diethyl malonate: a, nascent gel, b, the same gel after annealing for 5 min at 45°C (as indicated on the figure). Heating rate in both cases 20°C/min (with permission from Guenet *et al.*, 1985).

to the preparation concentration, C_{prep}, by the following relationship:

$$C_{eq} = G_{\infty}^{-1} C_{prep} \tag{1.3}$$

where the swelling ratio at equilibrium, G_{∞}, reads $G_{\infty} = P_{\infty}/P_0$.

Two regimes can be seen (Figure 1.9) on either side of a concentration, C_{γ} (in w/w), which differs in *cis*-decalin ($C_{\gamma} = 0.28$) and *trans*-decalin ($C_{\gamma} = 0.38$). Gels prepared below C_{γ} do not experience any significant swelling whereas those prepared above C_{γ} do. Interestingly enough, in the latter

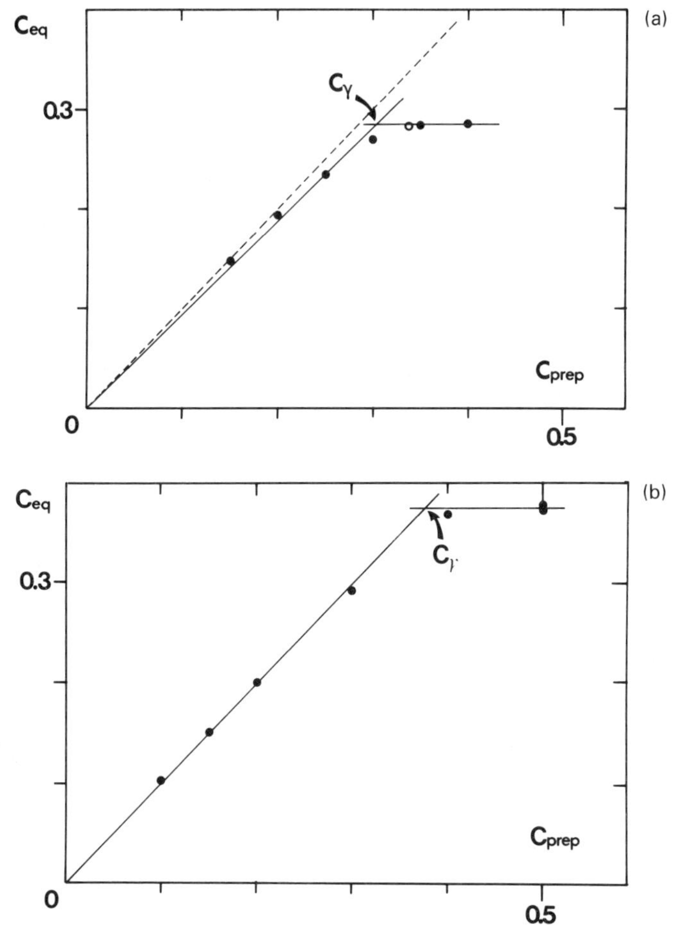

Figure 1.9 Equilibrium concentration, C_{eq}, *vs* preparation concentration, C_{prep}, determined after swelling gels to equilibrium in an excess of preparation solvent: (a) iPS–*cis*-decalin gels, •, FK1 fraction, ○, iPSG ($M_w = 2.12 \times 10^5$ with $M_w/M_n \simeq 2.6$); (b) iPS–*trans*-decalin (FK1). The dotted line stands for $C_{eq} = C_{prep}$ and C_{γ} represents the stoichiometric concentration (with permission from Klein *et al.*, 1990a).

case, the gels always swell at **the same equilibrium concentration** regardless of the preparation concentration. Klein *et al.* (1990*a*) conclude that this type of result is consistent with the existence of a polymer–solvent compound.

for $C_{prep} < C_{\gamma}$ the stoichiometry is fulfilled so that there is no need for the gel to absorb solvent. Then:

$$C_{eq} = C_{prep} \tag{1.4}$$

for $C_{prep} > C_{\gamma}$ the stoichiometry is no longer fulfilled, hence the need for taking the adequate amount of solvent to reach the stoichiometric concentration. Then:

$$G_{\infty} = C_{prep}/C_{\gamma} \tag{1.5}$$

which leads to:

$$C_{eq} = C_{\gamma} \tag{1.6}$$

The above analysis holds true in two cases: either there exists no disordered ('amorphous') material or, if disordered material is present, it does not absorb solvent on its own. The latter situation is probably the one that prevails in these gels. That the disordered material does not swell may be due to the fact that decalin is a θ solvent at room temperature but may also convey something deeper and not yet fully understood. The particular molecular structure (see Chapter 2) observed for these gels where the disordered material most probably possesses the same stoichiometry as the ordered material may be responsible for the phenomena involved. (See appendix 1.)

Not only is the swelling behaviour consistent with the existence of a polymer–solvent compound, but also the values of C_{γ} found under these conditions are in good agreement with those determined independently from thermal analyses.

To give additional support to their interpretation of the swelling behaviour, Klein *et al.* (1990*a*) refer to the same type of experiment as performed by He *et al.* (1987*a*) with a multiblock copolymer containing crystallizable sequences alternating with non-crystallizable ones (see the section devoted to this copolymer for further details (1.2.2.d)). The physical gel state can also be obtained from solutions in various solvents but no polymer–solvent compound is formed. In no case is the swelling behaviour of this multiblock copolymer similar to the one observed with polystyrene. Depending on the solvent quality, either C_{eq} increases simultaneously with C_{prep} (e.g. for *trans*-decalin) or $C_{eq} = C_{prep}$ over the whole range of concentrations investigated (up to 50%, e.g. for bromobenzene).

(ii) Atactic polystyrene

The reader might be surprised that both isotactic polystyrene and atactic polystyrene (aPS) gels are discussed in the same section. Indeed, since aPS is a non-crystallizable polymer one would expect a different gelation mechanism. It turns out that the gelation phenomenon is very similar in both systems.

Historically speaking, the first report on the physical gelation of aPS came from Wellinghoff *et al.* (1979) and Tan *et al.* (1983). The latter gave a list of solvents in which they claimed to be able to produce physical gels at low temperature. Here, by 'low temperatures' it is meant below 0°C. Most of their results were obtained by either the ball-drop method or tilting the test tube containing the solution to be gelled. Their results for the system aPS–carbon disulphide (aPS–CS_2) are given in Figure 1.10 together with those by Gan *et al.* (1985). It is worth emphasizing that this system gives gel melting temperatures close to room temperature unlike other aPS solutions for which the gel melting point is in the vicinity of −60°C.

These results were often greeted by disbelief, so great was the prejudice that physical gelation could only be due to a crystallization process.

While it turned out that some of the solvents listed by Tan *et al.* (1983) did not give true gels but highly viscous solutions (Plazek and Altares, 1986), at least one case, aPS–CS_2 was definitely shown by various rheological techniques to produce a genuine gel (Clark *et al.*, 1983; Gan *et al.*, 1985; Domzsy *et al.*, 1986; Plazek and In-Chul-Chag, 1991).

Before giving further details on the gelation behaviour of this polymer it is worth mentioning that Hyde and Taylor (1963), Benoit and Picot (1966) and Dautzenberg (1970) indirectly observed this phenomenon. By performing light-scattering experiments on concentrated solutions of aPS in several 'good' solvents they noticed anomalies in the scattering curve.

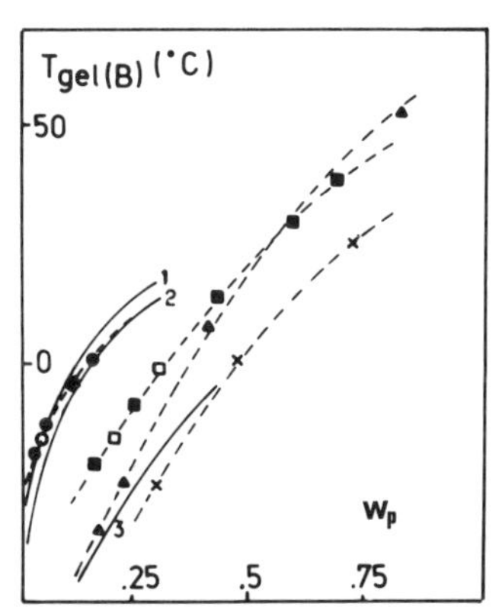

Figure 1.10 Determination of the gelation temperature as a function of concentration by different techniques for aPS–CS_2 gels. Solid squares by the ball drop method (solid lines stand for the results gathered by Tan *et al.*, 1983); (1) for aPS $M_w = 2 \times 10^6$, (2) for aPS $M_w = 10^6$ and (3) for aPS $M_w = 4 \times 10^3$); hollow squares = sphere rheometer (see for instance Gan *et al.*, 1985). ○, •, aPS $M_w = 1.7 \times 10^6$; □, ■, aPS $M_w = 1.8 \times 10^5$; ▲, aPS $M_w = 5 \times 10^3$; x, aPS $M_w = 6 \times 10^3$ (with permission from François *et al.*, 1986).

Theoretically, if the solution is homogeneous, the scattered intensity, $I(q)$ ($q = (4\pi/\lambda)\sin\theta/2$, θ being the scattering angle) should display a lorentzian form:

$$I^{-1}(q) \approx (q^2 + \xi^{-2}) \tag{1.7}$$

where ξ is the screening length defined by Edwards (1966).

Experimentally, a plot of $I^{-1}(q)$ vs q^2 that should have been linear showed a pronounced downturn at small angles (see Figure 1.11) sometimes referred to as enhanced low-angle scattering or ELAS for short. Once the results were shown to be genuine and not arising from dust, scientists were left with the problem of aggregates in aPS solution and, above all, in good

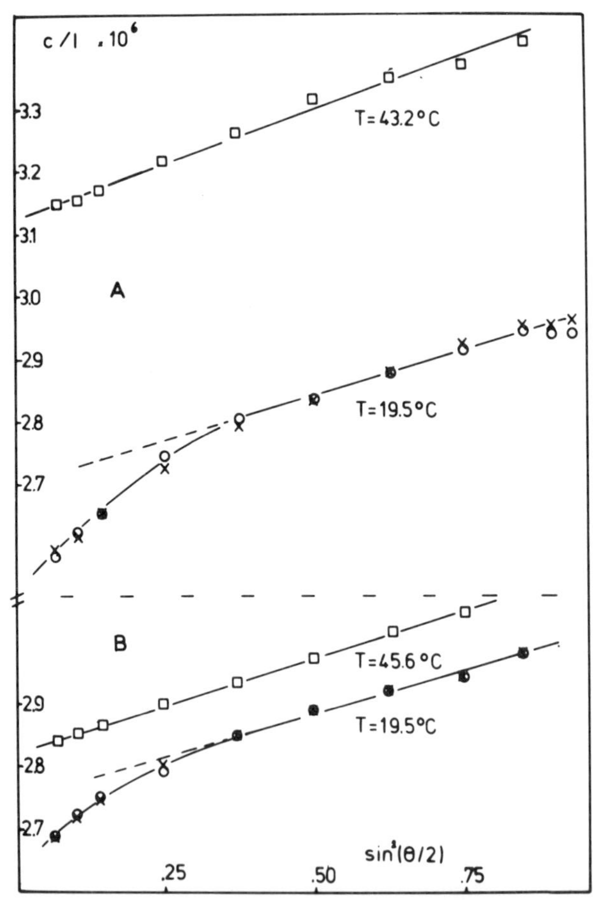

Figure 1.11 Zimm-plot (C/I *vs* $\sin^2\theta/2$, where θ is the scattering angle) for aPS in *para*-dioxane (A) and THF (B). $C = 5 \times 10^{-3}$ g/cm^3 at different temperatures (as indicated). ○ and x stand for the solution before heating and after cooling back, respectively, which shows the reversibility of the phenomenon of aggregation (with permission from Gan *et al.*, 1986).

solvents. At that time this did not make any sense: how could a good solvent lead to the formation of aggregates with an atactic polymer which had not shown the slightest trace of crystallinity in the bulk state?

A solution to this long-lasting question was proposed by Guenet *et al.* (1983). These authors noticed that the list of 'gelation solvents' given by

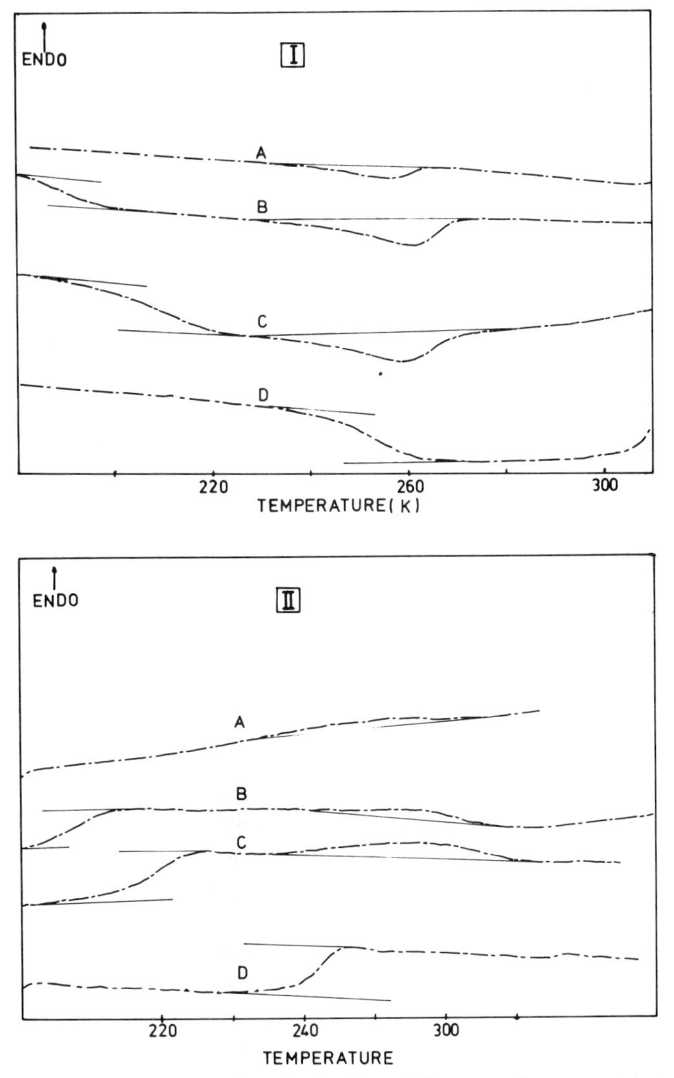

Figure 1.12 DSC thermograms obtained at different polymer concentration in aPS–CS_2 systems (I = on cooling and II on heating). aPS $M_w = 1.8 \times 10^5$. (A) $C_{pol} = 30.3\%$, (B) $C_{pol} = 47.8\%$, (C) $C_{pol} = 58.8\%$ and (D) $C_{pol} = 74.7\%$ (with permission from François *et al.*, 1986).

Tan *et al.* (1983) contained some of the solvents for which the ELAS effect was known to be observed. Conversely, Tan *et al.* (1983) also reported solvents in which no gelation could occur. Performing again the very same type of experiments as those described by Benoit and Picot (1966) they eventually showed that ELAS is absent in the solvents where gelation does not take place. They concluded that there was a direct correlation between the gelation phenomenon and the ELAS effect: ELAS is due to remnants of aggregates that are still present at room temperature. These results were confirmed by Gan *et al.* (1986) by means of a simple experiment which was not thought of 20 years ago, so much was the prejudice against permanent aggregation. It consists of heating the solution at a higher temperature (about 40°C) and **recording the scattering pattern at this temperature** (whereas former experiments consisted of heating the solution and then studying it once it had cooled down to room temperature). As expected the ELAS effect vanishes on heating and reappears reversibly on cooling to room temperature (Figure 1.11).

Further studies provided consistent explanations of why this effect, and correspondingly, of why gelation could occur in good solvents.

Careful differential scanning calorimetry experiments carried out by François *et al.* (1986) showed clearly the occurrence of gel formation exotherms and gel melting endotherms, thus implying the creation of order during the gelation process (Figure 1.12). This conclusion was in disagreement with the interpretation of Boyer *et al.* (1985) which consisted of considering the gelation phenomenon as being due to liquid-like association between chains and related to a second-order transition T_{ll}. This kind of transition should have given a discontinuity in the thermal behaviour, as a glass transition does, but no exothermic or endothermic signals. Of further note is the quite clear distinction between the glass transition and the gel melting (Figure 1.12).

When extrapolated to zero heating rate, the gel formation and the gel melting exhibit a hysteresis unlike the glass transition (Figure 1.13). This hysteresis is compatible with a nucleation and growth process, characteristic of first-order transitions, but not with a second-order transition for which it is absent, as demonstrated in this case by the glass transition.

It must be stressed that Boyer *et al.*'s interpretation relies upon the idea that CS_2 is a poor solvent of polystyrene. This arises mainly from the report by Tan *et al.* (1983) on the presence of a miscibility gap at low temperature from turbidity measurements leading them to determine a θ temperature of -73°C for aPS in CS_2. Theories were put forward on the basis of tricritical phenomena (Tanaka, 1989) to account for aPS gelation on the basis of these findings. Not only is CS_2 a good solvent at room temperature where concentrated solutions gel, but also, Gan *et al.* (1986) have questioned the validity of Tan *et al.*'s observation. They have pointed out that solutions prepared

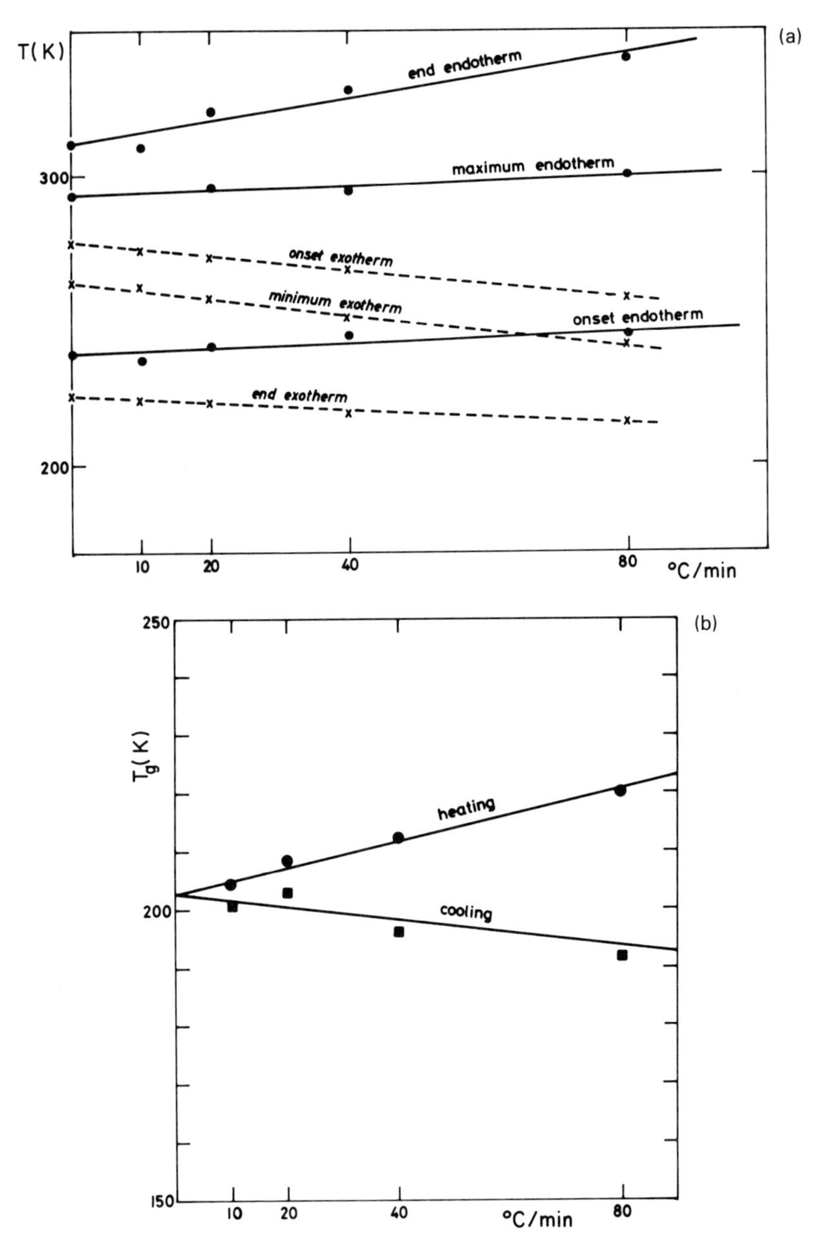

Figure 1.13 Variation of the gel formation temperature, gel melting temperature and glass transition temperature with heating and cooling rates in aPS–CS_2 (with permission from François *et al.*, 1986). (a) Dotted lines stand for the formation exotherm (onset, minimum and end) and the full line for the melting endotherm (onset, maximum and end). $M_w = 1.7 \times 10^6$ and $C_{pol} = 33.6\%$. Extrapolations to zero rate do not yield the same values on heating or on cooling. (b) Variation of T_g (glass transition) with heating and cooling rate. Here extrapolation to zero rate gives the same value. $M_w = 1.8 \times 10^5$ and $C_{pol} = 50.9\%$.

from carefully dried CS_2 do not display any turbidity down to the crystallization point of the solvent. Moreover, by measuring the intrinsic viscosity from room temperature down to $-50°C$ they have shown that CS_2 still remains a good solvent at very low temperatures where gelation takes place.

The phase diagrams obtained by François *et al.* (1986) on either cooling or heating reveal two features (Figure 1.14).

(1) Unlike the ball-drop method, DSC experiments do not show significant differences between the different molecular weights. This may be a result of the fact that the ball-drop results may be biased by solution viscosity particularly at high concentrations.

(2) Also, unlike the ball-drop method, the phase diagrams display a maximum in the vicinity of $C_{pol} = 50\%$ (w/w). According to François *et al.* (1986) this strongly suggests the formation of a polymer–solvent compound. The variation of either formation or melting enthalpies as a function of polymer concentration supports their statement. After carefully checking, by using lower and lower cooling rates, that the low melting enthalpies found at high concentrations did not arise from kinetic effects,

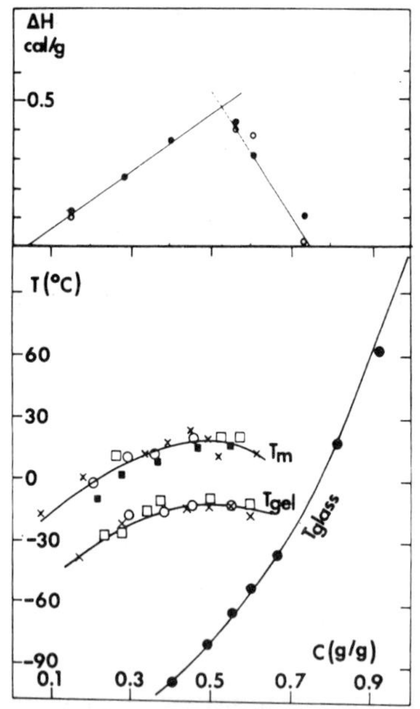

Figure 1.14 Gel melting points (upper curve) and gelation temperatures (lower curve) in aPS–CS_2 as determined by taking the maximum of the DSC endotherm or exotherm for polymer samples of different molecular weights: x, $M_w = 1.8 \times 10^5$; □, $M_w = 1.7 \times 10^6$; ■, $M_w = 6 \times 10^3$, ○, $M_w = 5.6 \times 10^6$. The glass transition line (•, T_{glass}) correspond to $M_w = 1.8 \times 10^5$. The variation of the formation enthalpy (•) and the melting enthalpy (○) for $M_w = 1.8 \times 10^5$ is given above. Solid lines are given only as a guide (data from François *et al.*, 1986).

they obtained the type of plot given in the inset of Figure 1.14 which shows the occurrence of a maximum, also near $C_{pol} = 50\%$ (w/w).

The fact that the gelation phenomenon, particularly the melting point, is so sensitive to the solvent type (for solvents of nearly identical χ_1 parameter) is consistent with the presence of polymer–solvent compounds in the physical junctions and is obviously reminiscent of what has been found with isotactic polystyrene. In particular, aPS gels produced in CS_2, which is probably the best solvent known for polystyrene ($\chi_1 = 0.4$), should have been characterized by low melting points had the gelation mechanism been related to any solid-solution-type crystallization or aggregation.

Here, it seems appropriate to mention results obtained by the solvent crystallization method which show the similitude between isotactic polystyrene and atactic polystyrene. This method consists of measuring in any polymer–solvent system (either solutions or gels) the amount of 'free solvent' (Yasuda *et al.*, 1979; Gan *et al.*, 1986; Guenet, 1986; Klein and Guenet, 1989).

This amount is determined by measuring the solvent melting enthalpy, ΔH_s, in the system under study. The enthalpy ΔH_s reads as a function of the melting enthalpy of the pure solvent ΔH_{os} and the polymer concentration x_p (w/w):

$$\Delta H_s = \Delta H_{os}\{1 - x_p[1 + \alpha(m_s/m_p)]\} \tag{1.8}$$

where m_s and m_p are the solvent and the monomer molecular weights respectively.

A plot ΔH_s vs x_p is linear and gives, once extrapolated to $\Delta H_s = 0$, a value x_p^0 from which the parameter α can be derived:

$$\alpha = \frac{(1 - x_p^0)m_p}{x_p^0 m_s} \tag{1.9}$$

α stands for the number of solvent molecules per monomer 'bound' to the polymer. The quotation marks are used on purpose in order to emphasize that the binding is not necessarily strong but sufficiently efficient to prevent these molecules from crystallizing with the 'free' ones.

Derivation of relation 1.8 rests upon the use of the lever rule (see Appendix 1). If a two-phase system is constituted of free solvent, on the one hand, and polymer + solvent (solid–solution, compound...) on the other hand, then the enthalpy of melting of the free solvent, ΔH_{fs}, reads:

$$\Delta H_{fs} = x_{fs}\Delta H_0$$

in which x_{fs} is the proportion of free solvent (w/w) and ΔH_0 the melting enthalpy of the pure solvent. The proportion of free solvent in turn reads:

$$x_{fs} = \frac{N_{fs}m_s}{N_p m_p + N_p \alpha m_s + N_{fs} m_s}$$

in which N_p and N_{fs} are the number of monomers and of free solvent molecules in the system, m_s and m_p the molecular weight of the solvent and of the monomer molecule and α the number of solvent molecules 'bound' per monomer to the polymer, respectively. The last two terms of the sum in the denominator correspond to the total amount of solvent. By expressing the proportion of polymer and combining the result with the above equation one finally ends up with:

$$x_{fs} = 1 - x_p[1 + \alpha(m_s/m_p)]$$

Introducing this relation in the first equation gives 1.8.

It is worth noting that the variation of the solvent melting enthalpy with polymer concentration must be linear. If this is not the case experimentally then it indicates some problems such as incomplete crystallization of the solvent. Deviations from linearity may appear close to the polymer concentration x_p^0 where almost all of the solvent is 'bound'.

It is worth mentioning that independent work was carried out by Errede (1989, 1990, 1991) via another method (determination of the amount of free solvent by an evaporation method) which seems to give similar results in the sense that the parameter α_G defined by Errede (note the coincidence in the designation of the parameter) is proportional to α as determined by Gan *et al.* (1986), Guenet (1987) and Klein and Guenet (1989).

In Figure 1.15 are reported the values of α as a function of the solvent molar volume V_m obtained from iPS gels in different solvents and aPS

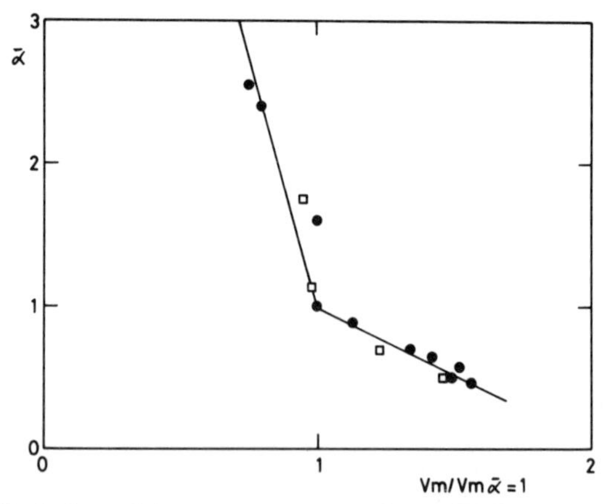

Figure 1.15 Variation of the parameter α as determined from the crystallization method in different solvents as a function of the ratio solvent molar volume/solvent molar volume determined at $\alpha = 1$ ($V_m/V_{m_{\alpha=1}}$); • = aPS and □ = iPS (with permission from Klein and Guenet, 1989).

solutions or gels. As can be seen both polymers display the same type of variation of α with V_m. This suggests a kind of universality of the system. Only the values are different: solvents of smaller molecular size are needed in aPS to obtain values of α larger than 1 unlike iPS. Also, when α is smaller than 1 either the gels are unstable in their nascent state (iPS) and tend to transform into the 3_1 form or gels do not form in the solvent (aPS). This suggests an effect of the solvent size which will be understood by elucidating the molecular structure (see Chapter 2).

Suffice it to say at this stage of the discussion that either polymer will probably form a gel by some common mechanism which is most probably not crystallization.

Needless to say, gelation of syndiotactic polystyrene, a newly synthesized polymer (Immirzi *et al.*, 1988), will be of prime interest, hopefully shedding further light on the gelation mechanism. This is especially so because François *et al.* (1988), on the basis of experiments carried out with epimerized isotactic samples (this chemical method (Shepherd *et al.*, 1979) allows one to go from the 100% isotactic form to the atactic one), suggest that aPS gelation is probably due to the syndiotactic sequences. The availability of the syndiotactic version of this polymer should then be helpful.

(b) Poly(methyl methacrylates)

Poly(methyl methacrylate) (PMMA) possesses the following chemical structure:

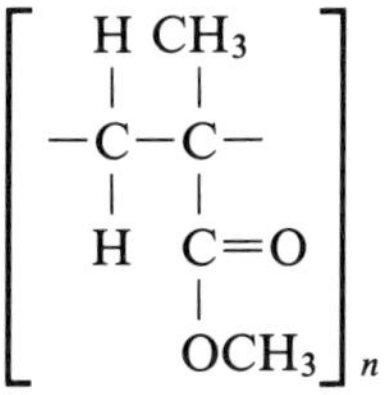

$$\left[\begin{array}{cc} \mathrm{H} & \mathrm{CH_3} \\ | & | \\ -\mathrm{C}- & \mathrm{C}- \\ | & | \\ \mathrm{H} & \mathrm{C{=}O} \\ & | \\ & \mathrm{OCH_3} \end{array} \right]_n$$

PMMA comes in three forms: atactic PMMA (aPMMA), isotactic PMMA (iPMMA) and syndioactic PMMA (sPMMA). Unlike for polystyrene, the glass transition temperature is markedly dependent on tacticity varying from 40°C for isotactic samples to above 120°C for syndiotactic samples.

Also, unlike isotactic or syndiotactic polystyrene, which both crystallize readily from the bulk state under appropriate thermal conditions (see for instance Keith *et al.* (1970) for iPS and Immirzi *et al.* (1988) for sPS), PMMAs, both syndiotactic and isotactic, do not exhibit strong tendencies to crystallization. Crystallization rates are different by several orders of magnitude. According to Kusuyama *et al.* (1983), crystallization of sPMMA cannot be achieved by thermal treatment alone but requires solvent induction. While thermoreversible gels do not display crystallinity, crystals can be produced from this state after subsequent drying (see Chapter 2 for further details).

Although there is no explicit statement that gels from PMMA solutions are solvent-induced, many results point towards this mechanism, as will be seen in what follows. Moreover, as aforementioned, the well-established fact that crystallization of sPMMA from the bulk state is only solvent-induced (Kusuyama *et al.*, 1983) allows one to anticipate what is actually taking place in the gel state.

As with polystyrene, all the existing versions of PMMAs (syndiotactic, isotactic and atactic) have been studied. It is worth mentioning here that blends of isotactic and syndiotactic PMMAs form a so-called stereocomplex which possesses gelation properties of its own. Here, all the cases will be considered.

(i) Gelation of any of the tactic versions

A large body of information on the formation of aggregates formed below the critical gel concentration is available in the literature for both iPMMA and sPMMA; Spěvaček and coworkers have contributed largely to this by means of ^{1}H-NMR, infrared spectroscopy, light scattering, osmometry and viscosity (see for instance Spěvaček and Schneider, 1987).

For their NMR studies, they have used the method of integrated band intensities. In high-resolution ^{1}H-NMR the intensity, I, is given by:

$$I = KNT^{-1}(1 + \gamma^2 H_{1ef}^2 T_1 T_2)^{-1/2} \tag{1.10}$$

where K is a constant, N is the number of protons per unit volume, T the absolute temperature, γ the gyromagnetic ratio, H_{1ef} the effective value of the radiofrequency magnetic field, T_1 and T_2 the spin–lattice and the spin–spin relaxation times respectively (the expression in parentheses is the saturation factor).

Proton mobility in monomer units involved in the physical links between chains is usually so low that the corresponding linewidths are of the order of several hundred Hertz thus escaping detection. As a result, in aggregated systems, only the 'non-aggregated protons' will contribute to the integrated band intensity. Thus the relative content of associated protons $p = N_a/N_{total}$ will be straightforwardly given by:

$$p = 1 - (I'/I_0) \tag{1.11}$$

where I' and I_0 are the integrated intensities of the partly aggregated system and the non-aggregated system respectively.

A typical plot of the variation of the integrated band intensity with temperature is given in Figure 1.16. Typical values of p obtained by Spěvaček

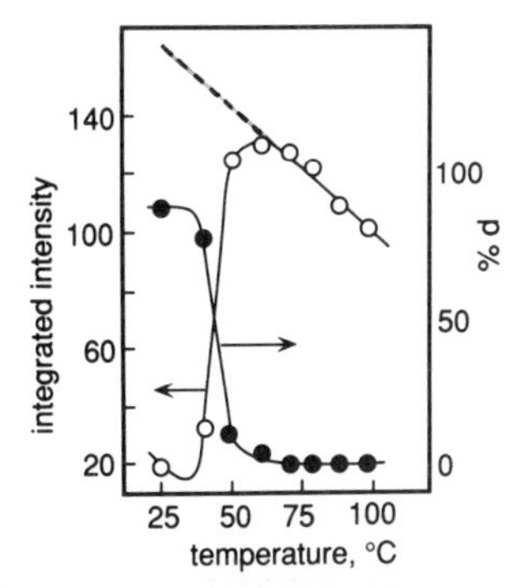

Figure 1.16 Temperature dependence of the integrated intensity of the OCH_3 proton band (○) and the fraction of associated monomer units (•) for sPMMA in toluene. The dotted line allows one to determine I_0 at the temperature where measurements are carried out (with permission from Spěvaček *et al.*, 1982)

and Schneider (1975) are reported in Table 1.1 together with values of the Flory interaction parameter, χ_1, determined for aPMMA.

This type of investigation shows that aggregation is dependent on both solvent type and stereoregularity. For instance, while strong aggregation occurs in toluene for a highly syndiotactic sample, this aggregation seems to be absent for a lower syndio tryad content and quite low for the isotactic sample. A similar trend is observed with *o*-dichlorobenzene. Conversely, noticeable aggregation is seen in acetonitrile for iPMMA but not for sPMMA. Acetonitrile is a θ-solvent for PMMA at 30°C. Interestingly and strangely enough, the aggregates of iPMMA formed in *o*-dichlorobenzene, although of lower content than in sPMMA, disappear at higher temperature.

Table 1.1 Values of associated units p at 27°C for solutions of isotactic PMMA (iPMMA, iso = 97%, hetero = 3%, syndio = 0%, M_n = 27 000) and syndiotactic PMMA (sPMMA1; iso = 2.5%, syndio = 88.5%, hetero = 9%, M_n = 170 000; sPMMA2; iso = 3.5%, syndio = 65%, hetero = 31.5%, M_n = 440 000). From Spěvaček and Schneider (1975).

	p (%)			
Solvent	iPMMA	sPMMA1	sPMMA2	χ_1
$CDCl_3$	2–12	0	0	0.377
C_6D_6	12–15	5	4	0.39
o-Dichlorobenzene	15	70–85	0	
CCl_4	20–24	(Insoluble)	40–48	
CD_3CN	18–23	5	—	

Spěvaček *et al.* (1982) also mention that aggregate stability depends upon both the length of time a sample is kept at a given temperature and the thermal history. The solution viscosity determined as a function of ageing time and concentration (see Figure 1.17) illustrates the kinetics of aggregate formation. While for the highest concentrations, η_{sp}/C eventually reaches a plateau, it keeps on increasing for the lowest concentrations. These authors do not comment on the maximum observed before the plateau regime.

Aggregate formation seems to proceed via two consecutive processes designated as primary and secondary. This can be demonstrated by using techniques that are sensitive to different phenomena such as NMR and turbidimetry. Sledaček *et al.* (1984) have shown that, while the fraction p of associated protons begins to level off, there is a sudden rise in turbidity of the sample. They account for these results by considering that intramolecular ordering takes place during the primary process and then aggregation of the ordered domains during the secondary process, causing the sharp increase in turbidity.

From light-scattering experiments (Mrkvickova *et al.*, 1983), it has also been pointed out that thermal history is important, particularly with iPMMA. For instance, in butyl acetate several heating and cooling cycles alter the scattering pattern changes, indicating better and better molecular dispersion which is retained, afterwards, at room temperature. They mention that only chloroform gives molecular dissolution without the need to go through heating and cooling cycles.

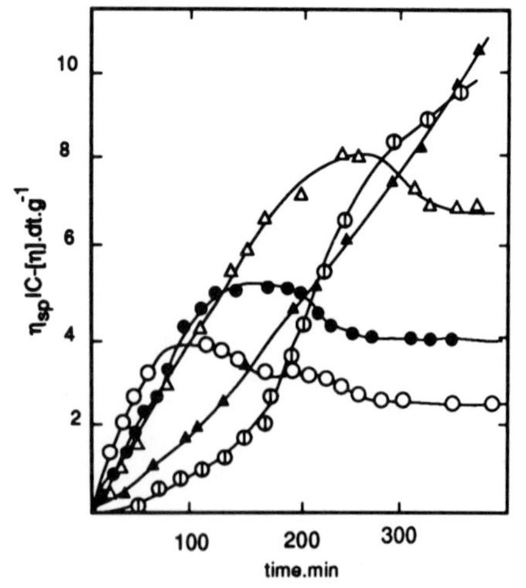

Figure 1.17 Dependence of η_{sp}/C-$[\eta]$ *vs* time ($[\eta] = 0.52$, determined at 60°C) for sample sPMMA-1 in toluene at 25°C: $C = 4.47$ (○), 2.79 (•), 2.03 (△), 1.4 (▲) and 0.86 (○) in g/l (with permission from Spěvaček *et al.*, 1982)

It is, however, worth emphasizing that, irrespective of the sample used, macroscopic dispersion readily occurs and only sophisticated techniques can reveal molecular aggregation. In addition, as explicitly mentioned by Spěvaček and Schneider (1975), the aggregation behaviour is not linked to solvent quality nor to solvent polarity (see for instance Table 1.1). For instance, aggregation can take place in a good solvent (good with respect to aPMMA) such as methyl ethyl ketone whereas the absence of aggregation is seen in a θ-solvent such as acetonitrile. This, as emphasized by the same authors, cannot be predicted by theoretical conformational analysis as performed by Sundararajan and Flory (1974) and Sundararajan (1977). It is also worth mentioning that Benoit and Picot (1966) observed the enhanced low-angle scattering in aPMMA solutions in methyl ethyl ketone, the same phenomenon as has been detected in aPS.

The absence of direct correlation between aggregation and solvent quality is most reminiscent of the situation encountered with polystyrene physical gels and physical aggregates.

The physical gel formation of iPMMA and/or sPMMA solutions has been studied from a thermodynamic point of view by Könnecke and Rehage (1981, 1983) and Berghmans *et al.* (1987) in *o*-xylene. According to Könnecke and Rehage (1981, 1983) gelation of iPMMA in this solvent is unusually slow. Hours of annealing at room temperature and below are required to obtain a gel for a concentration as high as 40% in polymer.

In contrast, gelation is much faster for sPMMA since a gel forms immediately on cooling as noticed by Berghmans *et al.* (1987). The gelation phenomenon can then easily be followed by DSC. In Figure 1.18, a typical get formation exotherm and gel melting endotherm are given for 10% solutions. They are quite similar to those obtained with polystyrene solutions. For more concentrated solutions Könnecke and Rehage (1983) report on the appearance of a second endotherm at lower temperature. Interestingly enough, these authors mention that this endotherm is invariant with concentration yet its magnitude increases with increasing annealing time (see Figure 1.19). Könnecke and Rehage have interpreted the low-melting temperature peak as arising from smaller and less perfect crystallites. However, in the light of results gained by He *et al.* (1987*a*) on block copolymers (see Section 1.2.2.d), it might be relevant and more appropriate to regard this peak as being due to the co-occurrence of a liquid–liquid phase separation during gel formation. In fact crystallization within a miscibility gap is known from the phase rule to give temperature-invariant transitions. The fact that clear gels are formed, which appears to be contradictory to a liquid–liquid phase separation, can be easily accounted for by the similar values of the solvent refractive index ($n_D = 1.505$) and the polymer refractive index ($n_D = 1.492$).

Berghmans *et al.* (1987) also show that the heat of gel formation and, correspondingly, the heat of gel melting are dependent on the sPMMA

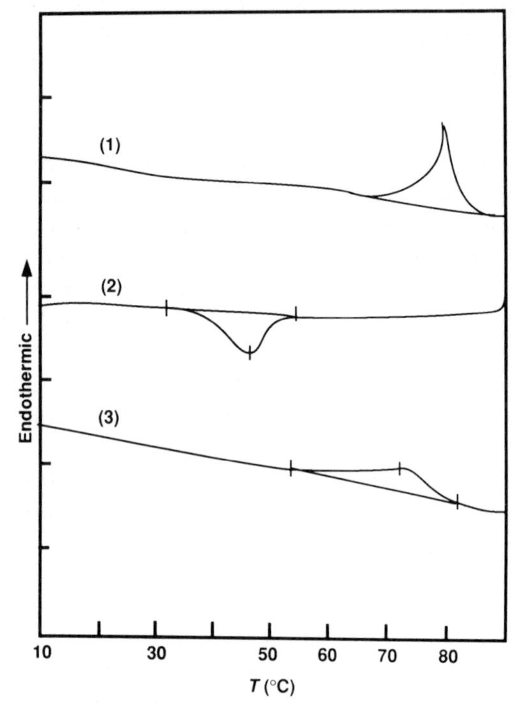

Figure 1.18 DSC thermograms of gels from sPMMA–*ortho*-xylene: (1) on heating, (2) on cooling and (3) on reheating (2) (with permission from Berghmans *et al.*, 1987, and the publishers Butterworth-Heinemann).

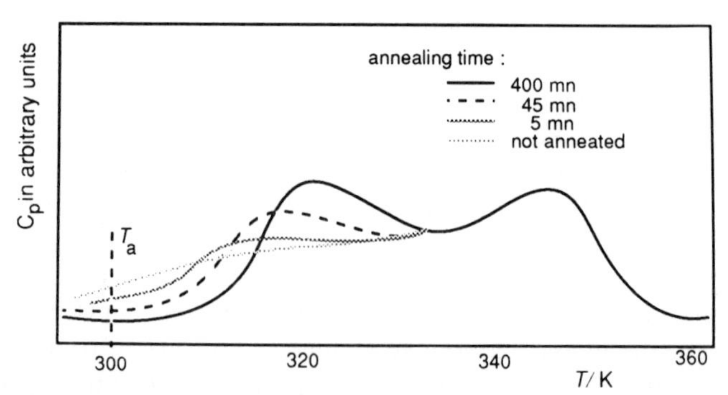

Figure 1.19 DSC trace of sPMMA–*ortho*-xylene gel after various annealing times at 300 K as indicated. Weight fraction of sPMMA = 0.4 (C_{pol} = 40%) (with permission from Könnecke and Rehage, 1983).

sample tacticity and vary from 6.4 J/g (0.15 cal/g) of gel for a sample with 67.5% of syndiotactic dyads up to 17.8 J/g (0.42 cal/g) of gel for a sample with 91.5% syndiotactic dyads. In contrast, the corresponding formation and melting temperatures are not significantly affected by tacticity (for a 10% solution: $T_{gel} = 50 \pm 3^\circ C$ and $T_m = 72.5 \pm 2.5^\circ C$).

The latter result is unexpected in the sense that the mean syndiotactic sequence length as calculated by Spěvaček *et al.* (1982) drops from about 22 tryads down to 5. In classical polymer crystallization such a discrepancy of the crystallizable sequence length should have altered the crystal melting temperature quite significantly. Different considerations, which will be developed in Chapter 2, led Berghmans *et al.* (1987) to disregard a mechanism of crystallization for sPMMA gelation in this solvent.

There are no complete phase diagrams for PMMAs. Tentative phase diagrams have been established by Könnecke and Rehage (1983) but they only show the high-melting temperature endotherm, especially at high concentrations (see for instance Figure 1.20). Interestingly enough, however, the plateau reached by the liquidus line at lower concentrations may suggest some invariant arising, as mentioned above for liquid–liquid phase separation.

Berghmans *et al.* (1987) also report on annealing effects on the sPMMA–*o*-xylene gels. Two cases are examined: (i) annealing just below and (ii) annealing just above the onset of the melting endotherm. In the former case annealing does not alter the melting behaviour significantly, while in the latter case one observes for the melting endotherm a sharpening together with a slight shift towards higher temperature (about 80°C against 72.5°C). Also, the annealing in the latter case gives enhanced melting enthalpies which suggests higher ordering.

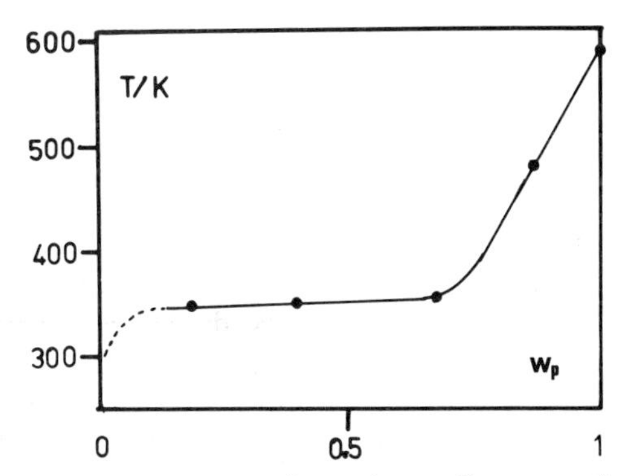

Figure 1.20 Temperature–concentration phase diagram of sPMMA–*ortho*-xylene (with permission from Könnecke and Rehage, 1983).

(ii) Gelation of the stereocomplex

In the early sixties Watanabe *et al.* (1961) found that mixtures of iPMMA and sPMMA in solution leads to the formation of a gel. Surprisingly, gelation occurs for solutions of iPMMA–sPMMA blends at room temperature in solvents, such as acetonitrile, in which gelation of either tactic PMMA is absent under the same conditions. Thus the mixture of sPMMA + iPMMA does not give additive properties but behaves as a new polymer. Liquori *et al.* (1965) investigated the aggregate formation by turbidimetry and optical density in the far-ultraviolet spectrum. They came to the conclusion that the most efficient composition occurs at ratio of iso/syndio of 1:2.

Further investigations were carried out by Biros *et al.* (1974) on the heat of formation of the complex as a function of the ratio iso to syndio in 1% dimethylformamide (DMF) solutions. By means of the Tamman's diagram, they came to the conclusion that the stoichiometry lies somewhere between iso/syndio = 1.5 and 2.

Some authors have reported a ratio of iso/syndio of 1:1 (Miyamoto and Inagaky, 1970; Borchard *et al.*, 1971; Kusakov and Mekenitskaya, 1973; de Boer and Challa, 1976; Shachovskaya and Kraeva, 1976). However, the more recent papers seem to agree in locating the stoichiometry of iso/syndio to 1.5–2 for highly tactic samples. For instance, Vorenkamp *et al.* (1979) eventually came to the conclusion by means of various methods that only one type of complex exists with a stoichiometry of iso/syndio of 1:2. Such a ratio is nicely accounted for by the complex molecular structure (see Section 2.3.1.c).

Concerning the solvent type, Liquori *et al.* (1965) also concluded that stereocomplex formation required polar solvents. This statement was first invalidated by Liu and Liu (1968) who demonstrated the possibility of preparing the stereocomplex in benzene. In addition, they observed by means of viscometry measurements that, most strikingly, complex formation occurs in both the non-polar benzene and the polar acetonitrile, but not in certain mixtures of these two solvents. For instance, a cosolvent system containing 80% benzene and 20% acetonitrile completely prevents complex formation. They also comment on the solvation power of the solvents. They emphasize that pure benzene, in which the complex is stable, solvates PMMA more strongly than a mixture of equal volumes of benzene and acetonitrile, in which the complex dissociates. This might be an indication of a polymer–polymer–solvent complex, a case not taken into consideration so far.

An additional study by Spěvaček and Schneider (1974) has clearly demonstrated that the extent of complex formation is not governed at all by solvent polarity or solvent quality. In Table 1.2 are given the content p of

Table 1.2 Values of the fraction p of associated units, the minimum length of associated sequences q_{min} (both measured at 25°C) and the temperature of total dis-aggregation T_m of stereocomplex aggregates in solvents of various dipole moments and solubility parameters. Mixtures of iso/syndio = 1:2 with iPMMA (iso = 97%, syndio = 0% and hetero = 3%) and sPMMA (iso = 3%, syndio = 66% and hetero = 31%). From Spěvaček and Schneider (1987).

Solvent	Debye	δ (cal$^{1/2}$/cm$^{3/2}$)	p (%)	q_{min}	T_m (°C)
CCl_4	0	8.6	95	3	95
CH_3CN	3.4	11.9	95	3	70
DMF	3.8	12.1	90	3	—
Toluene	0.37	8.9	80	4–5	85
o-Dichlorobenzene	2.5	10.0	38	9–10	70
Benzene	0	9.2	35–40	8–10	60
Benzaldehyde	2.8	9.4	25	12–13	60
$CHCl_3$	1.05	9.3	0	—	—

PMMA stereocomplexes formed in various solvents, as determined by NMR, together with the solubility parameter and the dipole moment. As can be seen, both acetonitrile (CH_3CN) and carbon tetrachloride (CCl_4) give the same values of p whereas they differ considerably with regard to their dipole moments. One is very polar (acetonitrile) while the other is not (carbon tetrachloride).

As with sPMMA, Spěvaček and Schneider (1987) found that a minimum length q_{min} for the syndiotactic sequences is required, which is about ten monomer units in aromatic solvents and about three in carbon tetrachloride or acetonitrile. Using terminology introduced originally by Challa *et al.* (1976), these authors consider that solvents for which $q_{min} = 3$ may be designated as strongly complexing, those for which $q_{min} = 10$ as weakly complexing and chloroform as non-complexing. They, however, hasten to add that this classification is not necessarily identical whatever the techniques used. For example, whereas toluene is regarded as a strongly complexing solvent from the NMR viewpoint, it appears to be weakly complexing from the calorimetry viewpoint (Biros *et al.*, 1974) where the heat of complex formation is found to be the lowest of all the systems investigated.

These results mainly concern the dilute systems where no macroscopic networks are formed. It would be of interest to examine the complexing power of a solvent in the gels.

Gel formation kinetics have been investigated by Katime and Quintana (1988*a*) for aggregates and gels in DMF. They have essentially used light scattering as a tool to follow the aggregation process by assuming direct proportionality between the scattered intensity extrapolated at zero angle and the amount of aggregates formed.

They have used two different syndiotactic samples that mainly differ through their tacticity (SW $M_w = 1.65 \times 10^5$, $s = 73\%$, $h = 26\%$, $i = 1\%$ and PO $M_w = 1.5 \times 10^5$, $s = 55\%$, $h = 40\%$, $i = 5\%$). As can be seen in Figure 1.21 the gelation kinetics, which are given by the derivative of these curves, are not very sensitive to temperature and to tacticity of the syndiotactic component. Yet, the magnitude of the zero-angle scattered intensity is larger with the syndiotactic sample of highest syndiotactic tryad content. This most probably indicates the amount of physical junctions to be larger with the syndiotactic sample for which the syndiotactic sequences

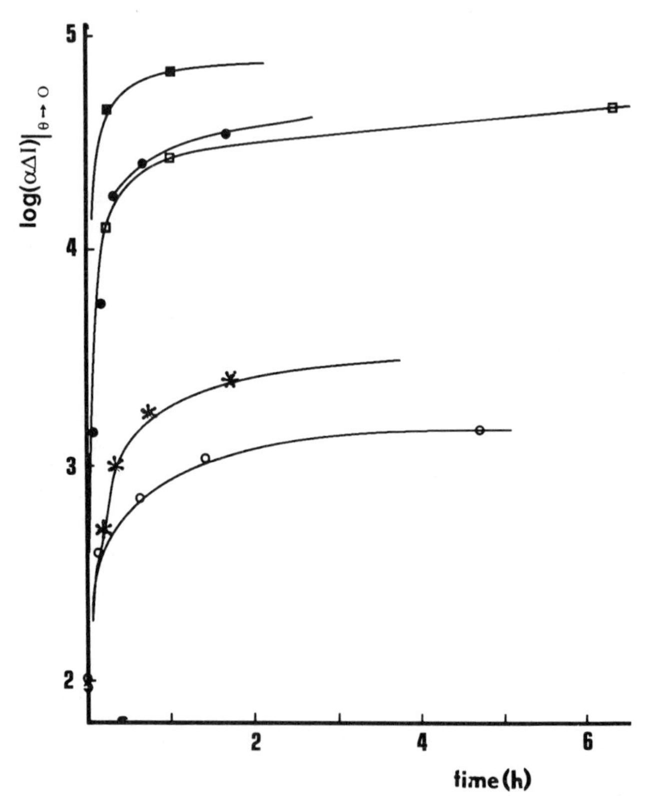

Figure 1.21 Time dependence of the scattered intensity extrapolated to zero angle as a function of tacticity and temperature of gel formation for sPMMA–iPMMA stereocomplex (ratio 2:1) in DMF for $C = 2 \times 10^{-2}$ g/cm^3. JR/SW at 273 K (■), 298 K (•) and 323 K (*); JR/PO at 273 K (□) and 298 K (○).

PMMA	M_w	M_w/M_n	i	h	s
JR	4×10^5	1.34	90%	7%	3%
SW	1.65×10^5	1.23	1%	26%	73%
PO	1.5×10^5	1.52	5%	40%	55%

(with permission from Katime and Quintana, 1988a).

are the longest, a statement in agreement with the conclusions derived by Spěvaček and his co-workers (see for instance Spěvaček and Schneider, 1987).

(c) Miscellaneous poly(alkyl acrylates)

In a series of papers, Wolf *et al.* (Jelich *et al.*, 1987; Nunes *et al.*, 1987; Nunes and Wolf, 1987) have reported on the gelation of preferentially syndiotactic (syndio = 60% and iso = 4%) poly(*n*-butyl methacrylate) (PBMA) samples in poor solvents (propane-2-ol and ethanol) for which the θ-temperature lies near or slightly below room temperature. By means of

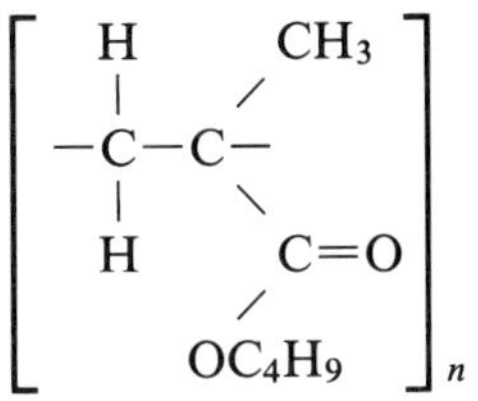

$$\left[-\underset{\substack{|\\ H}}{\overset{\substack{H\\ |}}{C}}-\underset{\substack{\diagdown \\ C=O \\ \diagup \\ OC_4H_9}}{\overset{\substack{CH_3 \\ \diagup}}{C}}- \right]_n$$

different techniques, they established the temperature–concentration phase diagram in propane-2-ol (Figure 1.22). From the measurement of the apparent viscosity, η_a, as a function of temperature, they determined the cloud point curve and, by extrapolation to zero of the velocity $v_{(T)}$ of a falling body, the gel melting curve. Details on the gel formation are, however, not provided so that information about the magnitude of the gel formation–gel fusion hysteresis, if any, is not available. They further mention that no melting endotherms can be detected from DSC investigations which led them to the conclusion that no crystallization mechanism is involved in the gelation process of this polymer, at least under these conditions. As will be discussed in Section 1.2.2.b, it is worth stressing that failure to detect a melting endotherm is by no means a sufficient criterion to state beyond doubt that crystallization is not involved.

Nunes *et al.* (1987) have determined the solution properties, i.e. the Flory interaction parameter χ_1, of their gelling systems and compared them with those of aPS–cyclohexane for which no gelation occurs. By means of light scattering in dilute solutions, vapour pressure measurements in concentrated solutions and inverse gas chromatography at the limit of the pure polymer, the variation of χ was established for the whole range of concentrations. Then the enthalpic and entropic contributions to χ were derived by the following two equations:

$$\chi = \chi_H + \chi_S \tag{1.12}$$

$$\chi_H = \Delta H_1 / RT\varphi^2 = -T(\partial\chi/\partial T) \tag{1.13}$$

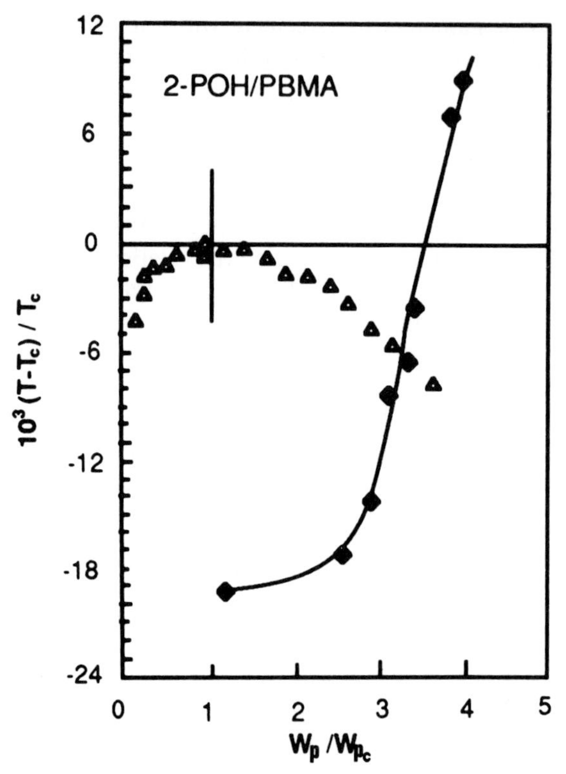

Figure 1.22 Phase diagram of poly (*n*-butyl methacrylate)/2-propanol reduced to the thermodynamic critical conditions for two different polymer fractions [T_c and w_{2c} being determined at the maximum of the cloud point curves, $T_c = 22.3°C$ and $w_{2c} = 0.082$ for $M_w = 5.2 \times 10^5$ ($M_w/M_n = 1.17$), while $T_c = 21.6°C$ and $w_{2c} = 0.086$ for $M_w = 4.7 \times 10^5$ ($M_w/M_n = 1.11$)]. △ stands for the cloud point curve, the gelation curve (solid line) stems from sedimentation measurements (with permission from Jelich *et al.*, 1987).

in which ΔH_1 is the enthalpy of dilution and φ the polymer volume fraction.

Whereas for aPS–cyclohexane solutions χ_H and χ_S are always positive and increase monotonically with polymer concentration, for propane-2-ol–PBMA and ethanol–PBMA the variation of χ_S as a function of concentration is seen to display a maximum, then to assume very negative values. Nunes *et al.* (1987) accordingly conclude that there is most probably cointercalation of the solvent between polymer chains which is another way of describing a polymer–solvent compound.

Gelation of long-side-chain poly(alkyl acrylates) has been investigated by Shibayev *et al.* (Plate and Shibayev, 1971, 1974; Tal'roze *et al.*, 1974; Borisova *et al.*, 1980, 1984). Special attention has been paid to poly(cetyl acrylate) (PA-16), poly(octadecyl acrylate) (PA-18) and poly(docosyl acrylate) (PA-22). The solvents used were hydrocarbons (C_nH_{2n+2}) and aliphatic alcohols ($C_nH_{2n+2}O$) of different numbers of carbon atoms.

$$\left[\begin{array}{llll} & \mathrm{H} & \mathrm{H} & \\ & | & | & \\ - & \mathrm{C} & -\mathrm{C}- & \\ & | & \quad\backslash & \\ & \mathrm{H} & & \mathrm{C{=}O} \\ & & & / \\ & & & \mathrm{OC_nH_{2n+1}} \end{array}\right]_p \qquad n = 16, 18, 22$$

Tal'roze *et al.* (1974) have found that the gel melting point increases with the length of the hydrocarbon solvent, whereas it is virtually constant in the case of the aliphatic alcohols (see Figure 1.23*a*). They have also determined the melting enthalpies as a function of polymer concentration (Figure 1.23*b*). The melting enthalpies display the same linear variation with polymer concentration C_{pol} regardless of the number of carbon atoms in the alcohol molecule and, once extrapolated to $C_{pol} = 1$, yield a value of 19 cal/g which is very close to that measured in bulk-crystallized samples ($\Delta H_{cry} = 20$ cal/g). Conversely, the variation, although linear, is different in cetane and yields $\Delta H = 27$ cal/g, a value much higher than the crystal melting enthalpy. The possibility of the existence of polymer–solvent compounds was evoked some years later by Borisova *et al.* (1980) when they discovered by means of NMR spectroscopy the existence of two populations of solvent molecules with definite distinct mobilities: the one occluded in the crystal structure possessing limited mobility and the other one located outside the junctions with normal mobility.

(d) Poly(4-methylpentene-1)

According to Charlet *et al.* (1984) the gelation conditions of poly(4-methylpentene-1) (P4MP1) depend strongly on the thermal history of the sample and particularly whether a nascent polymer (as-received) or a once-molten polymer is employed. A striking example of the thermal history is

$$\left[\begin{array}{lllll} & \mathrm{H} & \mathrm{H} & & \\ & | & | & & \\ - & \mathrm{C} & -\mathrm{C}- & & \\ & | & \quad\backslash & & \\ & \mathrm{H} & & \mathrm{CH_2} & \\ & & & / & \\ & & \mathrm{CH} & & \\ & & / \ \backslash & & \\ & \mathrm{CH_3} & & \mathrm{CH_3} & \end{array}\right]_n$$

the following: on heating an as-received sample (N) in cyclohexane a gel is formed at the dissolution temperature T_D whereas, under the same

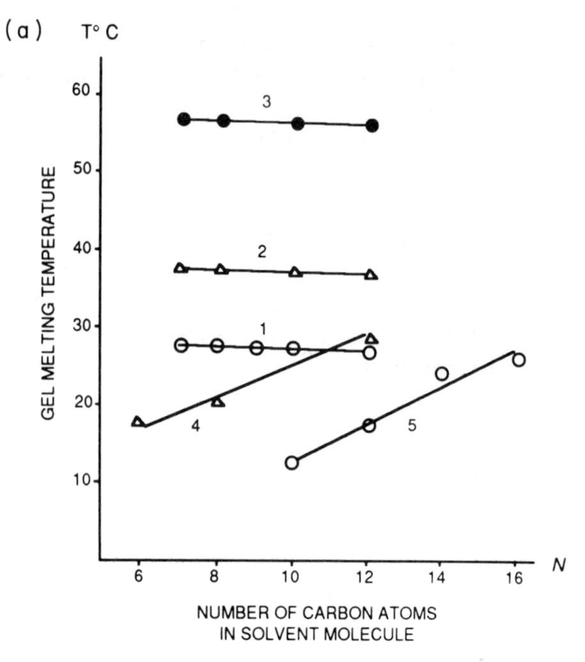

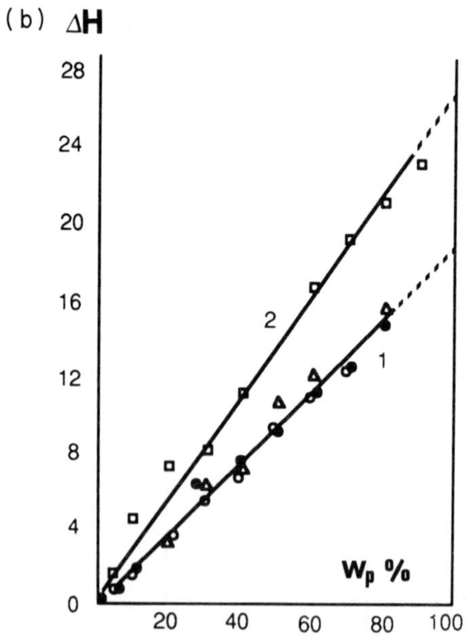

Figure 1.23 (a) Melting points for PA-16 (lines 1 and 5), PA-18 (lines 2 and 4) and PA-22 (line 3) gels in aliphatic alcohols (1, 2, 3) and hydrocarbons (4, 5) as a function of the number of carbons in the solvent molecule. (b) Melting enthalpy of PA-16 gels in alcohols and cetane as a function of the polymer concentration (• octanol, ○ decanol, △ dodecanol and □ cetane) (with permission from Tal'roze *et al.*, 1974).

conditions of temperature and concentration, a once-molten sample (M) gives a fluid solution. They also came to the conclusion that the capability of forming a gel at the polymer dissolution temperature is a special feature of P4MP1 unlike many other polyolefins for which supercoolings are usually required (that is, the gel usually forms at a lower temperature than the chain-folded crystals). Results of Charlet *et al.* (1984) reported in Figure 1.24 lead to some interesting conclusions: (i) the gels are remarkably stable with respect to the solvent boiling points (experiments were carried out in sealed tubes) and (ii) while the gelation temperature is lower for the once-molten sample (M), the gel melting temperature is clearly not affected by the thermal treatment applied to the sample.

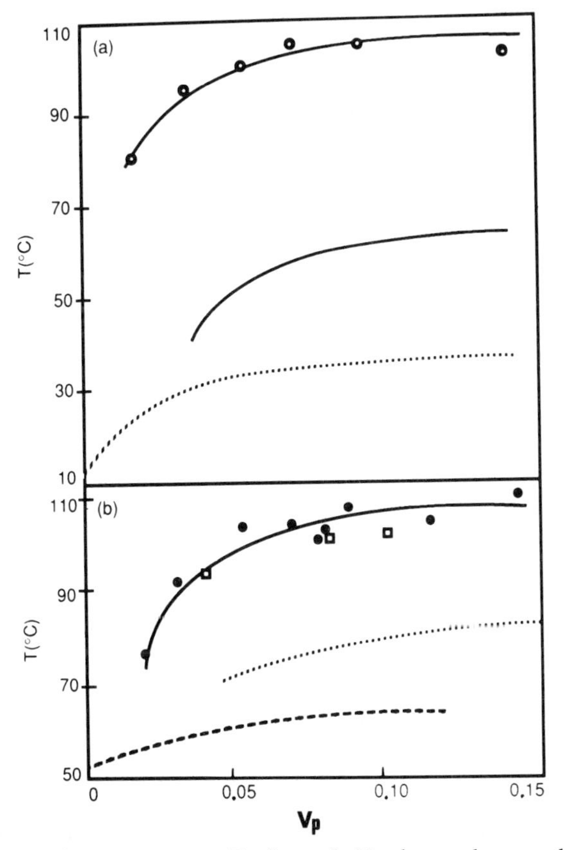

Figure 1.24 Solation temperature T_s of sample N gels *vs* polymer volume fraction v_p in *cyclo*-pentane (a) and *cyclo*-hexane (b). The dissolution temperature T_D of the chain-folded crystals of samples M (...) and N (– – –) is also plotted *vs* v_p. □ represents solution temperatures from protogel obtained from M (with permission from Charlet *et al.*, 1984).

However, the same authors hasten to add that heating (above gel melting and cooling (25°C)) cycles strongly influence the gelation habit. This habit is particularly sensitive to the solvent type. The use of cyclopentane instead of cyclohexane brings about dramatic changes.

Figure 1.25 gives the formation and melting temperatures after a second heating and cooling cycle in cyclohexane. As can be seen, the melting temperatures have noticeably dropped by about 30°C. In addition, the resulting gel is rather weak and is easily destroyed by simple shaking which eventually leads to polymer deposition. Charlet *et al.* (1984) conclude that the temperatures observed may represent the crystallization and not the gelation of the polymer. It is worth mentioning that cooling at 50°C instead of 25°C leads again to the formation of a rigid gel. There seems to be competition between chain-folded crystal growth and gelation in this solvent, the gel forming at higher temperature unlike in polystyrene systems for instance.

Conversely, the gelation habit turns out to be different in cyclopentane. Gel melting temperatures also decrease but not so dramatically as with cyclohexane. Besides, these temperatures are virtually the same independent of whether N or M samples have been used. Moreover, the gels still possess high rigidity, do not crumble upon shaking and are produced with the same or nearly the same characteristics independent of the cooling temperature

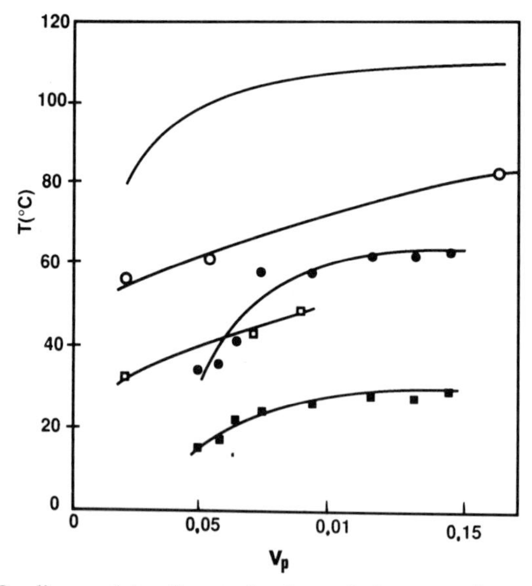

Figure 1.25 Cooling and heating cycles in *cyclo*-hexane mixtures. The gelation (□) and second solation (T_{S2}, ○) temperatures are plotted *vs* v_p for N (hollow) and M (solid symbols). For M systems, the second gelation and third solation temperatures are indistinguishable from those above (with permission from Charlet *et al.*, 1984).

(25°C or 50°C). In this solvent no competition between gelation and crystal growth occurs.

These results led Charlet *et al.* (1984) to infer that solvent is probably occluded in the physical junctions which means the existence of a polymer–solvent compound.

Later, Phuong-Nguyen and Delmas (1985) studied the heat of formation of the gels and pregels by calorimetry. Basically, the heat of formation, h_f per g of polymer can be calculated as:

$$h_f = h_{melting} + h_{mixing} \quad (1.14)$$

in which $h_{melting}$ is the heat of fusion of the crystals and h_{mixing} the heat of mixing of the amorphous polymer with the solvent. Introducing the heat of interaction h^E_{int}, which is the parameter of interest, one ends up with:

$$h^E_{int} = h_f - d_p(1 - v_p)^{-1} h_{melting} \quad (1.15)$$

in which v_p and d_p are the volume fraction and the density of the polymer respectively.

Typical results obtained as a function of temperature and concentration are given in Figure 1.26. As can be seen, there is a definite jump near $v_p = 0.02$ which turns out to be the critical gelation concentration. If the heat of gel formation and aggregate formation were to display the same variation it should be expressed theoretically by (Phuong-Nguyen *et al.*, 1982):

$$h^E_{int} = A v_p (B + v_p)^{-1} \quad (1.16)$$

in which A and B are constants. A, which is negative, is the heat of maximum interaction while B, which is positive, defines the value of v_p for which $h^E_{int} = A/2$. The steep decrease in h^E_{int} as a function of v_p,

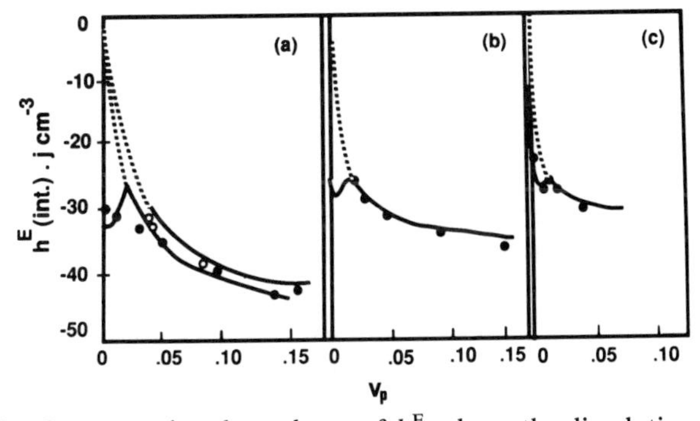

Figure 1.26 Concentration dependence of h^E_{int} above the dissolution temperature T_D of chain-folded crystals in *cyclo*-pentane at 25°C (a), 40°C (b) and 45°C (c). h^E_{int} determined in the gel region at 30°C has been added to (a) (hollow circles). The dotted lines are calculated from Figure 1.16 to fit the curves in the gel region ($v_p > 0.02$) (with permission from Phuong-Nguyen and Delmas, 1985).

Phuong-Nguyen and Delmas insist, is indicative of strong polymer–solvent and polymer–polymer–solvent interactions.

According to these authors the sharp change in h^{E}_{int} which occurs near the critical gel concentration is due to a change in quality of the junctions from the dilute to the gel phase. The P4MP1 gels would then not result from the appearance in solution of a larger number of cohesive junctions but would occur through a change in the state of aggregation. They further conclude that the large exothermic heats of interaction and their dependence on polymer concentration supports quite definitely the solvent-induced chain association.

(e) Poly(γ-benzyl-L-glutamate)

Poly(γ-benzyl-L-glutamate) (PBLG) is a synthetic polypeptide possessing the following chemical structure:

$$\left[\begin{array}{c} \mathrm{O} \\ \| \\ -\mathrm{NH}-\mathrm{CH_2}-\mathrm{C}- \\ | \\ [\mathrm{CH_2}]_2 \\ | \\ \mathrm{C}=\mathrm{O} \\ | \\ \mathrm{O} \\ | \\ \mathrm{C_6H_5} \end{array}\right]_n$$

Unlike the polymers discussed so far, PBLG is not a flexible chain but a fairly rigid one. This is why liquid crystalline phases can be obtained with this polymer.

This polypeptide is known to form gels in some solvents whereas it gives at least three different crystalline forms in others or liquid crystalline phases (nematic, cholesteric) depending upon concentration and temperature. Incidentally, these different crystalline forms depend strongly upon the solvent type (see for instance Watanabe *et al.*, 1981).

Doty *et al.* (1956) have shown that associations in dilute solutions depend upon the solvent type: solvents with very low or zero hydrogen-bonding potential (chloroform, dioxane, etc.) lead to aggregation while hydrogen-bonding solvents (e.g. dichloroacetic acid) prevent it.

The study of the gelation phenomenon has been restricted to a few solvents so far. Information is available for gels prepared in DMF (Miller *et al.*, 1978; Ginzburg *et al.*, 1985), benzyl alcohol (Sasaki, S. *et al.*, 1982, 1983; Hill and Donald, 1988) and toluene (Miller *et al.* 1978; Russo *et al.*, 1987).

A temperature–concentration phase diagram of the system PBLG–benzyl alcohol has been established by Sasaki, S. *et al.* (1983) on the basis of DSC experiments (Figure 1.27). Two types of structure are obtained: A-type gel and B-type gel. The A-type gel is obtained by quenching the solution between 60 and 48°C. The melting points of A-type gels fall on the liquidus line labelled a. Now, if the solution of concentration larger than v_2^{**} is quenched repeatedly from 61–62°C to below 48°C, B-type gels are formed whose melting temperatures are given by liquidus line b (the two types of gel have different diffraction patterns, the more ordered being B-type gels; see Chapter 2). This is an unexpected result if one considers that a gel quenched to lower temperature is liable to produce less stable structures of lower melting point.

Sasaki, S. *et al.* (1983) account for this unexpected result by invoking the presence of a miscibility gap, as described by Flory (1956) at low temperature. Suppose a solution of concentration v_0 is quenched to 55°C. According to these authors a complex-phase, polymer + solvent compound, will form. This is substantiated by the polymer volume fraction in the phase which is 0.76 independent of the value of v_0. Alternatively, if the system is quenched at lower temperature so as to enter the miscibility gap (they estimate this temperature as 48°C), then a liquid crystalline phase v_{lc}^* is formed which is rapidly transformed into a crystalline phase. Measurements of the polymer volume fraction give 0.95 which explains the difference in melting point between A-type and B-type gels in terms of solvation.

Hill and Donald (1988) report on the observation of three peaks in DSC experiments. They then consider that the biphasic region of the Flory-type phase diagram should be regarded as triphasic. This hypothesis implies Gibbs phase rule to be broken. In fact, in a two-component system the domain where three phases coexist is reduced to a point (fixed temperature and concentration, see Appendix 1). Gibbs phase rule can only be temporarily broken if one phase is metastable and spontaneously transforms into one of the two other phases.

The observation of three thermal events may actually hint at the occurrence of some solid–solid transformation, as has been already seen in concentrated solutions (see for instance McKinnon and Tobolsky, 1968) (for instance a metatectic melting is liable to give three melting endotherms; see Appendix 1 for further details).

Other studies on PBLG in DMF and toluene by Miller and coworkers (Miller *et al.*, 1978; Russo *et al.*, 1987) led to the conclusion that liquid–liquid phase separation was involved. Yet Russo *et al.* (1987) have not confirmed early assumptions that the mechanism of liquid–liquid phase separation occurs through spinodal decomposition, as was formerly suggested by Tohyama and Miller (1981). Here also a triphasic region is

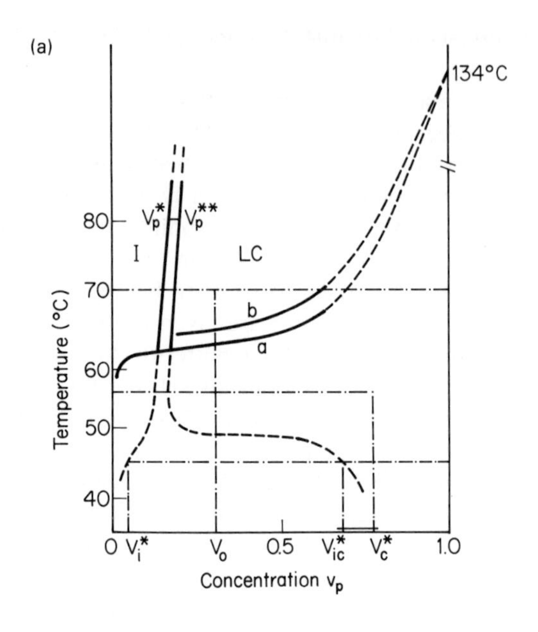

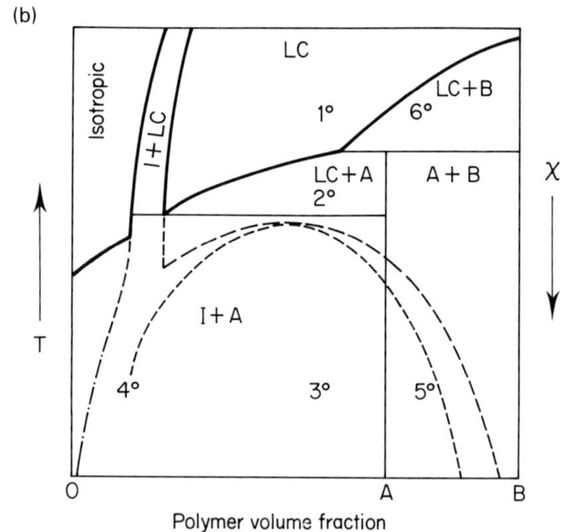

Figure 1.27 Temperature-concentration phase diagrams for poly [γ-benzyl-L-glutamate] in benzyl alcohol (I = isotropic region and LC = liquid-crystalline region: (a) from Saski *et al.* (with permission from Sasaki *et al.*, 1983); (b) from Cohen *et al.* According to these authors, the diagram is in accord with the observed phase transitions. Interestingly, it takes into account the formation of polymer-solvent compounds. A = complex A, and B = complex B. The numbered points correspond to: 1, the liquid-crystalline phase at 80°C; 2, complex B appearing by gelation at 50°C, quenching from 80°C (point 1) to 20°C (point 3) results in a liquid–liquid phase separation with a metastable liquid-crystalline phase (point 5) and a dilute isotropic phase (point 4). Further annealing at 70°–80°C results in the appearance of complex B (point 6) (with permission from Cohen *et al.*, 1991).

considered by Ginzburg *et al.* (1985) the relevancy of which should also be examined cautiously in light of Gibbs phase rule.

The work of Sasaki *et al.* (1983) suggests that gelation of PBLG is solvent-induced in the sense that a polymer–solvent compound is formed. This point is less clear with DMF and toluene for which no thermal analysis data are available. In particular, knowledge of the high-polymer concentration side of the phase diagram would be invaluable in throwing light on this question.

It should be noted that the temperature–concentration phase diagrams established so far are incomplete in the sense that no liquidus line related to the gel formation has been tentatively drawn (liquid–solid transition). Accordingly, whether there is liquid–liquid phase separation at the early stage might be important but this does not throw light on the way this phase separation is eventually frozen in at some later stage by some liquid–solid phase separation. Only investigations by thermal analysis, such as those made on polystyrenes or block copolymers, could provide some answers to these questions, keeping in mind that Gibbs phase rule must be obeyed even for systems formed out of equilibrium (Koeningsveld *et al.*, 1990). Recently, Dagan and co-workers (Dagan *et al.*, 1991) have proposed to draw some liquidus lines that are presented on Figure 1.27b. This phase diagram qualitatively accounts for most of the experimental results on the PBLG/benzyl alcohol system.*

1.2.2 Crystallization-induced gels

As already mentioned in the introduction of this chapter, polymers such as polyethylene or poly(vinyl chloride) fall into this category. Here, gel formation is very little sensitive to the solvent type, at least in the early stage, and is mainly controlled through crystallization.

As a rule, polymers characterized by high crystallization rates and high degrees of crystallinity in the bulk state will be found here. Concerning atactic polymers or copolymers, the crystallizable sequence involved in the gel junction formation obeys this definition. For instance, 100% syndiotactic PVC is a highly crystalline polymer, as is syndiotactic poly(vinyl alcohol).

(a) Polyethylene

Polyethylene possesses the simplest chemical structure ($[CH_2]_n$). This polymer crystallizes readily with crystallinity rates reaching 90%. The

---

* I only became aware of this work, which answers my concern, after submitting the manuscript of this book.

melting point lies in the range 100–140°C depending upon crystallization conditions. The glass transition temperature for an amorphous sample is $T_g = -85 \pm 5$°C.

Gelation of polyethylene occurs under special conditions. Pennings (1977) mentioned that physical gels can be obtained by stirring 1% solutions in *p*-xylene. Later, Barham *et al.* (1980) showed that gelation can also be obtained under quiescent conditions provided that the solution be stirred at high temperature and then cooled down to lower temperature. Figure 1.28 shows the maximum temperature at which the solution can be stirred to produce a gel as a function of polymer concentration. Stirring beyond $T \simeq 127$°C, however, no longer produces gelation. Narh *et al.* (1982) studied in more detail the effect of solution stirring by investigating the apparent viscosity measured in a Couette apparatus as a function of stirring time. As already reported by Pennings (1977), the apparent viscosity as a function of stirring time (or shear stress as a function of stirring time in the case of Pennings) may exhibit a maximum (Figure 1.29a). This effect was recognized in the early sixties by Lodge (1961), Peterlin and Turner (1963, 1965) and Peterlin *et al.* (1965) and are thought by these authors and Narh *et al.* (1982) to arise from an increase in the number of 'entanglements': Narh *et al.* (1982) use quotation marks to emphasize that these 'entanglements' are certainly not of the usual type.

Two types of effects on the variation of apparent viscosity *vs* stirring time were investigated by Narh *et al.* (1982): effect of temperature and effect of solution concentration. All the experiments were carried out with high molecular weight polyethylene ($M_w = 1.5 \times 10^6$) which is a prerequisite to achieve gelation. Figure 1.29(b) shows the apparent viscosity as a function

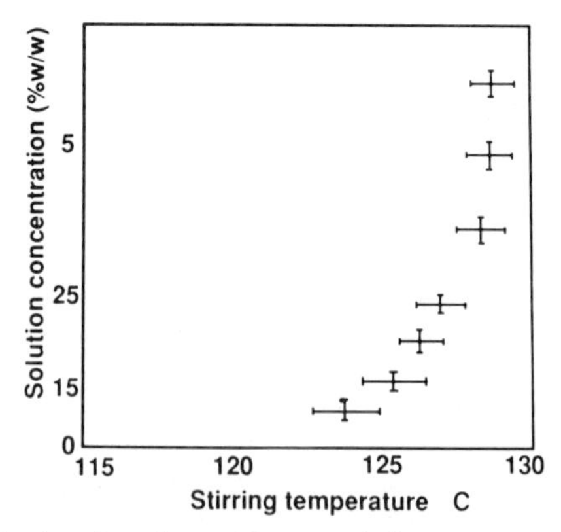

Figure 1.28 Graph showing the maximum stirring temperature from which solutions will form gels on cooling as a function of the polymer concentration. Error bars indicate the uncertainty in determining the final state of the product and the actual polymer concentration (with permission from Barham *et al.*, 1980).

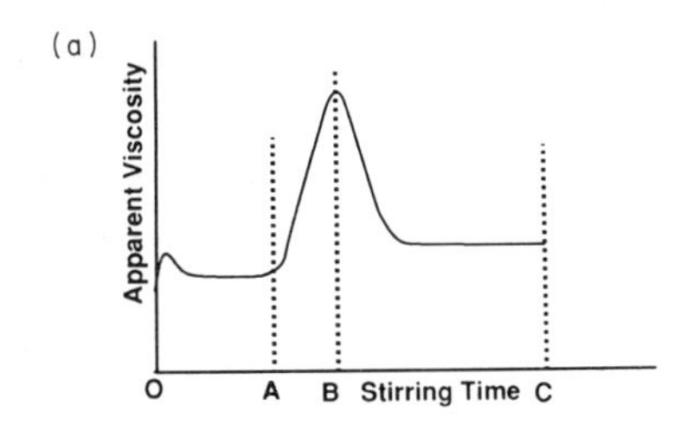

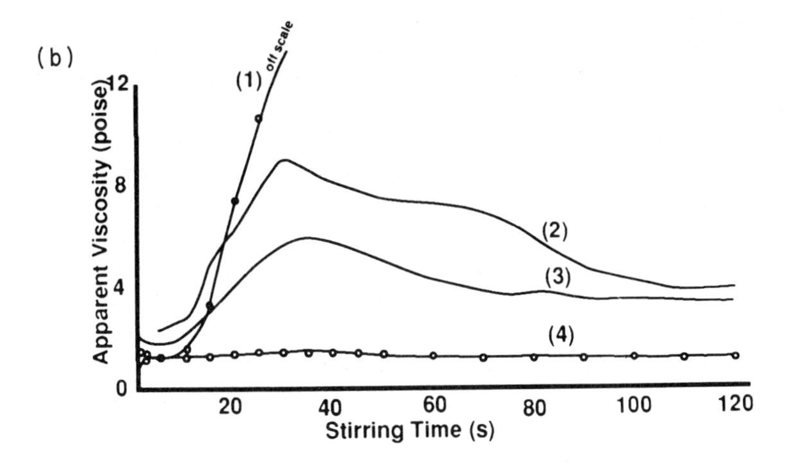

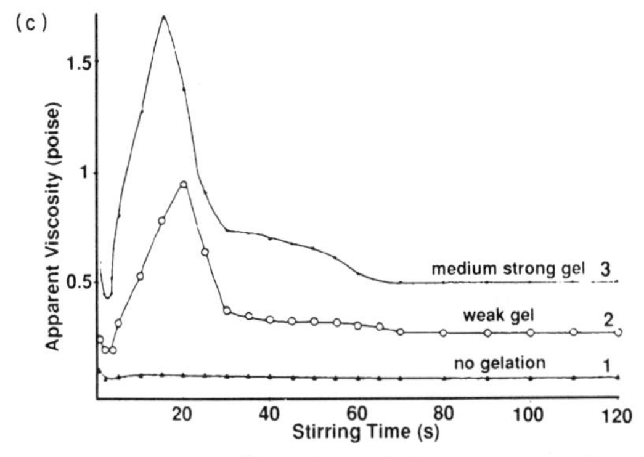

Figure 1.29 (a) Schematic representation of the apparent viscosity *vs* time. (b) Apparent viscosity as a function of stirring time for a 0.6% (w/w) solution of polyethylene ($M_w = 1.5 \times 10^6$) in decalin for a series of stirring temperatures, T_s: (○) = 115 °C, (△) = 120.8 °C, (x) = 124 °C and (□) = 140 °C. (c) Apparent viscosity as a function of stirring time for the same polyethylene sample at 124 °C but different concentrations (w/w): (•) = 0.4%, (○) = 0.2%, (△) = 0.1% (with permission from Narh *et al.*, 1982).

of stirring time for a concentration of 0.6% at different temperatures in decalin (earlier experiments by Pennings (1977) and Barham *et al.* (1980) were performed in xylene). This experiment shows clearly that the absence of a maximum is correlated with the failure to produce a gel. The same conclusion can be drawn from experiments carried out as a function of concentration at a given temperature (Figure 1.29c). Below a given concentration, no peak is seen and, correspondingly, no gelation occurs whatsoever. Narh *et al.* (1982) also found that, depending upon the moment at which stirring was stopped, a gel was formed or not. As shown in Figure 1.29(c), they define three domains: OA, AB and BC. They found that gelation occurs in regions AB and BC, the gels formed in BC being stronger than those formed in AB. Conversely, in domain OA only single crystals are obtained on cooling the solutions to room temperature.

Narh *et al.* (1982) attribute the onset of the increase in the apparent viscosity to the build up of network particles and the maximum to the break up of this network once it has grown right across the gap of the Couette apparatus.

Another interesting phenomenon reported by the same authors is the persistence of the gelation capability of the solution, once properly stirred, even if kept at high temperature. Figure 1.30 shows the time required to destroy the solution memory as a function of the temperature at which the solution has been stirred. This maximum memory time can be a matter of hours and is therefore considerably longer than would be expected with normal relaxation rates at such concentrations and temperatures.

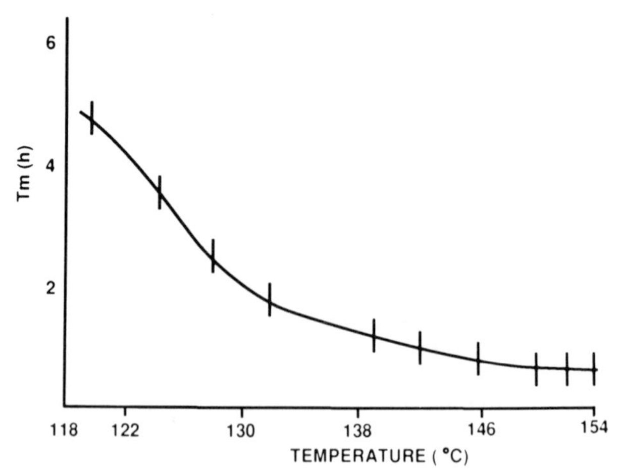

Figure 1.30 Time, τ_m, required to destroy memory of stirring effect as a function of solution temperature. High molecular weight polyethylene $C_{pol} = 0.6\%$ (w/w) (with permission from Narh *et al.*, 1982).

Smith and Lemstra (1980*a*) studied the thermal properties of gels formed by pumping a 2% (w/w) solution through a capillary into a cold water bath. Typical DSC thermograms are drawn in Figure 1.31 for different heating rates. Depending upon the heating rate, two maxima can be seen. The higher-melting peak grows at the expense of the lower-melting peak as the heating rate is decreased. Smith and Lemstra (1980*a*) concluded from this type of evolution that recrystallization was taking place, indicating a low polymer crystallinity in these gels.

Narh *et al.* (1982) obtained basically the same type of results in the temperature range investigated by Smith and Lemstra (1980*a*) but observed a broader endotherm at higher temperature (starting typically near 110°C and ending near 150°C) (Figure 1.32). They therefore suggest that this broad endotherm does correspond to gel melting whereas the sharp ones at lower temperatures correspond to the melting of chain-folded crystals undergoing some kind of fusion–recrystallization to account for the double peak. They also show that keeping the gel at 150°C for 2 h leads only to the formation of chain-folded crystals.

X-ray diffraction carried out on the wet gels by Smith and Lemstra (1980*a*) revealed some weak (as expected from the low crystallinity) yet sharp reflections which were identified with the 110 and 200 reflections of the orthorhombic crystal structure of polyethylene.

So far the gel junctions are thought to be of crystalline nature, although, to my knowledge, there have been no experiments carried out at 110°C, a temperature at which no chain-folded crystals are supposed to be left. In

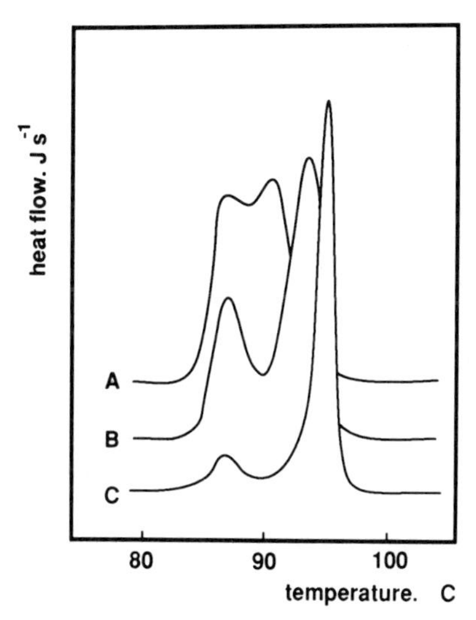

Figure 1.31 DSC endotherms of polyethylene–decalin gel. Heating rate: A = 40°C/min, B = 20°C/min and C = 5°C/min (with permission from Smith and Lemstra, 1980).

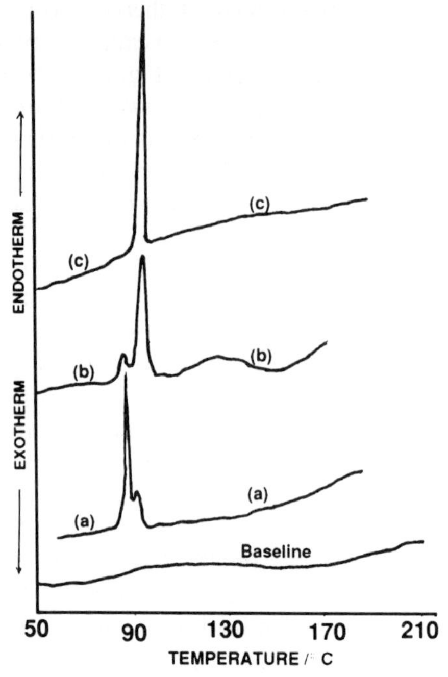

Figure 1.32 DSC endotherms for both polyethylene single crystals and gels in decalin: (a) single crystals suspension, (b) gel in decalin and (c) same as (b) but kept at 150°C for 2 h before cooling. Heating rate = 5°C/min (with permission from Narh *et al.*, 1982).

fact it is not yet known to what extent the diffraction pattern is due to the chain-folded crystals.

(b) Poly(vinyl chloride)

Poly(vinyl chloride) (PVC: $[C_2H_3Cl]_n$) is mainly obtained in the atactic form, although some highly syndiotactic material can be prepared at low temperatures (only short chains with a syndiotacticity close to 100% have been synthesized so far). Depending upon tacticity, the melting point of the crystals formed in the bulk state vary from 250°C to 300°C. It should be emphasized that, at such temperatures, melting is followed by the onset of chemical decomposition. The glass transition temperature of the atactic version is located near 80°C.

PVC usually forms physical gels in a large variety of solvents. So great is this capability that it is in fact difficult to find solvents in which molecular dispersion can be achieved. Tetrahydrofuran and cyclohexanone are two, at least at low concentrations.

As will be seen below, the early stage of gelation is little dependent upon solvent type although the later stage is. It should be emphasized that being discussed here is the gelation ability of the so-called 'atactic' version of PVC. NMR determination of tacticity gives the following figures for the tryads: iso = 0.18, syndio = 0.33 and hetero = 0.49.

It is difficult to trace back the discovery of the thermoreversible gelation property of PVC solutions. The first studies that are systematically quoted were carried out by Stein and Tobolsky (1948) and Aiken *et al.* (1947, 1949). From these studies crystallites were identified as constituting the gel junctions. As will be seen in Chapter 2, many confirmations of this statement have been obtained by many types of experiment and technique.

Curiously enough, although many investigations on the mechanical properties of the gel, such as compression modulus (Walter, 1954), have been carried out, detailed information on gel formation conditions have only been obtained during the past few years.

PVC gels are to some extent unique in the sense that, whereas the formation conditions are important, the ageing process after formation is as important. This process turns out to be very slow and can exceed days or even months. Such behaviour has not been reported for other polymers, at least to the same extent as for PVC. Only gelatin, a biopolymer, shows a comparable ageing phenomenon. This section will be subdivided into two subsections, the former dealing with the early stage of gelation and the latter with ageing.

(i) Early stage of gelation

Unlike other polymers, the gelation of PVC solutions cannot be straightforwardly followed by thermal analysis. In fact, no gel formation exotherm can be observed on cooling a solution nor can any melting endotherm be seen on heating a gel freshly prepared in the DSC sample pan (the situation encountered here is different from that seen with isotactic polystyrene for which a formation exotherm can always be observed). These observations were made by several authors on different PVC–solvent systems (Dorrestijn *et al.*, 1981; Yang and Geil, 1983; Mutin and Guenet, 1989). Mutin and Guenet (1989) further report that a high melting endotherm can be observed from a piece of gel obtained in a separate container, transferred into the DSC pan and scanned as usual. Yet, if a gel sample, once molten and cooled to room temperature in the DSC pan, is rescanned, no melting endotherm can be seen. Conversely, if this piece of gel is removed from the DSC pan, cut into several pieces, replaced into another DSC pan and then rescanned, the high melting endotherm reappears. Mutin and Guenet (1989) interpret these phenomena by considering a sort of constraint release effect. While the gel melting point is given within a few degrees to the maximum of the high melting endotherm, the corresponding enthalpy is artifactual and may result from the constraint on the bottom and the side of the pan as the piece of gel turns into a fluid liquid. Once the gel is molten and re-formed in the DSC pan, this constraint release effect is minimized, hence the absence of artifactual enthalpy (for further details, see Fig.1.34.).

That the enthalpy is artifactual has been further demonstrated by operating at different heating rates with pieces of gel prepared out of the DSC pans: extrapolation to zero heating rate yields an equilibrium melting enthalpy indiscernible from zero (Mutin and Guenet, 1989).

Whereas DSC can be a convenient tool for determining the gel melting point with, however, limited accuracy, it cannot give the gel melting enthalpy. Attempts have been made to measure this value from the gel melting point by using the Ferry–Eldridge theory (Harrison *et al.*, 1972; Takahashi *et al.*, 1972; Yang and Geil, 1983). This theory is, however, only valid for systems undergoing chemical equilibrium, that is transitions of second or higher orders. As a result, this theory implies point-like junctions (see Appendix 1 for more details) which is certainly not the case for PVC gels, the physical junctions of which are known to be made up of crystallites, as ascertained by many different techniques. The improper use of this theory leads to inconsistent results. For instance, Yang and Geil (1983) found, using this approach, that the junctions consisted of about two chains, which does not make sense for a crystallite.

Since, beyond doubt, physical junctions are composed of small crystallites, the absence of both a measurable melting enthalpy and a measurable formation enthalpy is puzzling. The easiest way to account for the 'missing' enthalpy is to consider a large range of crystal perfection which is liable to give a broad spectrum of crystal melting points. Another explanation has been suggested by Mutin and Guenet (1989), who point out that the crystallites are probably of very small size and may not therefore be regarded as three-dimensional objects (which means that their fractal dimension is lower than 3). According to a general theory on melting processes developed by Halperin and Nelson (1978) low-dimensionality systems may not give off any latent heat on melting. A detailed knowledge of the molecular structure will be needed to answer this question together with the use of more sensitive calorimeters.

The effect of the solvent type has been assessed by Harrison *et al.* (1972). For this purpose they have considered the approximation of the relation derived by Flory (relation 1.2) for crystalline systems and accordingly plotted $(1-\chi_1)/V_s$ vs T_m^{-1}. Within experimental uncertainties, they have obtained a straight line which indicates direct correlation of the gel melting behaviour with solvent quality. The results of Takahashi *et al.* (1972) as well as those of Yang and Geil (1983) point towards the same conclusion: the liquidus curves are virtually identical for solvents possessing similar solubility parameters and are shifted to lower temperatures with better solvents and to higher temperatures with poorer ones.

The effect of molecular weight has also been examined by Yang and Geil (1983), who found that the larger the molecular weight, the higher the gel melting point. They have interpreted this phenomenon by using the Eldrige–Ferry theory in which the melting point is predicted to vary with molecular weight. In fact, this theory may be valid just at C^* but does not hold beyond, as above C^* the molecular weight is not supposed to play any role as far as the number of contacts between different chains is concerned (see Appendix 1 for further reading). For instance, the screening length is molecular weight independent. Another approach to this effect may be contemplated. Mutin (1986) has developed in his thesis a short theoretical derivation allowing calculation of the number and fraction of syndiotactic sequences of a given length as well as the number and fraction of syndiotactic sequences of size larger than a given size. Considering a bernouillian distribution, Mutin concludes that at constant syndio and iso placement probability these numbers are actually molecular weight dependent. This particularly means that there exists an increasing number of long syndiotactic sequences with increasing molecular weight. Correspondingly, one may expect the presence of 'crystals' of larger size which are known to possess higher stability together with a higher melting point (see Appendix 1).

Mutin's derivation can be summarized as follows: consider α and β as the probability of obtaining an isotactic or a syndiotactic dyad respectively (with $\beta = 1 - \alpha$). The probability of finding a syndiotactic sequence of N_s consecutive monomers is:

$$\mathrm{p}(N_s) = \alpha^2 \beta^{N_s - 1}$$

The average number of syndiotactic sequences of N_s monomers for a chain of N monomers and their weight fraction are respectively:

$$T_s(N_s) = (N - N_s + 1)\alpha^2 \beta^{N_s - 1} \qquad W_s(N_s) = N_s(N - N_s)\alpha^2 \beta^{N_s - 1}/N$$

While the weight fraction can be shown to be little dependent on molecular weight, which indicates that the global tacticity is molecular weight independent, as expected, the number T_s is. It approximately increases with the molecular weight. One may also calculate the total number of sequences Z_s larger than a given size:

$$T_s(Z_s \geqslant N_s) = [\alpha(N - N_s + 1) - \beta]\,\beta^{N_s - 1}$$

(ii) Gel ageing

Significant ageing effects in PVC gels were reported by te Nijenhuis and Dijkstra (1975) and Dorrestijn *et al.* (1981) from measurement of the storage modulus as a function of ageing in PVC–dioctyl phthalate systems.

Ageing also takes place in dilute solutions in which only PVC aggregates are present. Mutin *et al.* (1988) have investigated the evolution of the intensity scattered at a given angle (90°C) with time. As can be seen in Figure 1.33, it takes almost 6 days for a 0.5×10^{-2} g/cm^3 solution to reach a constant value whereas a 1.6×10^{-2} g/cm^3 solution is still evolving towards a higher aggregation state after this period of time.

More recently Mutin and Guenet (1989) have investigated the ageing process by DSC. For gels aged for a minimum of 24 h they have observed the appearance of a low-melting endotherm near 50°C (see Figure 1.34a). Interestingly, unlike the high-melting endotherm, the low-melting endotherm is always detectable even with a gel sample formed in the DSC pan (see Figure 1.34b).

Two endotherms had already been observed by Guerrero and Keller (1981), the low-melting endotherm being in the same range as the one determined by Mutin and Guenet (1989). Guerrero and Keller wrongly assigned the low-melting endotherm to the gel melting and the high-melting endotherm to the fusion of chain-folded crystals that were said to be independent of the network. Mutin and Guenet clearly showed by measuring the compression modulus that the low-melting endotherm is due to the appearance of a second type of physical junction which forms during ageing. In fact, the rise in the low-melting endotherm correlates with the increase in compression modulus. Similarly, heating at 65°C rejuvenates the gel, as the nascent gel modulus is found again. As to the high-melting endotherm, it

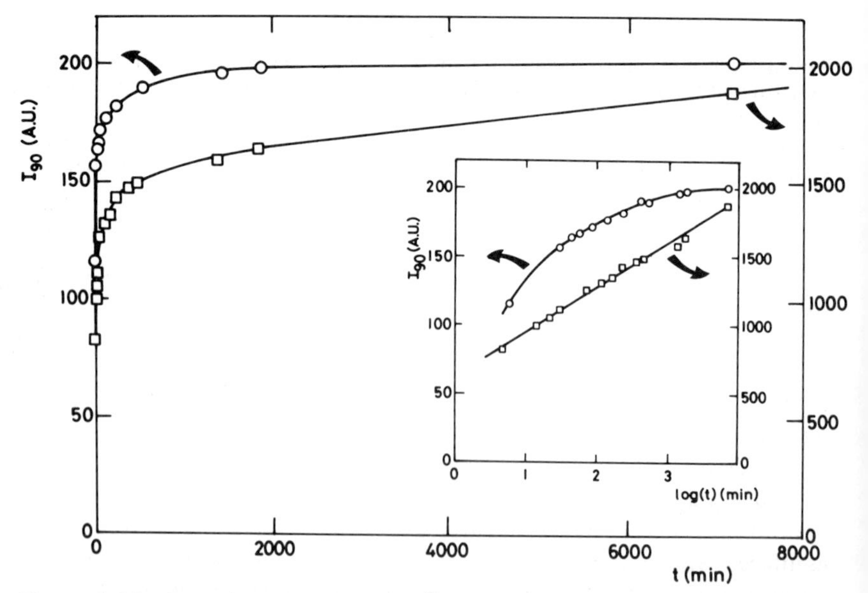

Figure 1.33 Intensity scattered at 90°C, I_{90}, as a function of ageing time. (○) $C = 5 \times 10^{-3}$ g/cm^3, (□) $C = 1.6 \times 10^{-2}$ g/cm^3. Inset: I_{90} as a function of log t (with permission from Mutin *et al.*, 1988, and the publishers Butterworth-Heinemann.).

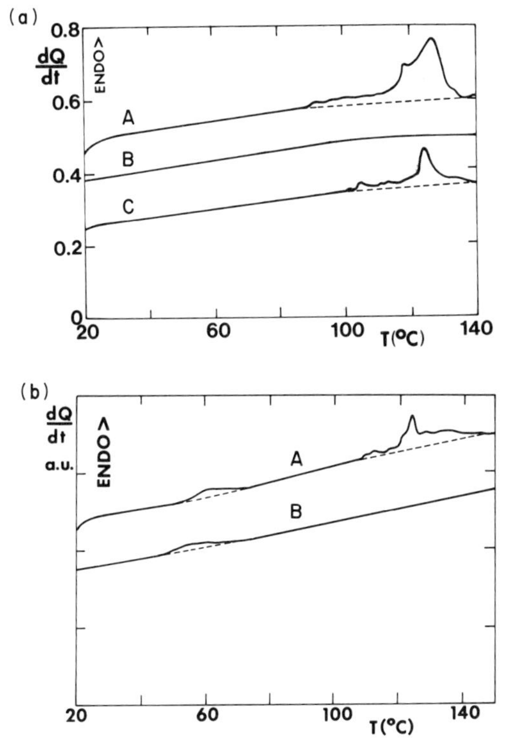

Figure 1.34 DSC thermograms on a PVC–diethyl malonate gel (atactic PVC synthesized at 50°C); ($C = 0.175$ g/cm^3). Upper figure: (A) first heating of a gel aged 30 min at 20°C; (B) second heating after the gel has been melted in the DSC pan to 20°C; (C) heating the same gel after cutting the gel into small pieces and placing them into a new DSC pan. The reappearance of the high-melting endotherm can be seen. This proves that its magnitude is artifactually enhanced by gel collapse in the DSC pan while melting. Lower figure: (A) first heating of a gel aged 24 hr at 20°C; (B) second heating of the same gel just after melting it in the DSC pan, cooling it to 20°C and ageing it at 20°C for 24 hr (with permission from Mutin and Guenet, 1989).

corresponds to the gel melting, as already shown above, and not to hypothetical chain-folded crystals. The misinterpretation of Guerrero and Keller arose from their belief, at that time, in the existence of a universal melting behaviour of thermoreversible gels based on the fact that iPS gels, once aged, also display two melting endotherms. They therefore regarded, as for iPS gels, the low-melting endotherm as the gel melting endotherm.

Mutin and Guenet (1989) also investigated the effect of polymer concentration. They found that the low-melting endotherm occurred at constant temperature whatever the concentration whereas the high-melting endotherm occurred at increasing temperature. The temperature–concentration phase diagram was accordingly established (Figure 1.35). The low-melting endotherm stands therefore as an invariant. This invariant represents a first-order transformation which has been designated by these authors as a transformation gel II ⇒ gel I. At this stage it is worth mentioning that similar endotherms also occurring near 50°C were observed by Leharne *et al.* (1979) on PVC plasticized by various phthalates (mass fraction of PVC between 0.45 and 0.71).

From their determination of the compression modulus, Mutin and Guenet (1989) concluded that two types of physical junction were present in the gel once aged. They noticed that the second type of junction, which disappeared above the transition, was solvent dependent. They further conjectured that the low-melting endotherms probably arise from junctions made up of the less regular tactic sequences together with incorporation of solvent molecules thus forming some kind of polymer–solvent compound. Although the idea about the ordering of the less tactic sequences had

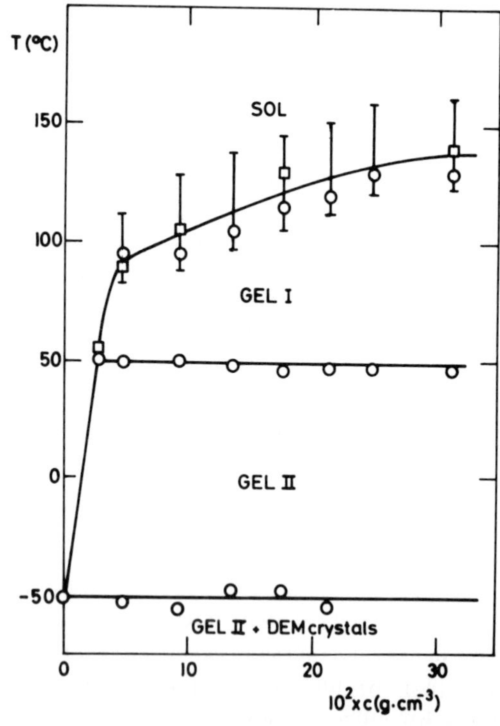

Figure 1.35 Temperature–concentration phase diagram from PVC–diethyl malonate gels aged 24 h at 20°C. (○) Temperatures determined from the first heating by DSC at a scan speed of 20°C/min (bars indicate the width of the melting endotherm) and (□) melting temperatures determined visually by the ball-drop method using a heating rate of 2°C/min (with permission from Mutin and Guenet, 1989).

already been suggested by Juijn *et al.* (1973), the participation of the solvent had not been considered. It should, however, be emphasized that Mutin and Guenet (1989) observed the low-melting endotherm with a large variety of solvents. It is probable that some solvents are more able to interact with PVC to eventually form a structure close to a polymer–solvent compound than others and, accordingly, one would not expect to observe the second type of physical link with all the solvents. This point is still not totally clear, although recent experiments carried out by Najeh *et al.* (1992) on the gel modulus with gels prepared from either monoesters or diesters support the concept of solvent participating in the second type of junction (see Chapter 3).

Studies on pregels, i.e. aggregates formed below the critical gel concentration, C_{gel}, by light scattering (Rayleigh scattering) (Mutin *et al.*, 1988) also demonstrated, admittedly indirectly, the existence of two types of junction. A detailed account of these results will be given in Chapter 2.

Two important questions deal with the number of phases and the type of transition between gel II and gel I. How can the gel consist of three phases at different temperatures and concentrations: two types of junction and the polymer-poor phase? Gibbs phase rules must be obeyed which implies that three phases can only exist at one temperature and one concentration with a two-component system (see Appendix 1). In the present case ought one to consider a three-component system: for instance, either a **mixture of highly syndiotactic and poorly syndiotactic chains** or **chains containing highly syndiotactic sequences alternating with less syndiotactic ones** as in a block copolymer? In the latter case block copolymers usually behave, as far as phase rules are concerned, as homopolymers. Yet, if both sequences of the copolymer can crystallize with different crystalline lattices and provided that the crystallization of one sequence does not hamper the other one from crystallizing, then one might consider the solutions of this polymer as a three-component system.

As for the type of transition from gel II to gel I the answer is still open to discussion with the available data. Nevertheless, two examples are discussed below for the sake of argument. Needless to say, only suitable experiments will give the final key to the puzzle.

As a first case one could envisage the formation of a peritectic which would imply the participation of the solvent but not to the extent of a polymer–solvent compound (see Appendix 1). This means that, above the transition, the polymer-rich phase (that is the network) should contain less solvent molecules than below this transition, hence an increase in its concentration. (See Appendix 1 for further details.)

As a second case one could consider the situation where the transition corresponds to the transformation of one solid solution into another due to a crystalline transformation in the crystalline part of the polymer-rich phase (metatectic transformation, see Appendix 1). Unlike the peritectic transformation, this type of transition should lead the polymer-rich phase to absorb solvent.

These considerations on the phase diagram outline the difficulty of elucidating the gel ageing mechanism and, correspondingly, the resulting molecular structure.

(c) Poly(vinyl alcohol)

Poly(vinyl alcohol) (PVA; $[-CH_2-CHOH-]_n$) comes in three varieties: atactic, isotactic and syndiotactic. Like PVC, atactic PVA displays a significant degree of crystallinity.

The physical gelation capability of the atactic and syndiotactic forms of this polymer have been known for a long time. As early as 1953 Sone *et al.* reported on the crystallinity of gel filaments spun from concentrated solutions.

Most studies have been carried out in water (see for instance Shibatani, 1970; Takahashi and Hiramitsu, 1974; Ogasawara *et al.*, 1975, 1976) but also in mixtures of water with another solvent such as acetone (Papkov *et al.*, 1966), aqueous 60% glycerol (Rogovina *et al.*, 1973) or ethylene glycol (Pines and Prins, 1973; Stoks *et al.*, 1988).

Shibatani (1970) and Takahashi and Hiramitsu (1974) have determined that the physical cross-links arise from the syndiotactic sequences. Ogasawara *et al.* (1975) have shown that the gel melting point, as determined macroscopically, is strongly sensitive to the syndiotactic content. For instance, the melting points increase from 40°C for a syndiotactic content of 58% up to 119°C for a content of 66%. In a following paper (Ogasawara *et al.*, 1976), they studied the gelation kinetics by dilatometry and found Avrami exponents (Avrami, 1939, 1940, 1941) close to 1 suggesting unidirectional crystallization, a result reminiscent of what is found for gelation (see Section 1.2.2.d).

The temperature–concentration phase diagram was established for concentrations up to 10% (w/w) (Figure 1.36). According to Ogasawara *et al.* (1976), the shape of this diagram is consistent with the co-occurrence of liquid–liquid demixtion: the liquidus line approaches a plateau between 5 and 10% as expected in such a case (monotectic transition, see Appendix 1).

Extensive investigation of the thermal behaviour of PVA–ethylene glycol systems as a function of the gelation temperature has been carried out by Stoks *et al.* (1988). They, however, restricted their study to only one polymer concentration (8.3%, w/w) and used an atactic PVA sample (triad content: iso = 0.2, hetero = 0.48 and syndio = 0.32). They emphasize that in this case the gel obtained may be described as paste-like rather than rigid-like, unlike gels formed from more syndiotactic samples.

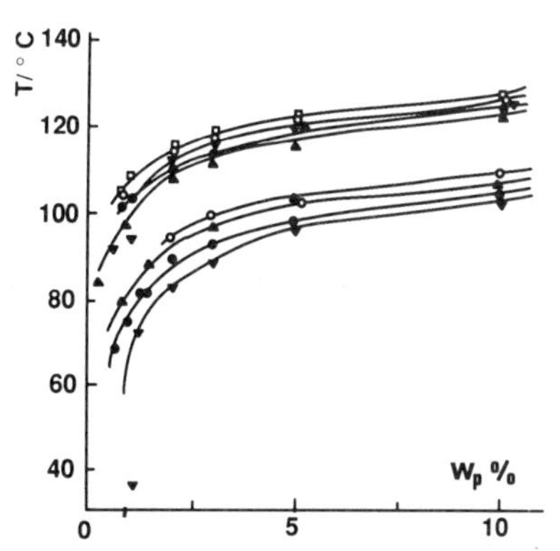

Figure 1.36 Temperature–concentration phase diagrams for poly[vinyl alcohol]/water gels. Upper curves: 64.3 mol% of syndiotactic dyads; lower curves: 58.4 mol%. Gelling temperatures of: (○) 0°C, (▽) 30°C, (•) 45°C, (△) 60°C and (□) 80°C (with permission from Ogasawara, Yuasa and Matsuzawa, 1976).

On heating a gel formed at room temperature, a large, broad endotherm is observed at high temperature (high-melting endotherm above 100°C) while another endotherm, of lower magnitude, appears on ageing at lower temperature (low-melting endotherm) (Figure 1.37). This low-melting endotherm is not seen for gels prepared above 100°C.

The variation of the position of the high- and low-melting endotherms as a function of gelation temperature shows that the high-melting endotherm

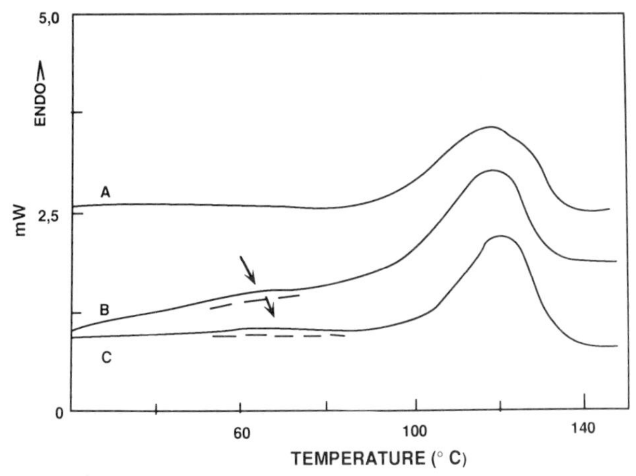

Figure 1.37 Melting endotherms of poly[vinyl alcohol]/ethylene glycol gels formed at room temperature at different ageing times: A, 0 h; B, 118 h and C, 447 h. Arrows indicate the low-melting endotherm (mW = milliWatts) (with permission from Stoks *et al.*, 1988).

remains constant up to a gelation temperature of 100°C while the low-melting endotherm increases linearly.

Stoks *et al.* (1988) consider that gelation occurring in ethylene glycol solutions quenched to below 100°C proceeds via liquid–liquid phase separation frozen in at its early stage by crystallization, an assumption already put forward by Pines and Prins (1973) for the system PVA–water–ethylene glycol. It is perhaps worth adding that the low-melting endotherm might represent the signature of a monotectic transition which is a first-order transition seen when crystallization has taken place within a miscibility gap. As a result, at constant quenching temperature the monotectic transition temperature is concentration invariant (see Section 1.2.2.d). Unfortunately, Stoks *et al.* have not collected any data as a function of polymer concentration that could have given further information.

Worth underlining is the tendency of these gels to display pronounced syneresis on ageing. This is usually attributed to continuous crystallization together with rearrangements at the molecular level.

(d) Multiblock copolymers

For chronological reasons the gelation of random copolymers should have been presented first. However, its mechanism will probably be better understood once the more complete study carried out on multiblock copolymers has been presented.

Hitherto, the only investigations on the physical gelation behaviour of multiblock copolymers, in which one sequence is crystallizable, have been carried out by He *et al.* (1987*a*, 1988, 1989). The multiblock copolymer studied consisted of alternating sequences of poly(dimethyl siloxane) (PDMS) and poly[(dimethyl silyl-1,(dimethylene)silyl – 4] benzene (COSP = crystalline organosilicic polymer):

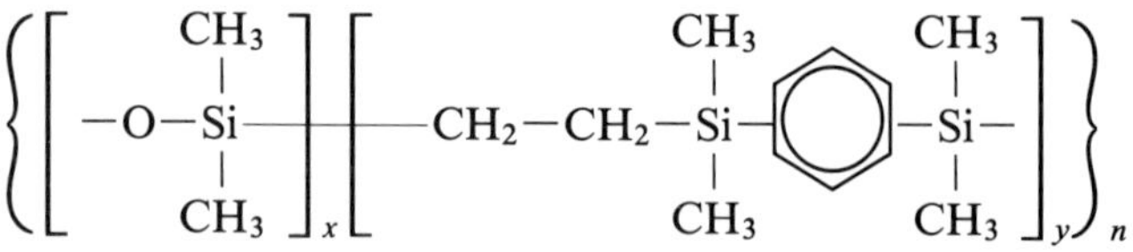

$$\left\{ \left[-O-\underset{CH_3}{\overset{CH_3}{\underset{|}{\overset{|}{Si}}}}- \right]_x \left[-CH_2-CH_2-\underset{CH_3}{\overset{CH_3}{\underset{|}{\overset{|}{Si}}}}-C_6H_4-\underset{CH_3}{\overset{CH_3}{\underset{|}{\overset{|}{Si}}}}- \right]_y \right\}_n$$

Concerning COSP, He *et al.* (1987*b*) report melting points ranging from 165°C to 189°C and glass transition temperature from −2°C to 24°C depending on the chain length. PDMS is characterized by a glass transition temperature of −127°C. PDMS can crystallize at low temperature but He *et al.*'s investigation was always carried out well above.

Of further note is the fact that PDMS and COSP are incompatible. COSP is insoluble in PDMS even at high temperatures.

The amount of crystalline sequence, X_{COSP}, was varied by changing the length of either sequence. Results reported by He *et al.* (1987*a*, 1988, 1989) deal with copolymer samples, the characteristics of which are gathered in Table 1.3.

He *et al.* (1987*a*) first investigated the gelation mechanism in *trans*-decalin. *Trans*-decalin turns out to be a good solvent for the PDMS sequence. Conversely, from investigations into the thermal properties of COSP–*trans*-decalin solutions quenched rapidly to low temperature, He *et al.* (1987*a*) conclude that *trans*-decalin is a poor solvent for the crystalline sequence. In fact, the temperature–concentration phase diagram displays an invariant at 50°C which He *et al.* (1987*a*) attribute to a monotectic transition. This transition occurs whenever crystallization takes place within a miscibility gap (see Appendix 1). In addition, the morphology of the COSP crystals obtained in dilute solutions with this thermal treatment is ball-shaped which, on the basis of the work by Schaaf *et al.* (1987), is a further indication of the occurrence of a liquid–liquid phase separation prior to crystallization.

To form a gel, He *et al.* (1987*a*) indicate that the copolymer solutions must be quenched rapidly to low temperature (– 18°C in the case of *trans*-decalin) whereas a slow cooling produces only a crumbling system. Figure 1.38, the temperature–concentration phase diagram, shows that the thermal behaviour of copolymer gels includes an invariant near 50°C (this temperature depends slightly on the size of the crystalline sequence). Whatever the copolymer composition, all the temperature–concentration phase diagrams in *trans*-decalin exhibit this monotectic transition. He *et al.* (1987*a*, 1988) conclude that, under these preparation conditions, the gelation mechanism proceeds via liquid–liquid phase separation frozen in at its early stage by crystallization. They also discuss the fact that the monotectic transition is visible well above the miscibility gap. They emphasize that this monotectic transition represents a memory effect of the crystallization conditions on to the melting behaviour which arises from the intrinsic ability of these solutions to reach high undercoolings before the onset of crystallization.

Table 1.3 Characteristics of the multiblock copolymers used by He *et al.* (1987*a*, 1988, 1989). M_{nCOSP} = number-averaged molecular weight of the crystalline sequence; M_{nPDMS} = number-averaged molecular weight of the amorphous sequence; X_{COSP} = weight fraction of the crystalline sequence; M_{wCOPO} and M_{nCOPO} the weight-averaged and number-averaged molecular weights of the copolymer as determined from light scattering and GPC.

Name	M_{nCOSP}	M_{nPDMS}	X_{COSP}	M_{wCOPO}	M_{nCOPO}	$M_{\mathrm{wCOPO}}/M_{\mathrm{nCOPO}}$
Copo10	2000	17 100	0.11	215 000	88 000	2.18
Copo20	4000	17 100	0.19	130 000	62 000	1.79
Copo50	9700	9 700	0.50	290 000	—	—

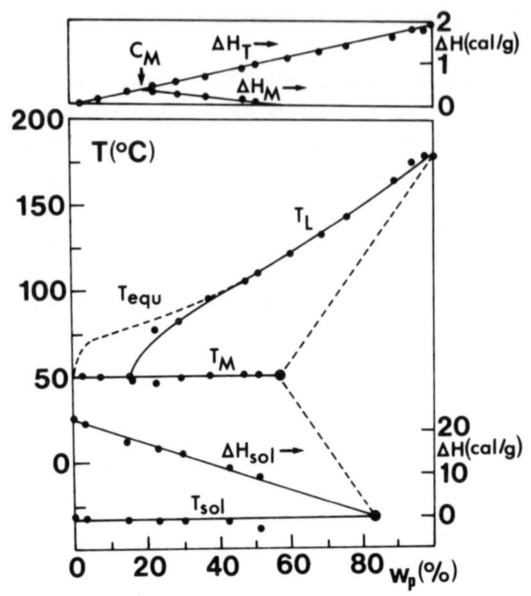

Figure 1.38 Temperature–concentration phase diagram for Copo20–*trans*-decalin gels (concentration in w/w). T_{sol} = solvent melting temperature crystallized after gel formation and ΔH_{sol} the associated enthalpy. T_{M} = monotectic transition; T_{L} = liquidus line for a gel prepared by a quench to -18°C (T_{eq} = equilibrium melting temperature (crystals) formed at very low cooling rate). Above, Tamman's diagram for the enthalpies associated to the melting endotherms: C_{M} = monotectic concentration, ΔH_{M} = enthalpy associated to the monotectic transition and ΔH_{T} = total melting enthalpy (monotectic + final melting endotherms) (with permission from He *et al.*, 1987a).

Also, the phase diagram indicates that the polymer-rich phase, which is the continuous phase, consists of a solid solution the composition of which, C_{α}, can be determined at the monotectic transition.

Here, it seems worth stressing that the occurrence of two endotherms does not necessarily imply the presence of two distinct structures, network and single crystals for instance. As will be seen in Chapter 3, **partial melting** of the gel structure takes place at T_{M} for $C > C_{\text{M}}$ rather than to the melting of one structure, while the other one would remain unaltered. Again, unlike iPS gels the low-melting endotherm does not correspond to the gel macroscopic melting.

The only major difference observed with copolymer composition is the value of the monotectic concentration, C_{M}, which decreases with increasing content of COSP sequences, X_{COSP} (33% for $X_{\text{COSP}} = 0.1$, 20% for $X_{\text{COSP}} = 0.2$ and 10% for $X_{\text{COSP}} = 0.5$). According to He *et al.* (1988), this may stem from kinetic effects. The monotectic concentration, they say, is probably equal or proportional to the concentration reached by the polymer-rich phase just before crystallization occurs. It therefore depends on how deep the solution can be quenched within the miscibility gap before crystallization takes over: the deeper the quench within the miscibility gap, the higher the polymer-rich phase concentration. For slow crystallization kinetics (low COSP content), the quench will be deeper than for rapid crystallization kinetics (high COSP content): hence a higher value for C_M in the former situation.

He *et al.* (1988) point out that the quality of the solvent towards the copolymer seems to be mainly determined by the presence of COSP sequences since liquid–liquid phase separation takes place in *trans*-decalin with a COSP content as low as 0.1. The fact that PDMS and COSP are totally incompatible, even above COSP melting point, may explain this.

The gel melting enthalpy variation as a function of concentration, once extrapolated to 100% copolymer (Figure 1.38), gives values that are very close to those determined in the bulk state for the crystallized COSP. He *et al.* (1987*a*, 1988) conclude that the COSP sequences probably crystallize in the same form as in the bulk state. Here, there is no evidence for the formation of a polymer–solvent compound.

To test this point further, He *et al.* (1989) have investigated the effect of the solvent type. Gels were prepared in bromobenzene and 1-phenyldodecane. Unlike *trans*-decalin, bromobenzene and 1-phenyldodecane are poor solvents for the PDMS sequence at room temperature. For instance, He *et al.* (1989) report that the critical temperature T_c for a PDMS sample with $M_w = 1.7 \times 10^4$ is 40.8 ± 0.5°C in bromobenzene and 85.2 ± 0.5°C in 1-phenyldodecane. Similarly, a thermal study of the COSP solutions shows that both solvents are poor for this sequence.

The temperature–concentration phase diagrams determined from gels prepared in bromobenzene (Figure 1.39) are very similar to those obtained in *trans*-decalin. As with *trans*-decalin there is an invariant at $T = 55$°C. Also, the monotectic concentration C_M increases with decreasing content of COSP sequence. Conversely, while the phase diagram's overall shape is similar in 1-phenyldodecane, all the lines are shifted to higher temperature (Figure 1.40). He *et al.* (1989) account for this shift by the fact that 1-phenyldodecane is the poorest of the three solvents used for both the PDMS and the COSP sequence. Flory's relation (see 1.1) for the melting point depression indicating that the poorer the solvent, the higher the melting point for a given concentration applies here quite well.

Also, as with *trans*-decalin, the melting enthalpies determined by extrapolating to 100% copolymer are virtually identical, which leads one to conclude that the crystalline structure in the gel is the same as in the bulk state and, therefore, that no polymer–solvent compounds are formed.

In all the cases, a gel is obtained thanks to the occurrence of a liquid–liquid phase separation frozen in at its early stage by crystallization.

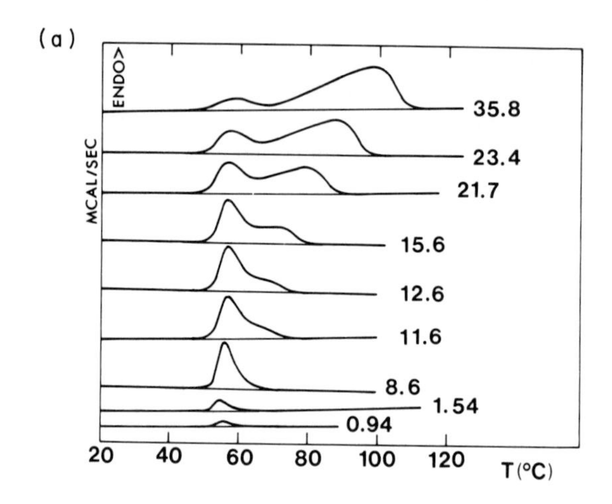

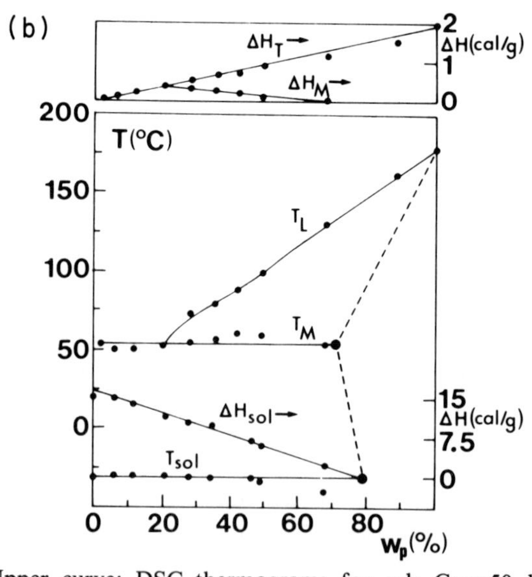

Figure 1.39 Upper curve: DSC thermograms for gels Copo50–bromobenzene gels prepared by a quench to -18°C (concentrations as indicated in w/w). T_M = monotectic transition and T_L = final melting. Lower curve: Temperature–concentration phase diagram for Copo20–bromobenzene gels. T_{sol} = melting temperature of the solvent crystallized after the gel has formed and ΔH_{sol} the associated enthalpy. ΔH_M and ΔH_T = melting enthalpies associated to the monotectic transition and to the total melting (monotectic + final melting) (with permission from He *et al.*, 1989).

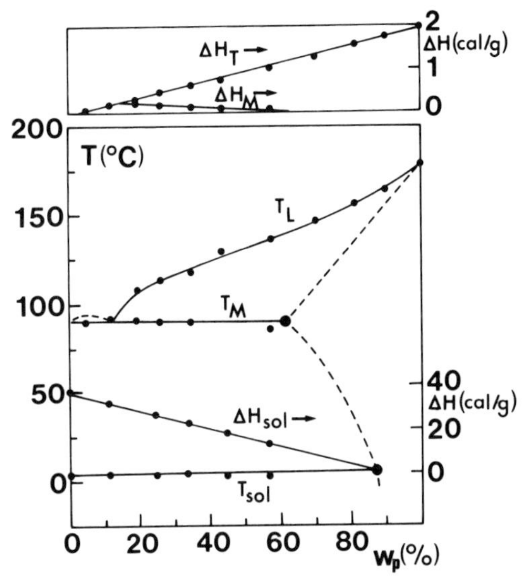

Figure 1.40 Temperature–concentration phase diagrams for Copo20–1-phenyl dodecane prepared through a quench to 20 °C (concentration in w/w) (with permission from He *et al.*, 1989).

This study shows also, as emphasized by Koeningsveld (1990), that phase rules are obeyed even for systems prepared out of equilibrium.

Gel swelling behaviour has also been studied by He *et al.* (1987*a*,*b*, 1988, 1989) in the light of the phase diagrams. Since the phase diagrams indicate that no solvent is occluded in the crystallized sequences, the only region that can absorb solvent is the amorphous region and it should do so for osmotic reasons. Under these conditions the equilibrium swelling ratio P_∞/P_0 (in w/w see Section 1.2.1.a) can be expressed as follows (He *et al.*, 1987*a*):

$$G_\infty = 1 - [1 - (C_\alpha/C_\beta)]\, X_\alpha \tag{1.17}$$

in which C_α and C_β are the concentration of the polymer-rich phase before and after swelling respectively and X_α, the amount of polymer-rich phase.

Relation 1.17 implicitly contains the hypothesis that the polymer-rich phase should always swell at the same concentration, C_β, independent of the preparation polymer concentration. Obviously, as the crystalline domains of the polymer-rich phase cannot incorporate solvent molecules, only the amorphous domain, i.e. the PDMS chains linking the crystalline domains, will. That these chains always swell up to the same concentration can be justified by assuming, as is done for chemical gels, that they tend to attain some C^* concentration that is only dependent upon the PDMS chain length (see de Gennes, 1979).

The amount of polymer-rich phase is given by:

$$X_\alpha = [C_{\text{prep}} - C_{\text{liq}}(T)] / [C_s(T) - C_{\text{liq}}(T)] \qquad (1.18)$$

in which C_{prep} is the preparation concentration, $C_{\text{liq}}(T)$ and $C_s(T)$ are the concentrations defined by the liquidus line and the solidus line respectively. From the experimental phase diagrams indicating that $C_{\text{liq}}(T=20^\circ\text{C}) \simeq 0$ and $C_s(T=20^\circ\text{C}) \simeq C_\alpha$, He *et al.* (1987) have approximated X_α at 20°C to:

$$X_\alpha \simeq C_{\text{prep}}/C_\alpha \qquad (1.19)$$

which finally gives:

$$G_\infty = 1 - [1 - (C_\alpha/C_\beta)]\, C_{\text{prep}}/C_\alpha \qquad (1.20)$$

The experimental variations are well reproduced with relation 1.20, as shown by the linear variation of G_∞ as a function of C_{prep}/C_α in the case of gels in *trans*-decalin (Figure 1.41). In bromobenzene and 1-phenyldodecane, no swelling occurs whatsoever, which is consistent with the fact that both solvents are poor solvents of the PDMS sequence. Obviously, under these conditions relation 1.20 is also satisfied. This shows the difference from iPS gels for which no swelling was seen below a given concentration and considerable swelling above. These simple swelling experiments underline the different natures of the two types of gel.

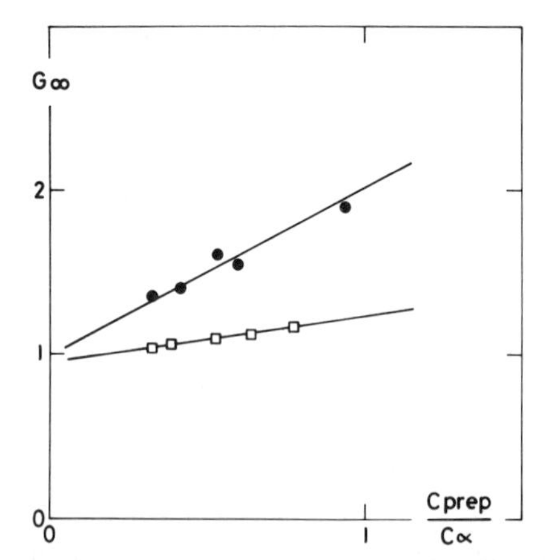

Figure 1.41 Equilibrium swelling degree, G_∞, as a function of the ratio C_{prep}/C_α (C_α being the concentration of the polymer-rich phase at the monotectic transition. It is the concentration where the monotectic endotherm vanishes): gels prepared in *trans*-decalin by a quench to −18°C then immersed in an excess of preparation solvent: (•) Copo20 and (□) Copo50 (with permission from He *et al.*, 1988).

(e) Random copolymers

Reports on the gelation of random copolymers are scarce. Paul (1967) has investigated acrylonitrile–vinyl acetate copolymers, Takahashi (1973) and Takahashi and Hiramitsu (1974) poly(ethylene–co-vinyl acetate) and Berghmans *et al.* (1979) poly(ethylene terephthalate–co-isophthalate).

It must be stressed that these authors consider that gelation has occurred when the solution can no longer flow. There is no determination of gel-like behaviour by rheological experiments except in Paul's investigation.

The investigation by Paul of acetonitrile–vinyl acetate random copolymers was focused on a sample containing 7.7% by weight of vinyl acetate (Paul, 1967). The chemical formula is:

$$\left[CH_2{-}\underset{\underset{C\equiv N}{|}}{CH} \right]_n \left[CH_2{-}\underset{\underset{\underset{\underset{CH_3}{|}}{\underset{CO}{|}}}{O}}{\underset{|}{CH}} \right]_m$$

Paul has determined by the **mercury-drop penetration method** the gel melting point as a function of copolymer volume fraction v_2 in dimethyl acetamide (DMAc) (Figure 1.42). The gel melting point increases very steeply within a limited range of concentration. The gel preparation temperature, 25°C or −78°C, does not alter the melting behaviour significantly while gelation is several orders of magnitude faster at −78°C. The gel melting point thus determined coincides quite nicely with the onset of a gel melting endotherm observed by differential thermal analysis (DTA), indicating that first-order transition is involved, most probably crystal melting.

The role of 'gelator' in this copolymer is undoubtedly played by the acrylonitrile sequences. This receives support from a study of the gel melting point in DMAc at constant polymer concentration but differing compositions. Pure acrylonitrile gels melt at 130°C whereas increasing content of vinyl acetate leads to a marked decrease in gel melting point (about 55°C for a sample containing 9% by weight of vinyl acetate monomers).

Paul has also investigated the effect of solvent type on the gel melting point. He has found by using the solubility parameter that the better the solvent the lower the gel melting point at a given polymer volume fraction which does suggest that gel formation is crystallization-induced.

From his work on the melting behaviour with concentration and solvent type on **poly(ethylene–co-vinyl acetate)**, Takahashi (1973) also comes to the conclusion that crystallization is involved in the gelation mechanism of this copolymer. Here it needs to be specified that the mole fraction of vinyl acetate in the copolymer used by Takahashi never exceeded 0.36. As a result,

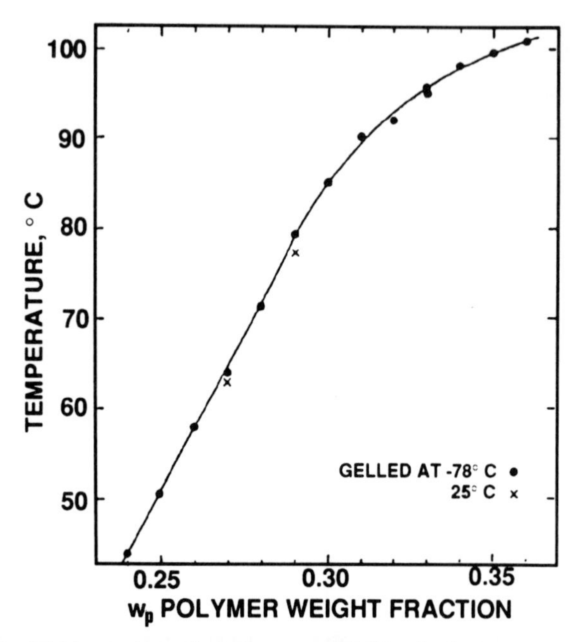

Figure 1.42 Melting points of random acetonitrile-vinyl acetate copolymer gels in dimethyl acetate as a function of copolymer weight fraction; gelled at (○) -78°C and (x) 25°C (with permission from Paul, 1967).

the average number of ethylene sequences in a row was always high enough to dominate the gelation phenomenon through their crystallization.

The chemical formula of **poly(terephthalate–co-isophthalate)** investigated by Berghmans *et al.* (1979) is:

$$\left[-C_6H_4-\overset{\overset{O}{\|}}{C}-O-CH_2-CH_2-O-\underset{\underset{O}{\|}}{C}- \right]_n \left[-C_6H_4-\underset{\underset{O}{\|}}{C}-O-CH_2-CH_2-O-\overset{\overset{O}{\|}}{C}- \right]_p$$

Berghmans *et al.* (1979) have determined the gelation induction time at 20°C in different solvents with solutions containing 20% (w/w) of a copolymer whose composition (w/w) was 60% terephthalate/40% isophthalate. Values are gathered in Table 1.4. These results suggest a direct correlation of the gelation behaviour with solvent quality. Although values of χ_1 are not available for this system, one can make a pretty good guess at the gelation mechanism by considering the theoretical expression relating

Table 1.4 Gelation induction times for a 60/40 copolymer at 20°C for various solvents (concentration 20%, w/w). From Berghmans *et al.* (1979).

Solvent	Induction time	Gel melting temperature (visual observation) (°C)
o-Dichlorobenzene	10 min	83
Anisole	30 min	78
Acetophenone	5 h	74
Tetrahydrofuran	5 h	74
Cyclohexanone	5.5 h	73
Dichloroethane	6.5 h	70
Nitrobenzene	9 h	69
Dioxane	5 weeks	55

crystal growth, G, to undercooling, ΔT:

$$G = G_0 \exp(-\Delta G_a/kT)\exp(-K_g/fT\Delta T) \tag{1.21}$$

where G_0 is a constant, the first exponential corresponds to a transport term, and the second exponential contains the effect of undercooling, ΔT, as well as a term, K_g, which depends upon the crystallization regime and a so-called correction factor f (for further reading, see Lauritzen and Hoffmann (1960) and Hoffmann and Lauritzen (1961)). As a first approximation, i.e. considering the transport term as similar in all the solvents investigated and the same crystallization regime, crystal growth and correspondingly gelation kinetics depend upon the undercooling ΔT: the smaller the undercooling, the slower the crystallization. Besides, Flory's relation indicates that, at fixed temperature and concentration, the smaller the undercooling, the better the solvent. Consequently, the better the solvent, the slower the crystallization or gelation rate and the lower the gel melting point, which is what Table 1.4 shows.

Berghmans *et al.* (1979) also conclude that gelation of this copolymer occurs via classical crystallization. From the experimental values of the melting enthalpy and using $\Delta H_m = 27$ cal/g for the melting enthalpy of PET (polyethylene terephthalate) they end up with a crystallinity of $x_c \simeq 0.06$ for the 60/40 sample in gels prepared from anisole. They, however, report that the gel thermal behaviour is complex and subject to quenching temperature as well as to annealing conditions. Throughout their study anisole was used as the solvent. In this solvent, gelation times are relatively short, as shown in Table 1.4. It turns out to be a poor solvent as hinted at by the significant discrepancy between the solubility parameters of the polymer ($\delta_{pol} = 10.53$) and the solvent ($\delta_{anisole} = 9.51$).

As expected, the gelation induction time decreases with decreasing temperature. Yet, for the copolymer of lowest ethylene terephthalate content there is a minimum near 0°C for which no explanation is provided.

The DSC melting traces reveal two endotherms whose position and magnitude depend strongly upon gelation temperature. For instance, Figure 1.43 shows that both endotherms are shifted towards higher temperature when the gelation temperature is increased. Also, the low-melting endotherm grows faster than the high-melting endotherm as a function of gelation time (Figure 1.44).

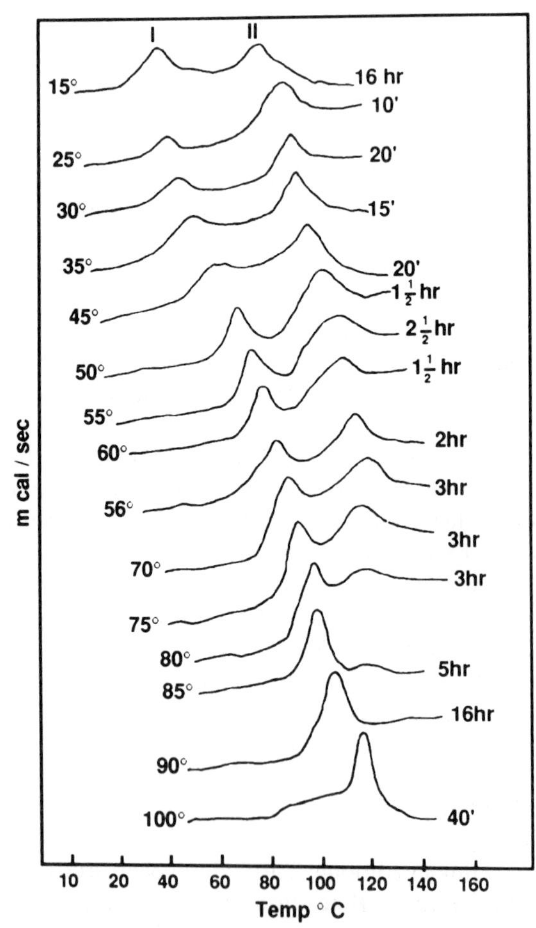

Figure 1.43 DSC endotherms of gels in anisole of a 70% terephthalate/30% isophthalate copolymer. Gelation temperatures and times as indicated (with permission from Berghmans *et al.*, 1979).

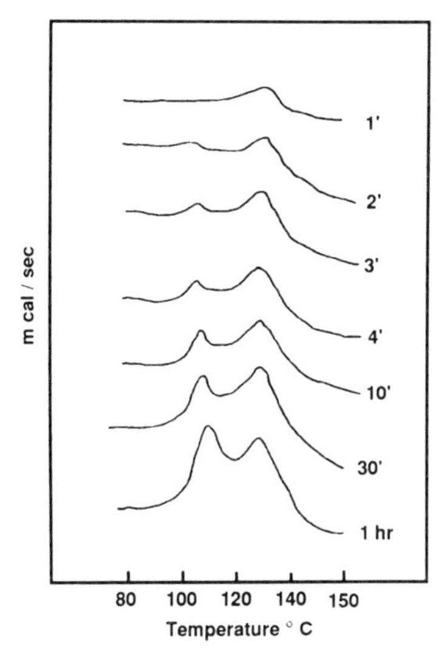

Figure 1.44 Melting endotherms of gels of copolymer 80% terephthalate/20% isophthalate as a function of gelation time ($T_{gel} = 90^{\circ}$C; heating rate 20 °C/min) (with permission from Berghmans *et al.*, 1979).

According to Berghmans *et al.* (1979), the low-melting endotherm relates to the gel melting and, correspondingly, to the fusion of fringed-micellar crystals, while the high-melting endotherm represents the melting of chain-folded crystals.

In the light of the results gained by He *et al.* (1987*a*, 1988, 1989) on multi-block copolymers one may, however, wonder whether the low-melting endotherm does not represent a monotectic transition arising from the interference of a liquid–liquid phase separation on crystallization. Since anisole happens to be a poor solvent, this mechanism is liable to occur. Unfortunately, no data are available as a function of concentration. As has been already discussed in Section 1.2.2.d, the monotectic transitions should entail no variation of the low-melting endotherm as a function of concentration at fixed gelation temperature.

1.3 BIOPOLYMERS

In this section two classes of biological macromolecules will be discussed in particular: **gelatin gels** and **polysaccharide gels**, including in the latter class gels obtained from man-made cellulose derivatives.

Water is usually the solvent used for preparing the gels, unlike synthetic polymers. On this account two classes may be distinguished: **neutral**

biopolymers such as gelatin or agarose and **charged biopolymers** such as carrageenans. Whereas the addition of inorganic salts to water plays only a secondary role in the former (mainly altering solvent quality), it becomes an essential parameter in the latter where polyelectrolyte effects come into play.

1.3.1 Neutral biopolymers

(a) Gelatin gels

Gelatin is a protein which is prepared from partial hydrolytic degradation of collagen, the latter being found in various animal tissues (skin), tendons and bones.

As is the case with every protein of natural origin, the chemical formula is complex and composed of several amino acids. Although the formula of gelatin depends upon its origin, it always contains large amounts of proline (Pro), hydroxyproline (Hyp) and glycine (Gly):

Proline Hydroxyproline 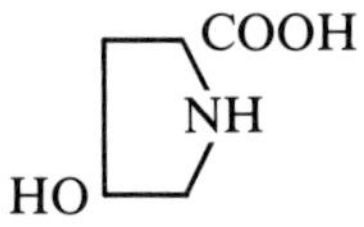 Glycine

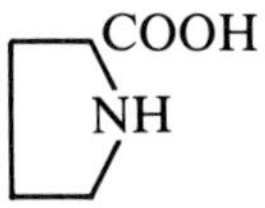

COOH COOH

NH NH NH_2-CH_2-COOH

HO

Sequences of (Gly-X-Pro) and (Gly-X-Hyp), in which X is another amino acid, are the commonest.

Common industrial gelatins are mixtures of various components, including α-gelatin (one chain), β-gelatin (two chains linked by covalent bonds) and γ-gelatin (as β but with three chains).

The chemical structure depends on the recipe used when processing collagen raw material. Processing with a weak acid (pH $\simeq$ 4) allows the amide groups to be preserved so that the global composition of amino acids is close to that of collagen. Alkaline processing induces hydrolysis of the amide groups of asparagine and glutamine which increases the proportion of COO^- groups. Finally, enzymatic processing (Pronase) allows one to obtain a low-polydispersed sample of high purity.

Results are obviously subject to the gelatin composition although in some cases investigations have been performed on fractionated and well-defined samples (for more details see for instance Ramachandran, 1967; Ward and Courts, 1977).

Gelation of gelatin solutions has been investigated by several techniques. Since there is little doubt that partial renaturation of the collagen helical structure occurs during gelation, optical rotation determination has been used as a main tool (see for instance Eagland *et al.*, 1974; Chatellier *et al.*,

1985; Djabourov *et al.*, 1988*a*). Other experimental approaches have made use of light scattering (see for instance Pines and Prins, 1973), NMR (see for instance Finer *et al.*, 1975), dilatometry (Flory and Garrett, 1958), DSC (Godard *et al.*, 1978) and the like.

Gel formation and melting behaviour are strongly dependent upon temperature, concentration and molecular composition of the gelatin sample (Flory and Weaver, 1960; Macsuga, 1972).

Harrington and Karr (1970) were able to point out the occurrence of a maximum in the crystallization half-time as a function of temperature for RCM-*Ascaris* (reduced carboxymethylated *Ascaris* cuticle collagen) and rat skin α_1-gelatin chains in 0.15M-NaCl–0.01M-acetate–50% ethylene glycol. From these results they have concluded that the rate of renaturation finds an explanation in theories developed for nucleation and growth of polymer crystallites.

Godard *et al.* (1978) have extensively studied by DSC gel formation, as well as its melting, as a function of quenching temperature and gelatin concentration. The gelatin they used contained 53% components of molecular weight lower than α, 18% of α chains, 16% of β chains and 13% of components of molecular weight higher than β. Typical melting endotherms obtained by these investigators are drawn in Figure 1.45. By measuring the gel melting enthalpy, ΔH_{gel}, as a function of time the kinetics of gelation have been established. The gelation isotherm exhibits a sigmoïdal shape due to primary crystallization which is followed by a linear variation which Godard *et al.* interpret as arising from secondary crystallization. Analysis of the gelation kinetics with Avrami's equations (Figure 1.46) (Avrami, 1939, 1940, 1941) leads to an exponent close to 1 (1.1 to 1.14 depending on

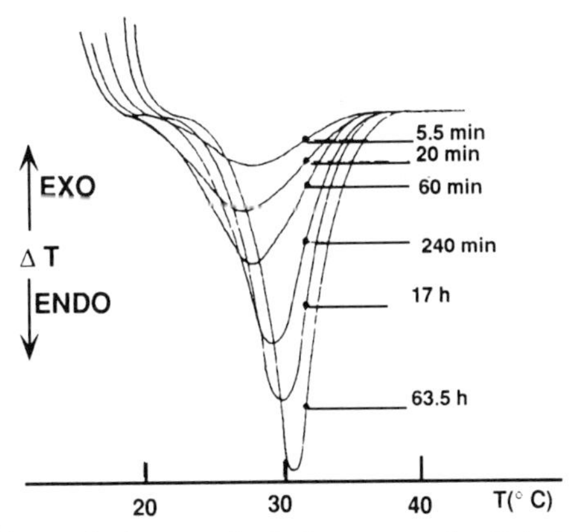

Figure 1.45 Melting endotherms of gelatin–water gels. Gelation time as indicated ($C_{pol} = 20\%$, gelation temperature $= 15^\circ C$, scanning rate $= 2.5^\circ C/min$) (with permission from Godard *et al.*, 1978).

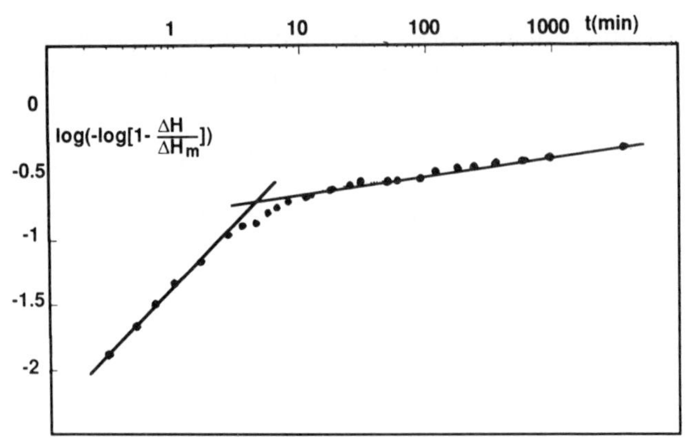

Figure 1.46 Avrami plot of the melting enthalpy $\Delta H/\Delta H_m$ *vs* time for gelatin gel $C = 20\%$, $T_{gel} = 15°C$. ΔH is the actual gel melting enthalpy while ΔH_m is the heat of fusion for an infinite crystallization time and taken to be 10 cal/g (with permission from Godard *et al.*, 1978).

the method used to separate the primary from the secondary crystallization effect). Godard *et al.* conclude from the value of this exponent that gelation of gelatin is a one-dimensional growth (crystallization) from predetermined nuclei. Such a result is in agreement with the actual gel fibre-like morphology (see Chapter 2).

Godard *et al.* have also examined the effect of gelatin concentration on the gel melting point (see Figure 1.47). The equilibrium melting points are determined by measuring the gel melting point *vs* crystallization temperature. They found that the variation from a 5.1% gel to a 31.5% gel is only 5.2°C (37 against 42.2°C). Similar results have been reported by several authors (Fouradier and Venet, 1950; Izmaïlova *et al.*, 1965; Godovskii *et al.*, 1971).

Godard *et al.* (1978) have found that secondary crystallization takes place on top of primary crystallization, as is demonstrated by the increase in the gel melting point beyond a given crystallization time (Figure 1.45). They also show that the experimental half-crystallization time, $t_{0.5}$, as determined by the method of Vidotto and Kovacs (1967), agrees with a theoretical expression derived by Pennings (1977):

$$\log(t_{0.5}/\Delta T^2) = A + (\pi\sigma_s^2 l T_m^0)/(2.303\Delta h_f kT\Delta T) \tag{1.22}$$

in which T_m^0 is the equilibrium melting point, $T = T_m^0 - T$ is the undercooling, σ_s is the lateral surface free energy per unit area, l the length of the secondary nucleus, Δh_f the enthalpy of fusion of the collagen crystal and A a term including the transport term $\exp(-\Delta G_a/kT)$ taken as constant in the temperature range investigated. Godard *et al.* (1980) have confirmed these results by optical rotation, as it has been shown by Macsuga (1972) that, to within good approximation, there is direct proportionality between $[\alpha]$ and ΔH_{gel}.

From the experimental results, they find $\sigma_s \simeq 7.5$ erg/cm^3 which allows them to calculate the diameter of the fibrils, D, through the following relations:

$$D = 2\gamma\sigma_s T_m^0/\Delta h_f \Delta T \qquad \text{with} \qquad T_m = T_m^0[1-(2/\gamma)] + (2/\gamma)T_c \quad (1.23)$$

The values of D computed by Godard *et al.* are reported in Table 1.5. These values show that the lower the undercooling, the larger the fibril diameter as expected. Also, it is revealed by the use of this theory that the fibril diameter decreases markedly with concentration, which is the direct consequence of the small melting point variation with concentration.

However, close inspection of the calculated data shows that the relative discrepancy between the diameter at $C = 5\%$ and $C = 31.5\%$ increases with decreasing undercooling which seems rather paradoxical. In fact, one would rather expect a smaller discrepancy of size as a function of concentration on approaching thermodynamic equilibrium where crystals are supposed to reach infinite size. All this went unnoticed in the paper of Godard *et al.* The considerations developed here suggest that the slight melting point variation as a function of concentration may convey something else.

It is worth emphasizing here that Godard *et al.*'s analysis does not consider the possibility of a gelatin–water complex. If one compares it with what is known for synthetic polymers, such a melting point variation is indeed small compared with the same range of concentration for block or random copolymers (see Sections 1.2.2.d and 1.2.2.e) and PVC (Section 1.2.2.b). Conversely, it is quite reminiscent of what has been found for isotactic polystyrene gels in *cis*-decalin. This suggests the occurrence of a gelation–water complex, a concept already put forward by Bear (1952) and Zaides (1954) and supported by others (Mrevlishvili and Sharimanov, 1978; Djabourov *et al.*, 1988*a*).

Table 1.5 Values of the fibril diameter, D, as calculated from relation 1.23 as a function of crystallization temperature (T_c) and for different concentrations (in w/w).

		D (nm)				
T_c (°C)	T_m (°C) (for 1%)	1% [(a)]	5% [(b)]	9.8% [(b)]	20% [(b)]	31.5% [(b)]
10	26	63	58	56	53	50
12.5	26.2	70				
15	26.5	78	72	68	64	59
17.5	27.7	88	81	76	71	65
20	28.5	102	93	86	79	72
22.5	29.7	121	109	101	91	81
25	30.8	148	131	119	106	93

[(a)] Results obtained from optical rotation and [(b)] from calorimetry. From Godard *et al.* (1980).

The study by Flory and Garrett (1958) of gelatin gels in ethylene glycol and its comparison with gelatin gels in water adds more information to the argument. They performed their measurements on gelatin, beef achilles tendon collagen and rat tail tendon, the last two being swollen in ethylene glycol. Despite the fact that gelatin possesses a lower degree of crystallinity in gelatin–ethylene glycol gels, these investigators found that the gel melting point is only 1 °C below that determined for collagen mixtures of the same concentration. Melting temperatures plotted against composition reveal that the melting behaviour can be well reproduced theoretically by Flory's relation (relation 1.1) (Flory, 1953) with $T_m^0 = 145\,°C$ and $\Delta H \simeq 24$ cal/g (Flory and Garrett, 1958).

They also point out that the virtual coincidence of the melting points for gelatin–ethylene glycol and collagen–ethylene glycol affords convincing evidence that the aggregation process in gelatin thermoreversible gels is none other than a reversion towards the native crystalline form of collagen.

Yet, Flory and Garrett (1958) have noticed, without actually much dwelling upon the point, that Flory's relation applied to gelatin–water gels does not yield the same melting point T_m^0. This can be easily demonstrated by considering a variant form of Flory's relation (1.1).

$$\frac{1}{T_{m_p}} = \frac{RV_p}{\Delta H_{0_p} V_s}(v_s - \chi_1 v_s^2) + \frac{1}{T_{m_p}^0} \tag{1.24}$$

A plot of $1/T_m$ *vs* v_s should then give a **downturn parabola** from which T_m^0 can be straightforwardly derived without making any further assumptions about the system. In Figure 1.47 are plotted Flory and Garrett's results (FG for short) and Godard *et al.*'s results (G for short). The melting point variation with concentration of gelatin–ethylene glycol gels is well reproduced with Flory's equation, through a fit with a polynome of degree 2 which eventually yields $T_m^0 = 141\,°C$ and $\Delta H = 27$ cal/g (since Flory's relation is only valid above C^*, here the value obtained by FG in very dilute solution has been excluded, hence the slight difference with FG theoretical fit).

Conversely, in the same representation gelatin–water gels display an **upturn parabola** instead and yield extrapolated melting points for the bulk state much lower than those found with the gelatin–ethylene glycol system ($T_m^0 = 30 \pm 2\,°C$ instead of 141 °C in ethylene glycol). This accordingly indicates that the original Flory's relation cannot describe the melting behaviour of gelatin–water gel as the bulk melting point is not obtained. Clearly, this analysis hints at the possibility of a gelatin–water complex being formed in which case a modified version of Flory's relation may appear more suited (see for instance Appendix 1 relation A1.10). Incidentally, although FG and G results are slightly different for gelatin–water gels, they lead to nearly the same extrapolated melting point.

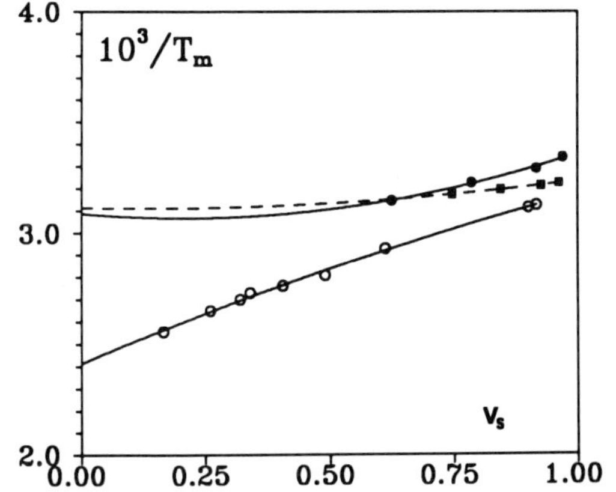

Figure 1.47 Inverse of the melting temperature *vs* solvent volume fraction. (•) Data taken from Flory and Garrett, 1958 (gelatin/water gels) and (■) Godard *et al.*, 1978 (gelatin–water gels) and (○) Flory and Garrett, 1958 (gelatin–diethyl glycol systems). The different lines are obtained from a best fit with Flory's relation (relation 1.24) by using a polynomial of second order which entails no assumption as to the values of the constants. Note the discrepancy between T_m^0 in aqueous gels and in ethylene glycol systems.

More details on the melting behaviour of gelatin–water gels has been gathered over the past few years. Djabourov *et al.* (1988*a*) have also investigated the aqueous gel melting behaviour. By using polarimetry and scanning the gel sample as a function of temperature with a heating rate of 0.05°C/min they have obtained the variation of $[\alpha]$ as a function of temperature. By taking the derivative $\mathrm{d}[\alpha]/\mathrm{d}T$ the resulting curves are supposed to be comparable with DSC curves according to these authors. From these experiments they observe two temperature regimes:

(1) Below $T = 20°\mathrm{C}$ the melting curve is composed of a single peak which depends upon the quenching temperature.

(2) For $20°\mathrm{C} \leqslant T \leqslant 28°\mathrm{C}$ the melting curve consists of two peaks, the high-melting peak being at 36°C, that is corresponding to the thermal stability of collagen. Two types of helices with different stabilities are therefore present.

According to Djabourov *et al.* (1988*a*), Godard *et al.*'s interpretation supposes that the extrapolated melting temperature of fully developed crystals ought to depart significantly from the melting temperature of a single triple helix, as association should give the system higher stability. Since their measurements show the network to possess a stability comparable with the collagen triple helix, Djabourov *et al.* conclude that helix renaturation is a conformational transformation which induces a three-chain association.

The thermal history of gelatin gels can have considerable influence on such parameters as the gelation time. For instance, Chatellier *et al.* (1985) report that SEC chromatograms determined for 11% gelatin–water samples that have never gelled, once gelled or twice gelled differ if the solution is kept at 45°C for 30 min. Conversely, the thermal history is erased once the systems are annealed for a minimum of 30 min at 50°C.

The swelling behaviour of gelatin–water gels has been investigated by Northrop (1927) and Northrop and Kunitz (1927, 1931). They determined the change in weight of gel samples immersed in an excess of water (pH 4.7, maintained at the isoelectric point). Below a concentration of 0.1 g/cm^3 the gel samples deswell while above this concentration they swell (Figure 1.48). Deswelling is a most remarkable feature which differentiates gelatin gels from chemical gels. Owing to the lack of data at higher concentrations, it cannot be decided whether some level-off is reached which would mean that the gels always swell at the same concentration. If such were the case, this would provide further support for the existence of a gelatin–water complex and show additional resemblance to iPS–*cis*-decalin gels.

(b) Neutral polysaccharide gels

Neutral polysaccharides occur naturally in algae, plants, bacteria and animals. The phenomenon of thermoreversible gelation of naturally occurring polysaccharides has been extensively studied on agarose, amylose and amylopectin (both components of starch). While cellulose does not possess gelation capability, man-made cellulose derivatives such as hydroxypropylcellulose and cellulose nitrate do.

(i) Agarose

Agarose is the gelling component of agar which is found in red seaweeds. Agarose is an alternating copolymer of 1,4-linked 3,6-anhydro-α-L-galactose and 1,3-linked β-D-galactose:

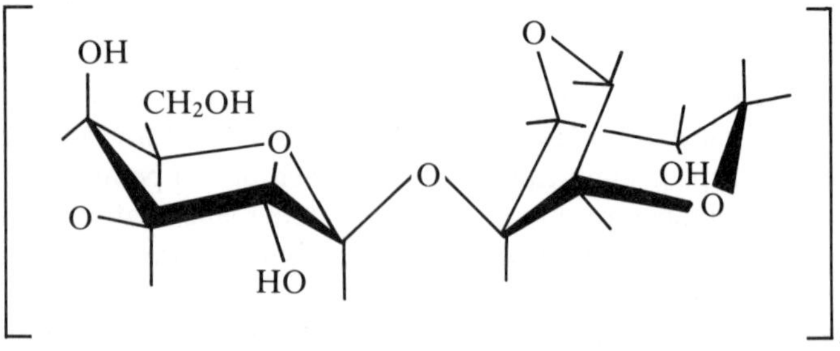

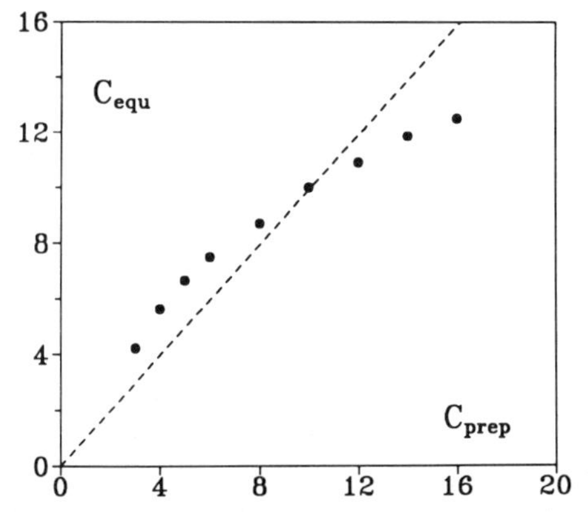

Figure 1.48 Swelling behaviour of gelatin–water gels. The equilibrium concentration, C_{eq}, reached after swelling to equilibrium is plotted as a function of the preparation concentration, C_{prep}. The dotted line represents $C_{eq} = C_{prep}$, concentrations in g/100 cc (data taken from Northrop and Kunitz, 1931).

Agar or agar-agar (a name of Malay origin) is extracted from red seaweeds of the types *Gelidium* or *Gracilia*. Agar consists of two fractions: a (virtually) neutral one, agarose, and a charged one with no gelling properties (agaropectin). Precipitation of agaropectin by means of cetylpyridinium chloride (Hjerten, 1962) or adsorption of agaropectin on to aluminium hydroxide (Barteling, 1969) are two of the various methods employed to recover agarose.

Impurities along the chain, such as galactose or galactose 6-sulphate instead of 3,6-anhydro-α-L-galactose, can also be found. Hickson and Polson (1968) have determined the molecular weight of agarose polymers by three methods and found a value of about 1.2×10^5. Djabourov *et al.* (1989) report weight-averaged molecular weights of about 5.12×10^5 with a polydispersity $M_w/M_n \simeq 3.3$. More precise and reliable determinations have been achieved by Rochas and Lahaye (Rochas and Lahaye, 1989) by means of an aqueous size exclusion chromatography coupled with a low-angle laser light scattering. According to these authors, the majority of the agarose samples from different commercial origins have molecular weights ranging from 8×10^4 to 1.4×10^5 with polydispersities lower than 1.7. From their measurements they have derived an experimental relation for the intrinsic viscosity *vs* molecular weight in 0.75M NaSCN aqueous solutions at 35°C: $[\eta] = 0.07\ M^{0.72}\ cm^3/g$. (Under these conditions no gelation takes place.)

Like synthetic polymers, the sol–gel and the gel–sol transitions in the system agarose–water display a strong hysteresis in temperature. For instance a 3 mg/cm^3 solution gels at $T \simeq 40$°C and the resulting gel melts at $T \simeq 80$°C (see for instance Rees *et al.*, 1969; Pines and Prins, 1973;

Hayashi *et al.*, 1977; Letherby and Young, 1981). Rees *et al.* (1969) have tentatively explained this hysteresis by considering two valleys in the representation of the free energy of the system as a function of entropy and temperature. However, one may wonder whether this is not so because gelation occurs through **homogeneous nucleation** only. Under these conditions all systems are expected to exhibit large undercoolings (see Appendix 1).

Feke and Prins (1974) have studied the kinetics of gel formation by elastic light scattering. By measuring the increase in scattered light at a given angle, these investigators come to the conclusion that there are two types of gel formation mechanism depending upon the quenching temperature. At low quenching temperature ($T_{quench} = 25.5^{\circ}C$) they claim that the temporal evolution of intensity is consistent with a spinodal decomposition. This mechanism which takes place in the so-called unstable region of the miscibility gap (see Appendix 1) involves only diffusion (Cahn and Hilliard, 1959, 1965). Under these conditions Cahn and Hilliard's theory predicts that the temporal evolution of intensity at the early stage of the spinodal decomposition should be:

$$I(q, t) = I(q, 0)\exp[2r(q)t] \tag{1.25}$$

in which $r(q)$ is the growth rate and q the momentum transfer ($q = (4\pi/\lambda)\sin(\theta/2)$). As can be seen in Figure 1.49, this behaviour can be found at the early stage of gelation after a so-called response time, t_b.

Conversely, at higher quenching temperatures ($T_{quench} = 40^{\circ}C$), the gelation occurs through a nucleated process as no exponential growth is seen.

As another test, Feke and Prins (1974) have found that the scattering pattern displays a maximum at different momentum transfers depending upon the quenching temperature, as expected from a spinodal decomposition (Figure 1.50). It is worth stressing, however, that the observation of a maximum is not a sufficient condition to conclude in favour of this mechanism. As long as periodicity appears, whatever the underlying mechanism, a Bragg-type maximum is liable to be observed.

Finally, as a last test, these investigators have determined the growth rate, $r(q)$, as a function of the momentum transfer. From Cahn's approach it should be given at a given quenching temperature as:

$$r(q) \simeq q^2(q_c^2 - q^2) \tag{1.26}$$

in which q_c is the critical momentum transfer. They did find that $r(q)/q^2$, as measured from the exponential increase of intensity, varies linearly with q^2.

Djabourov *et al.* (1989) tend to agree with the liquid–liquid phase separation mechanism although they hasten to add that little extra substantiation

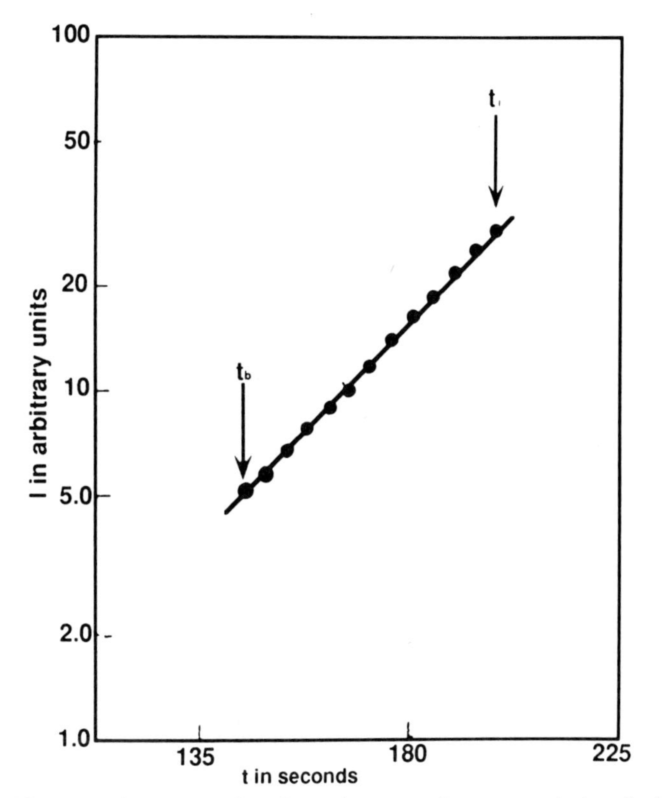

Figure 1.49 Intensity scattered at the early stage of agarose gelation (in logarithmic scale) plotted as a function of time showing that the relation $I(q, t) \approx \exp - (2r(q)t)$, wherein $r(q)$ is the growth rate of the spinodal decomposition, is obeyed (with permission from Feke and Prinz, 1974).

is provided by either electron micrographs as available or their own optical microscopy.

It also might be worth adding that one of the major problems in observing the spinodal decomposition mechanism lies in the difficulty of reaching the unstable domain in the miscibility gap without any alteration of the sample. This is supposed to be feasible with viscous systems for which the relaxation time or rather the response time is long enough compared with the quenching time (time needed to reach the desired temperature). The question then remains whether the experimental evidence given by Feke and Prins constitutes unshakable proof of spinodal decomposition being involved in agarose gel formation at low temperatures.

Still, there seems little doubt that gelation occurs through a liquid–liquid phase separation whatever its mechanism may be (diffusion or nucleation and growth). Data collected by Letherby and Young (1981) are very

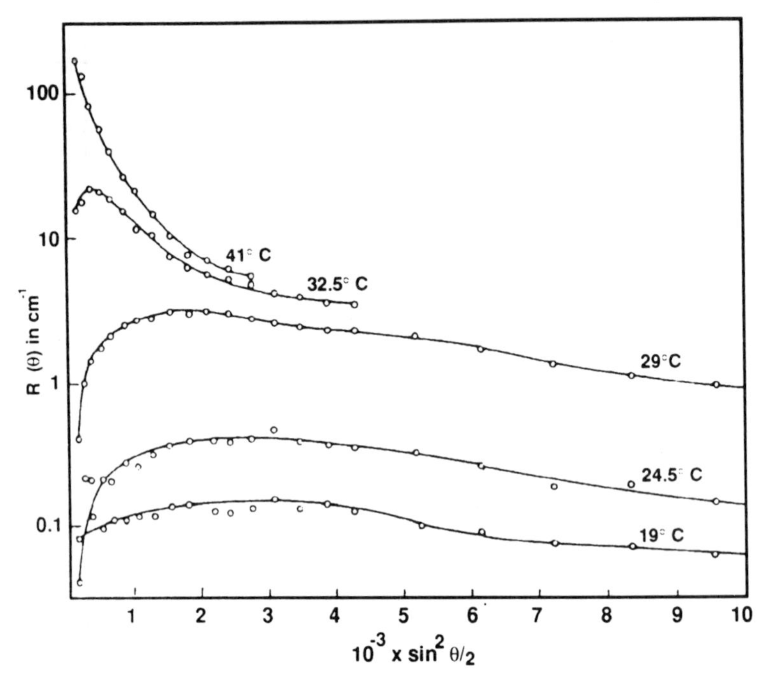

Figure 1.50 Logarithm of the equilibrium Rayleigh ratios (V_v polarization) as a function of the scattering angle for 1% (w/w) aqueous agarose gels quenched to the indicated temperatures from 82°C (with permission from Feke and Prinz, 1974).

enlightening on this point. First they agree to fix the θ-point of agarose in water at $T = 35°C$ which ultimately implies the existence of a miscibility gap below. Second, they have shown that, while gel formation temperature is concentration dependent, gel melting is not. In the concentration range investigated (up to 3%, w/w) the gel formation line strongly resembles a binodal (Figure 1.51). This resemblance and the fact that the gel melting point does not vary with concentration are most reminiscent of what has been found by He *et al.* (1987*a*, 1988) for multiblock copolymers (see Section 1.2.2.d). The melting point invariance may be indicative of a monotectic transition which agrees well with the occurrence of a liquid–liquid phase separation.

Further studies were carried out by Watase and Nishinari (1987) and Watase *et al.* (1989) on agarose samples of lower molecular weight ($[\eta] = 1.7$ dl/g in 0.01 mol/l aqueous sodium thiocyanate solution at 35°C) which allowed these investigators to study gels of higher concentration (up to 40%, w/w). They observed an exotherm near 30°C for gels heated up for the first time. They have attributed this exotherm to rearrangements. However, gels molten in the DSC pan then heated up do not show this

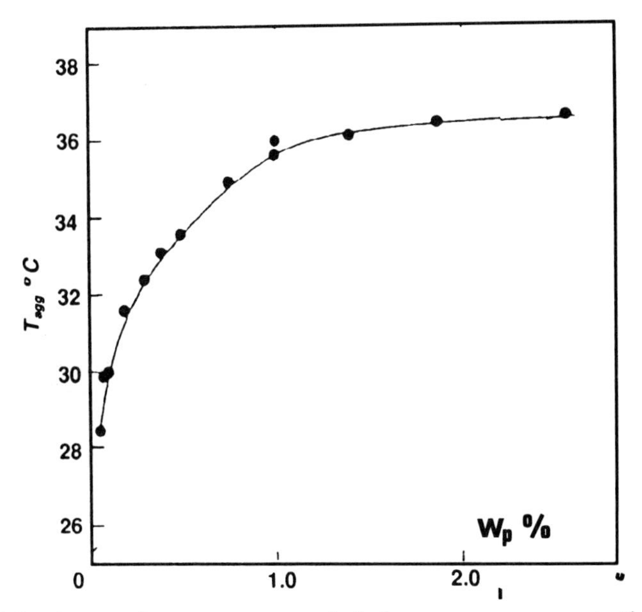

Figure 1.51 Aggregation temperature plotted *vs* agarose concentration (w/w) (data from Letherby and Young, 1981 with permission).

thermal event, which is in fact probably due to constraint release (see Section 1.2.2.b) and has, accordingly, nothing to do with anything taking place in the gel. At any rate, it should be emphasized that DSC curves are meaningful only if the system is first molten in the DSC pan then treated as required.

Quenching gels molten in the DSC pan and then reheating them gives curves such as those in Figure 1.52. As can be seen, there is a broad melting endotherm followed by a narrower one. Interestingly, the position of the maximum of the broad endotherm is virtually invariant with concentration whereas the second endotherm increases slightly (+4°C). Again, this is consistent with a monotectic transition arising from a liquid–liquid phase separation. These thermograms are very reminiscent of what has been observed with multiblock copolymers (Section 1.2.2.d) except that the low-melting endotherm is broader than the high-melting one.

In the same studies, use was made, with more or less success, of the Eldridge and Ferry (1954) theory. Applying this theory to the problem under consideration is irrelevant and provides meaningless data, as it deals with second-order transitions whereas agarose gel melting is typically a first-order transition (see Appendix 1).

Watase *et al.* (1989) also report that the melting enthalpy displays a linear dependence on polymer concentration. Yet, there is no deconvolution

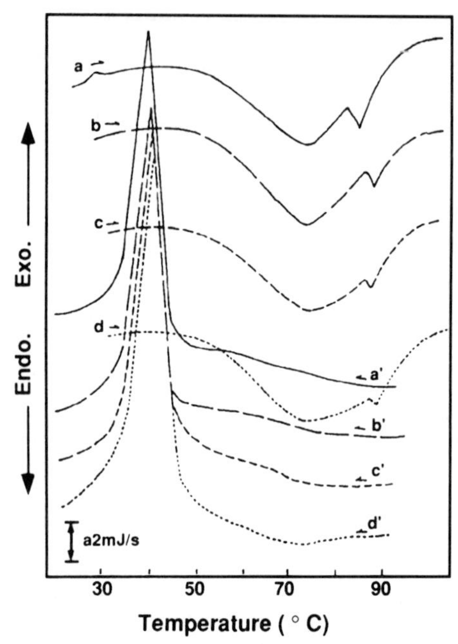

Figure 1.52 DSC curves of concentrated agarose–water gels. Heating and cooling rates 2°C/min. (a) Reheating and (a′) cooling for a 25% (w/w) gel; (b) reheating and (b′) cooling for a 30% gel; (c) reheating and (c′) cooling for a 35% gel; (d) reheating and (d′) cooling for a 40% gel (with permission from Watase *et al.*, 1989).

between the low- and the high-melting endotherm so that the monotectic concentration, if any, is not defined (incidentally Watase *et al.* do not provide any explanation for the temperature invariance).

Hayashi *et al.* (1980) also agree with the liquid–liquid phase separation being involved, but come to the additional conclusion that the temperature–concentration phase diagram must be similar to those described by Flory (1956) for rod-like polymers. These diagrams display a so-called chimney which Hayashi *et al.* (1980) locate between $C = 0.8$ g/dl and 1.4 g/dl by means of fluorescence depolarization experiments and the falling time of a steel ball at 45°C and 49°C, i.e. above the temperature of gel formation.

The gelation temperature has been found to be dependent upon the degree of methylation of agarose (in fact the CH_2OH group is sometimes replaced by a CH_2OCH_3 group or the OH group replaced on the other ring by a OCH_3 group). Depending on whether the methylation is of natural origin or carried out *in vitro*, the melting point increases (up to 45°C for 8% methylation) or decreases (down to 17°C for about 9% methylation) respectively (Guiseley, 1970).

Agarose in pure water has not been the only system investigated. Water containing soluble inorganic salts has also been used as a gelation medium. In spite of the neutral character of agarose, gelation properties are anion-sensitive, as reported some 40 years ago by McBain (1950) and more recently by Letherby and Young (1981) and Piculell and Nielsen (1989). For instance, addition of NaSCN to water reduces drastically the gel formation temperature from 40° ± 5°C to 2° ± 2°C for a 1.5M solution (Piculell and Nielsen, 1989).

It is currently believed that the effects observed arise from an increase in solvent quality with increasing salt content (see, for instance, Norton *et al.*, 1984). According to Piculell and Nilsson (1989), the enhancement of solvent quality results from preferential absorption which, they rightly emphasize, may take place even with an inert solute on account of the non-ideality of the solvent mixture.

Attempts to quantify the anion effect have been carried out by Piculell and Nilsson (1989) by an approach derived by Schellman (1987) for selective binding of denaturant molecules to macromolecules. These authors have worked out the following expression for the variation of chemical potential per repeating unit, $\Delta\mu_{\mathrm{hc}}$, at the coil–helix transition:

$$\Delta\mu_{\mathrm{hc}} = \Delta\mu^0_{\mathrm{hc}} - RT\Delta n \sum_i \ln\left[1 + (K_{+,i} + K_{-,i} - 2)x_s\right] \qquad (1.27)$$

where $\Delta\mu^0_{\mathrm{hc}}$ is the variation of chemical potential without added salt, x_s the mole fraction of the anion (or the cation), K equilibrium constants associated with the exchange between water molecules and denaturant molecules on a macromolecular site and Δn the number of new sites that are exposed as a result of the transition (assuming that all sites are equiprobable).

They have finally derived the onset temperature T_0 as a function of the partial molar enthalpy, Δh_{hc}, associated with the transition:

$$T_{\mathrm{m}} = \frac{\Delta h_{\mathrm{hc}}/\Delta n R}{\Delta h_{\mathrm{hc}}/(\Delta n R T^0_{\mathrm{m}}) + \ln\left[1 + (K_+ + K_- - 2)x_s\right]} \qquad (1.28)$$

By means of relation (1.28), Piculell and Nilsson (1989) produce reasonable fits to their experimental data.

(ii) Starch components: amylose and amylopectin

Amylose is a plant polysaccharide which is one of the two major components of starch (about 20–32%), the other one being amylopectin (see for instance French, 1984). Amylose is a linear chain constituted of only one residue (1 → 4)-α-D-glucan.

Amylopectin is chemically similar to amylose but it is highly branched and composed of short (1 → 4)-α-D-glucan chains which are linked together through α-(1 → 6) branchpoints.

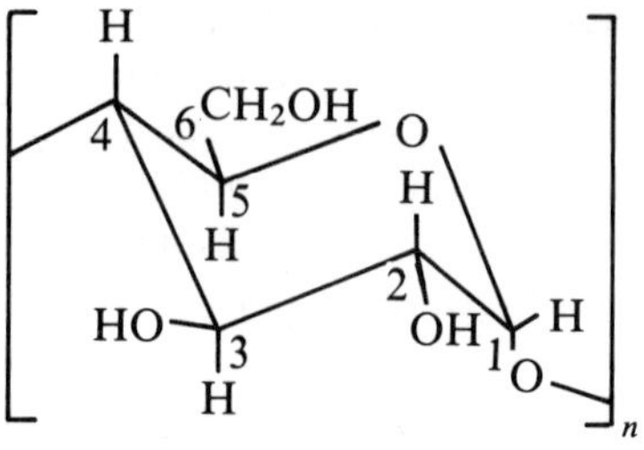

Amylose and amylopectin owe their names to the greek word for starch, 'αμυλον' (amylon) which can be literally translated by 'without millstone' thus indicating that the extraction process of starch from barley or from corn seeds did not require the use of a millstone. In many studies starch is obtained from smooth-seeded peas by the method of Adkins and Greenwood (1966) relying on aqueous extraction. The seeds are left overnight at 4°C in an NaCl solution (0.1M) containing 0.1M-$HgCl_2$, the latter being added to inhibit the action of the enzymes. The swollen seeds are eventually pounded and then passed through a sieve. Starch is finally recovered by decanting and washed with 0.1M-NaCl to eliminate all sorts of cytoplasmic proteins. Amylose is obtained from starch by leaching the starch granules. Amylopectin is usually obtained from waxy maize starch by dissolving the starch granules in aqueous 90% dimethyl sulphoxide at 95°C followed by continuous stirring at room temperature for 48 h and then precipitation in ethanol.

Amylose gelation is an important phenomenon since it is believed to be of considerable influence in the textural properties of starchy food products. (Amylose gelation in starch is sometimes also referred to as retrogradation.) Only recently have investigations been carried out on this subject which is, paradoxically, relatively new compared with the knowledge of amylose.

Dissolving amylose in water cannot be achieved straightforwardly. There are several methods: (i) preparation of amylose by heating at 95°C solutions from the complex formed with butan-1-ol after starch extraction and then removing the alcohol by a heated nitrogen stream (Miles *et al.*, 1985), (ii) heating amylose–water mixtures at 160–170 °C in a sealed vessel which allows clear solutions to be obtained (Gidley and Bulpin, 1989) and (iii) dissolution in alkaline solutions (amylose in 1M-KOH) (Kitamura *et al.*, 1984; Doublier and Choplin, 1989). Whereas in the first two cases the gel is obtained by cooling, in the latter it is formed by neutralization with HCl.

According to Miles *et al.* (1985) amylose gelation occurs via liquid–liquid phase separation (it should be underlined here that Miles *et al.* designate it ambiguously as phase separation in their paper). There is little in their paper to substantiate this view, i.e. that amylose gelation occurs via the liquid–liquid phase separation, except the increase in gel opacity with time (a clear gel is obtained first which becomes more and more opaque with time). As has been seen with isotactic polystyrene, such an increase in opacity may be due to the transformation of the original gel structure into another crystalline form. After all, amylose is a crystallizable polysaccharide unlike agarose, the latter giving only gels.

That there is competition between crystallization and gelation is clearly shown by the work of Gidley and Bulpin (1989). These investigators have studied the effect of amylose molecular weight on the gelation behaviour with enzymatically synthesized samples (polydispersities $M_w/M_n < 1.15$). They have found that low molecular weight samples (degree of polymerization, DP < 110) give only precipitates (most probably crystalline), middle molecular weight amyloses (250 < DP < 660) tend to give precipitates and gels and high molecular weight samples (DP > 1100) form only gels.

It is striking how similar the gelation of amylose is to the gelation of some synthetic polymers (competition between precipitation or crystallization and gelation as in isotactic polystyrene or the need for long chains as in polyethylene) and yet no comparison has been explicitly made so far.

Turbidity usually rises with the gel modulus G'. For instance, Doublier and Choplin (1989) found a direct correlation between the cloud time (defined arbitrarily as the departure of the turbidity curve from the baseline) and the gel time (defined as the point of separation between G' and G''). They further report that for a 1.6% amylose gel in 0.2M KCl above 42°C turbidity does not occur at all (they define this temperature as the cloud limiting temperature: c.l.t.). They note that this temperature varies markedly with concentration: from 35°C for 1% to 45°C for 1.8% solutions. They consider that the evolution of the c.l.t. with concentration defines the cloud point curve below which liquid–liquid phase separation occurs. As a result, cooling the alkaline solution to 25°C (where it is still in a good solvent) and altering the pH from 14 to 7 brings the system below the miscibility gap. They conclude that gels are formed through liquid–liquid phase separation frozen in at some stage by crystallization.

It is worthwhile to mention, however, that experimental results seem to be very sensitive to the preparation method. Doublier and Choplin (1989) emphasize that recovering amylose from the amylose–butanol complex or heating mixtures with water up to 160–170°C produces results which are dependent on the thermal history.

Despite the suffix pectin, **amylopectin** is chemically different from pectin and gives gels that are thermoreversible. As mentioned above, amylopectin is a polysaccharide extracted from starch and constitutes its major component.

The gelation capability of amylopectin was reported by Ring (1985). Gels can be formed by quenching at low temperature (usually around 1°C) (Ring *et al.*, 1987). Gelation kinetics are exceedingly slow, probably the slowest, to our present knowledge, for thermoreversible gels, since weeks rather than minutes are usually required. In addition, gelation occurs at concentrations higher than 10%.

A broad gel melting endotherm can be observed in the range 40–70°C with a maximum at 58 ± 4°C for 20 or 25% solutions. Ring *et al.* (1987) have shown that the heat of melting increases slowly with the gelation time. It takes approximately 40 days to reach the maximum value.

(iii) Cellulose derivatives

Cellulose derivatives possessing gelling properties are ester or ether derivatives such as cellulose nitrate (nitrocellulose), cellulose butyrate, methylcellulose or hydroxypropylmethylcellulose. Cellulose has the following ideal chemical structure:

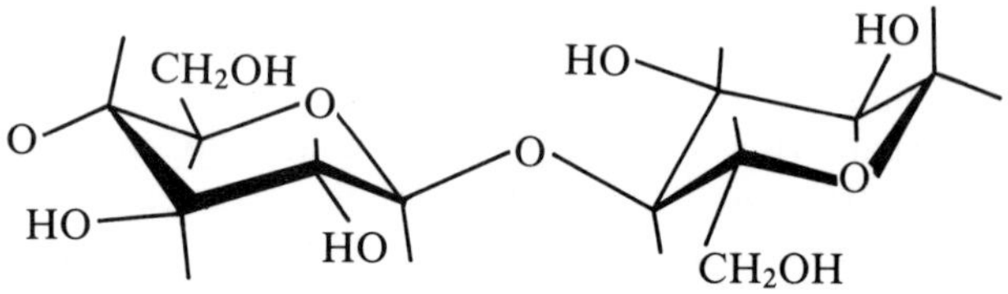

Substitution occurs on the OH groups by replacement of the hydrogen atom. In practice cellulose is only partly substituted so that the chemical structure of highly substituted nitrocellulose, for instance, is typically somewhere between dinitro and trinitrocellulose.

These derivatives display very special gelation properties in that **they form gels upon heating which reversibly melt upon cooling**. This is the only known case so far. Although the gelation phenomenon has been known for a long time (Heymann, 1935; Vacher, 1940), fundamental studies on the gelation phenomenon are relatively scarce. Conversely, practical applications are numerous, from the food industry to printing technology.

Early reports on the gelation behaviour of cellulose nitrate in various solvents were made by Doolittle (1946). More extensive studies on the gelation conditions were carried out by Newman *et al.* (1956). These authors established a temperature–concentration phase diagram in ethanol (see Figure 1.53). It is known that these systems exhibit a low critical-solubility temperature, i.e. there exists a miscibility gap at high temperature leading to liquid–liquid phase separation (see for instance Newman *et al.*, 1956; Sarkar, 1979; Conio *et al.*, 1983). According to earlier views put forward by Rees (1972), these gels were termed 'micellar gels' which means that gel junctions would be analogous to non-ionic surfactants and would, therefore, not display any ordering. Yet, the tendency of these derivatives to form mesophase (Werbowyj and Gray, 1976, 1980; Conio *et al.*, 1983) points towards some kind of organization.

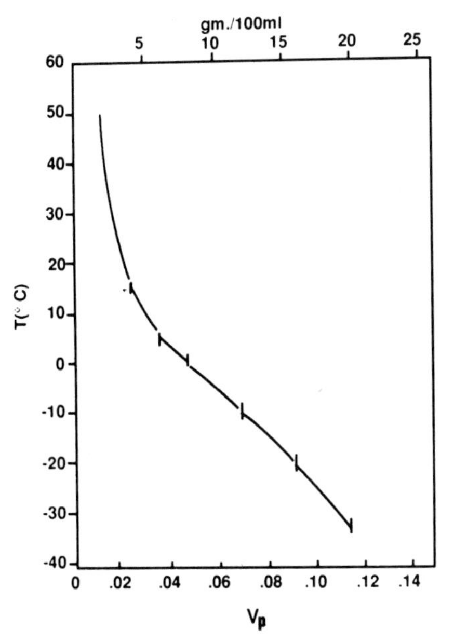

Figure 1.53 Visual melting or liquefaction temperatures for cellulose nitrate–alcohol system as a function of the polymer volume fraction (with permission from Newman *et al.*, 1958).

The reasons why these derivatives form thermoreversible gels upon heating is thus linked to the fact that the solvent becomes poorer at high temperature which favours chain–chain interactions. On the other hand, the type of substitution plays an important role. For instance, solutions of derivatives such as sulphoethylcellulose and hydroxyethylcellulose do not produce gels. Hydroxypropylcellulose, although it precipitates upon heating, does not form a gel at low concentration as nitrocellulose does. Sarkar (1979) mentions that solutions of hydroxyethylmethylcellulose, containing a high degree of methoxyl and a low degree of hydroxyethyl substitution gel, whereas the polymer containing a high degree of hydroxyethyl and a low degree of methoxyl substitution does not form gels (Sarkar, 1979).

1.3.2 Charged biopolymers

(a) Carrageenans

Some of these polysaccharides can produce gels that are formed irreversibly (such as alginates or pectins for instance). They are beyond the scope of the present monograph and, therefore, will not be discussed here.

Of main interest are carageenan gels. Carrageenans are electrolytc polysaccharides containing ester sulphate groups. They are copolymers of alternating 1,4-linked α-D-galactose and 1,3-linked β-D-galactose. Carrageenans are extracted, as in agar, from red seaweeds and come in three major types designated by means of greek letters ϰ, ι, λ. Only ι-carrageenan and ϰ-carrageenan possess gelling capability. Their general chemical structures are as follows:

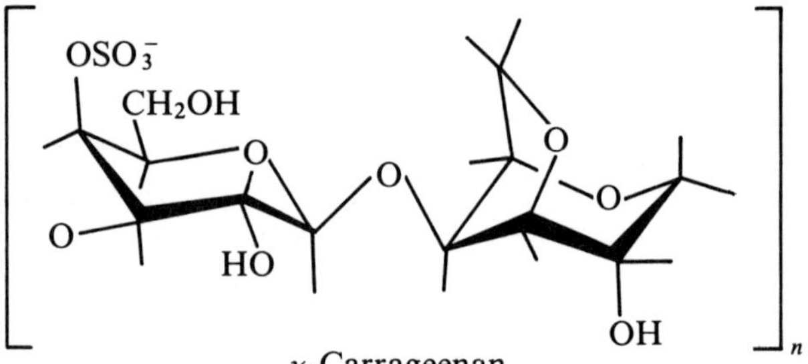

ϰ-Carrageenan

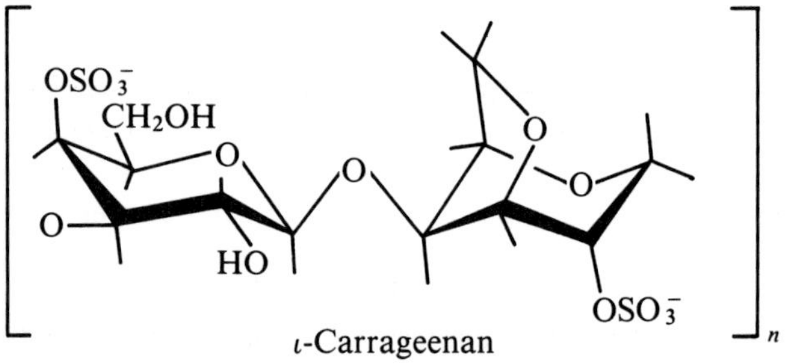

ι-Carrageenan

Carrageenans owe their name to an Irish town Carragheen, near Waterford, where red seaweeds of carrageenan type (*Condrus crispus*) are abundant.

As can be seen, the only difference between ι-carrageenan and ϰ-carrageenan lies in the presence of an additional sulphate group in the former on the 1,4-linked α-D-galactose residue. Despite the care taken to obtain both types in the purest form, the actual chemical structure always differs slightly from the ideal given above. The 1,4-linked α-D-galactose residue turns out to be partly replaced by 1,4-linked galactose 6-sulphate and 2,6-disulphate.

As expected for polyelectrolytes, the counterion plays a major role in solution properties and, correspondingly, in the gelation process. For instance, it is traditionally mentioned that ι-carrageenan is calcium-'sensitive' while ϰ-carrageenan is more potassium-'sensitive'.

Like most of the thermoreversible gels known to date, carrageenan gels display hysteresis in their formation and melting temperature. However, ι-carrageenan gels form at about 65–68°C and melt 2–5°C above, which is probably the smallest hysteresis reported so far (including that for synthetic polymers). In contrast, $\varkappa$-carrageenan gels are characterized by a range in gelation temperature, melting temperature and hysteresis width (see, for instance, Dea *et al.*, 1972). Clearly, gel formation and gel melting temperatures are strongly dependent upon the nature of the counterion. Conversely, gel formation temperature and gel melting temperature, although strongly dependent upon the salt content, are little sensitive to the polymer concentration at least in the range investigated.

Extensive studies by optical rotation of the effect of both polymer and cation concentrations on $\varkappa$-carrageenan gelation have been carried out by Rochas and Rinaudo (1980). These investigators came to the conclusion that the melting temperature T_m is dependent on both the polymer concentration, C_p, and the ionic concentration, C_s. Yet, T_m is independent of polymer concentration if the total ionic concentration of free counterions, C_T, is taken into account instead. This concentration is:

$$C_T = \bar{\gamma} C_p + C_s \tag{1.29}$$

in which $\bar{\gamma}$ is the average activity coefficient of the cations present on the polymer since $\varkappa$-carrageenan is in the form of a salt the activity coefficient, $\bar{\gamma}$, of which depends upon the conformation. The value of $\bar{\gamma}$ is then calculated from:

$$\bar{\gamma} = \tfrac{1}{2}(\gamma_{\text{coil}} + \gamma_{\text{helix}}) \tag{1.30}$$

On establishing the phase diagram of T_m as a function of C_T, both for cooling and for heating, these authors have observed two regimes (Figure 1.54). Below $C_T = C^* = 7 \times 10^{-3}$ eq/l, the formation and melting process takes place at the same temperature which corresponds to a second- or higher-order transition. This is illustrated by polarimetry experiments for which no hysteresis is detected upon cooling or upon reheating (see Figure 1.55A). Conversely, above $C_T = C^* = 7 \times 10^{-3}$ eq/l, these processes are differentiated (Figure 1.55B) which is a characteristic of first-order transitions involving a nucleation and growth process. This is correlated with the gel formation which occurs only for $C_T > 7 \times 10^{-3}$ eq/l.

It might be of interest to point out that such behaviour at $C_T = C^*$ is reminiscent of what is expected for a tricritical point (as is the θ-point, for instance (see Daoud and Jannink, 1976)) where a first-order line and a second-order line meet.

Rochas and Rinaudo (1980) have analysed these data by using Manning's theory (Manning, 1972). This theory, which makes use of the Van't Hoff equation, considers a second-order transition. It gives the following

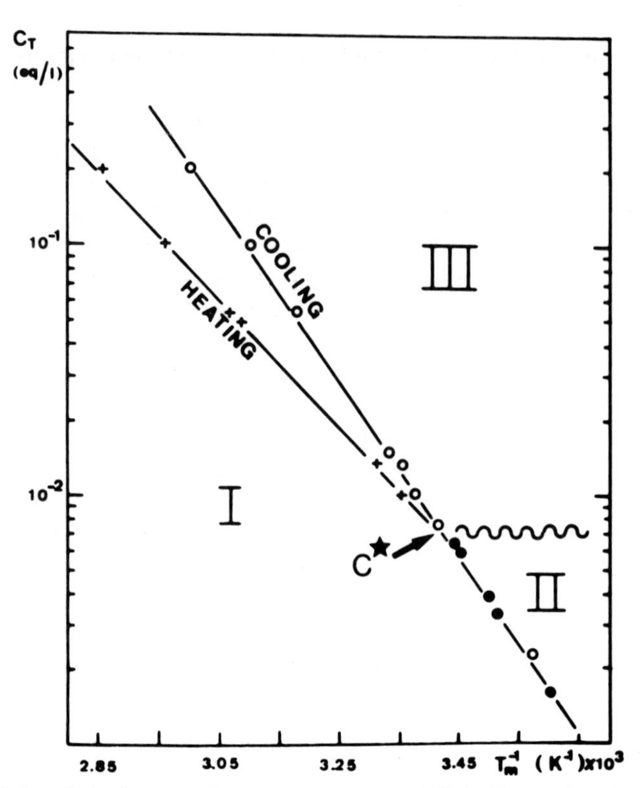

Figure 1.54 Phase diagram for ϰ-carrageenan established by optical rotation representing the inverse of the helix-coil transition temperature, T_m^{-1}, as a function of the total free potassium concentration. Above C* the curve splits in two: (x) transition determined on heating and (○) transition determined on cooling. Region I = coil state, region II = dimer of helices (no macroscopic gel formed) and region III = gel. The limit between II and III is somewhat diffuse as indicated by the oscillating line (with permission from Rochas and Rinaudo, 1984).

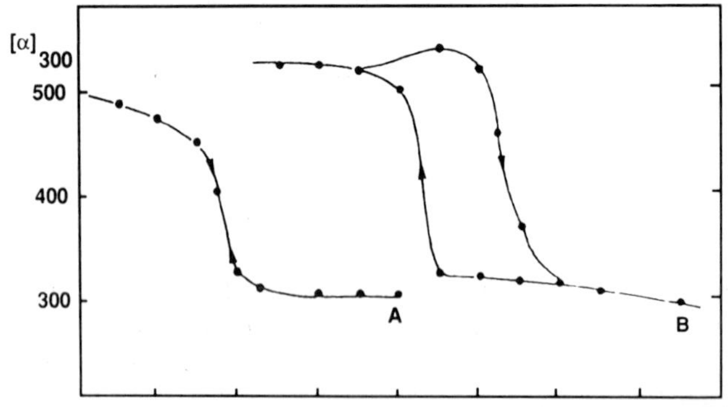

Figure 1.55 Dependence of the specific rotation $[\alpha]_{300}$ *vs* temperature in potassium–ϰ-carrageenan systems for two different salt and polymer contents: (concentrations in 10^2 eq/l) A: $C_p = 1.2$, $C_{salt} = 0$ and $C_T = 0.64$; B: $C_P = 0.13$, $C_{salt} = 5$ and $C_T = 5.07$ (with permission from Rochas and Rinaudo, 1980).

relationship between T_m and C_T:

$$\partial(1/T_m) = -(2.3R/\Delta H_{melting})(\partial \log C_T) \qquad (1.31)$$

Provided that the enthalpy associated with the melting process, $\Delta H_{melting}$, is a constant over the investigated range of temperature, this relationship indicates a linear variation between $\log(1/T_m)$ and $\log(C_T)$ which is effectively what Rochas and Rinaudo found. They also observed a similar variation in dimethyl sulphoxide and dimethylformamide (Rochas and Rinaudo, 1982*b*), although no gels are formed in these solvents in the absense of added salt.

It should be noted here that a theory based on a **second-order transition** is used to fit data above $C_T = C^* = 7 \times 10^{-3}$ eq/l that manifestedly represent a **first-order transition**. That a first-order transition is involved seems to be substantiated by calorimetric measurements (Rochas and Rinaudo, 1982*b*) that display a melting endotherm. Clearly, there might be something to eludicate here.

Rochas and Rinaudo (1980) have further shown that, while the variation of T_m *vs* C_T always obeys linearity in a log–log representation, there is a

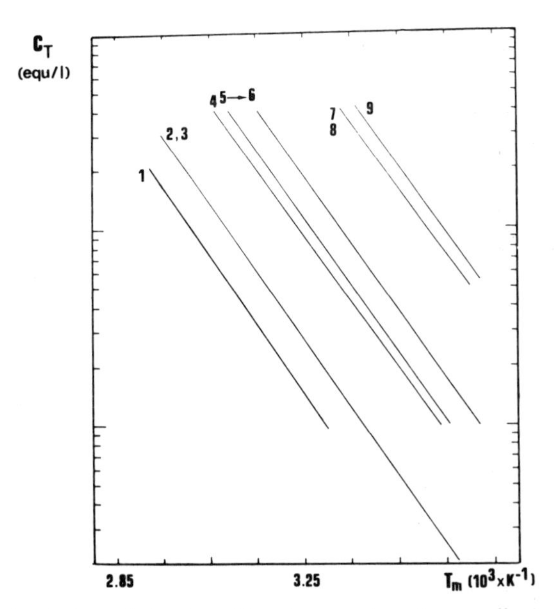

Figure 1.56 Variation of the transition temperature (on cooling) with the total ionic concentration, C_T, for various cations: curves 1 to 4 and 7 to 9, monovalent cations (Rb^+, K^+, Cs^+, NH_4^+, $N^+(CH_3)_4$ and Li^+, respectively); 5 to 6 range for Ba^{2+}, Ca^{2+}, Sr^{2+}, Mg^{2+}, Zn^{2+} and Co^{2+} (with permission from Rochas and Rinaudo, 1980).

significant shift of the curves with the nature of the cation. For instance, at constant ionic concentration, Rb^+ give a gel with a higher melting point ($T_m \simeq 60°C$) than Li^+ or sodium Na^+ cations ($T_m \simeq 10°C$) (see Figure 1.56).

Monovalent and divalent cations do not behave exactly the same. T_m plotted as a function of the partial ionic volume goes through a maximum for monovalent cations whereas it increases gradually with divalent cations.

Conversely, T_m is not very sensitive to the nature of the anion and its partial ionic volume (typically $\pm 3°C$).

(b) Gellans

Gellan (trade mark from Merck and Co.) is a weak polyelectrolyte extracellular polysaccharide produced from a bacterium (*Pseudomonas elodea*). The chemical structure is composed of a tetrasaccharide unit which consists of two β-D-glucose, one β-D-glucuronic acid and one α-L-rhamnose residue arranged together as follows:

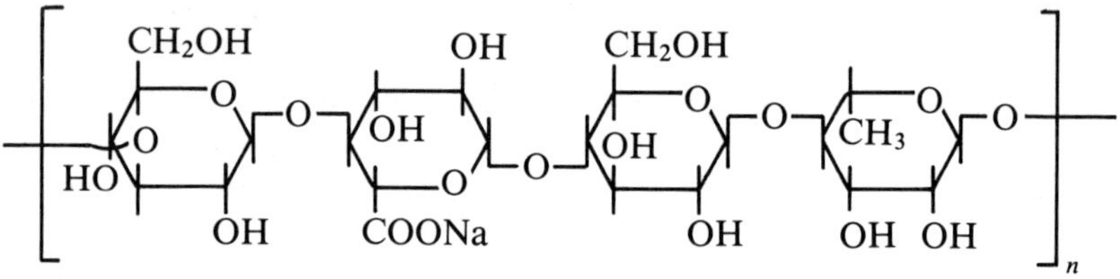

CH_2OH OH CH_2OH
O O O O O O O O
HO OH OH CH_3
OH COONa OH OH OH
n

The native macromolecule contains about 6% by weight of *O*-acetyl groups but can also be obtained in the fully deacetylated form. In both cases, thermoreversible gels can be produced but gels are weaker in the native form. Apart from conformational investigations, there is little information available so far on the gelation behaviour with temperature and concentration of this polymer, although it is currently being intensively studied. Indeed, gellan might be a serious contender as a replacement for gelling agents such as carrageenans since its production does not rely upon uncontrollable factors such as weather conditions.

Gel formation and gel melting show a pronounced hysteresis (40–60°C) (Kang and Veeder, 1982). As with carrageenans, gelation is strongly dependent upon the nature of the cation and its concentration (Grasdalen and Smidsrød, 1987).

1.3.3 Blends of biopolymers

Blends of biopolymers, particularly of polysaccharides present interesting features. For instance two non-gelling polysaccharides such as galactomannan and xanthan form gels once mixed under appropriate conditions.

In fact galactomannan can form gels under particular conditions. Dea *et al.* (1986*a*) have shown that by freezing the aqueous galactomannan solution a gel can be obtained after reheating to room temperature. The term non-gelling polysaccharides is therefore somewhat abusive.

Galactomannans are naturally occurring copolysaccharides sometimes referred to as carob gum or guar gum. They consist of linear chains of $1 \rightarrow 4$ linked b-D-mannopyranosyl residues with $1 \rightarrow 6$ linked a-D-galactopyranosyl side groups. The amount of galactopyranose units varies from 15% to 49% depending upon the species out of which the gum is extracted. Xanthan is a bacterial polysaccharide produced by *Xhantomonas campestris*. It consists of D-glucose monomers linked in β $(1 \rightarrow 4)$ that bear every two monomers a lateral chain made up with D-mannose β $(1 \rightarrow 4)$, D-glucuronique acid β $(1 \rightarrow 2)$ and α-D-mannose.

Cairns *et al.* (1986*b*) have shown that mixing xanthan and galactomannan at room temperature does not lead to gelation. Gelation only occurs when the mixture is heated above 95°C which corresponds to the helix-coil transition of xanthan.

In other cases, adding a non-gelling polysaccharide to a gelling one enhances the gel properties, particularly the mechanical properties. For instance, it has been known for a long time that addition of galactomannan to gelling polysaccharides produces gels stronger and less brittle (see for instance Baker *et al.*, 1949). Yet systematic investigations of these gels are quite recent. As a rule preparation of these gels is achieved by mixing appropriate volumes of aqueous solutions heated up to 95°C that are cooled down to room temperature. So far no ternary phase diagrams have been established for these systems.

Investigations into blends of gelatin + polysaccharides have also been recently carried out by Clark *et al.* (1983).

2 Gel morphology and molecular structure: gelation mechanisms

'It is often necessary to forget one's doubts and to follow the consequences of one's assumptions wherever they may lead.'

Steven Weinberg
in The First Three Minutes

2.1 INTRODUCTION

The determination of both gel morphology and molecular structure certainly constitutes the most controversial area of thermoreversible gelation. This is probably so because this type of investigation usually requires that the gel sample be 'tampered' with. In other words, there are few non-destructive techniques that allow direct access to structural data. Also, in many cases the amount of information is not sufficient to decide unambiguously between opposite interpretations.

Studies of **gel morphology** necessitates removal of the solvent, as the network can only be observed in most cases by means of electron microscopy. Although this can be done by freeze-drying techniques, there is no certainty that what is eventually seen corresponds to the actual gel morphology in the wet state. It is always difficult to assess to what extent the preparation method has introduced artifacts. Even if the gel sample possesses a mesh size large enough to be examined by optical microscopy, there still might be some questions as to whether placing the preparation between glass slides has brought about unwanted effects such as orientation by shearing.

The same types of problem occur when it comes to elucidating the **short-range molecular structure**. When a sample contains 99% solvent, it is usually difficult to produce X-ray diffraction patterns that are not totally 'smeared' out by the solvent halo, hence the need for drying. Again, how can we assess the effect of drying the sample on the original gel structure? For instance, phase diagrams presented in Chapter 1 show that in some cases polymer–solvent compounds are formed. There is little doubt that the drying process will considerably alter the initial structure here.

In addition, it turns out that the short-range structure is mostly poorly ordered. To improve the signal further, dried gels are also unidirectionally stretched. Although through this procedure diffractograms may eventually display well-defined reflections, the question still remains as to whether the data obtained under these conditions are representative of the original gel state (for further details on the determination of helical structures see Appendix 2).

This pessimistic view should, however, be toned down. There do exist some non-destructive spectroscopic techniques that allow investigation of the short-range structure. Among them are NMR and FTIR (Fourier transform infra-red) spectroscopy. However, since it is difficult to calculate what a spectra should be for a given structure, these techniques rely in most cases on comparison: comparing the gel state, on the one hand, and the dried-and-stretched state for which diffraction data are available, on the other hand.

A direct technique of structural determination, particularly useful when it comes to studying polymer-solvent compounds, is **neutron diffraction**. Here one of the gel components is replaced by its deuterium labelled counterpart, which, by virtue of the difference in scattering amplitudes, allows enhancement of its signal with respect to the unlabelled component. As a result, drying is not so 'desperately' important, since the polymer diffraction power can be amplified with respect to that of the solvent. Obviously, the relevancy of this diffraction technique rests upon the absence of significant alteration of the original structure by isotopic substitution. In addition to the diffraction investigations which are carried out at large angles, neutron scattering experiments performed at small angles can supply information on the chain structure within the gel. This can be of great help for understanding the gelation mechanism since it becomes possible to discover directly what conformational changes occur during the sol–gel transition.

Finally, it must be stressed that instead of studying the gel itself, a study of the aggregates (sometimes referred to as pregels) formed below the critical gel concentration, C_{gel}, can provide useful information on the gel state. Thanks to the submicronic size of aggregates, these investigations can be achieved by means of other non-destructive techniques such as light scattering. A typical example of the relevance of using such a technique is the variation in the radius of gyration during transition from the sol to the gel state. For instance, if this parameter increases significantly, then it can be inferred that gelation probably occurs via three-dimensional growth. Conversely, if it does not increase, then this suggests that aggregation proceeds via one-dimensional growth. A thorough investigation of the pregel state must also include the determination of the chain conformation just above the aggregation temperature, T_{agg}, where chains are still isolated, to discover whether there are any particular chain statistics that favour gelation.

Studies on gels and pregels have naturally led scientists to the question: how does the critical gel concentration, C_{gel}, relate to the well-known chain overlapping concentration C^*? As a reminder, C^* is given by:

$$C^* \approx M/(4N_A R^3) \tag{2.1}$$

in which R is the chain radius of gyration, M its molecular weight and N_A the Avogadro's number. The experimental value of C^* is usually determined from the cross-over apparent in the variation of the intrinsic viscosity $[\eta]$ as a function of concentration (Graessley, 1974). The question as to whether C^* and C gel ought to be proportional, or even equal, has been particularly discussed in the case of biopolymers and becomes quite interesting when $C_{gel} \ll C^*$ or in other words when the gel mesh size is larger than the chain dimension.

A knowledge of morphology and molecular structure together with phase diagrams should be now sufficient to elucidate, either partly or totally, the gelation mechanism. This is why discussion of the latter has been postponed to this chapter. As in Chapter 1, Chapter 2 will be subdivided into two main sections: **biopolymers** and **synthetic polymers**. Much thorough work was carried out on biopolymer morphology and molecular structure before systematic investigations began on synthetic polymers. As a result it seems judicious to tackle the presentation of this chapter in chronological order and, accordingly, discuss biopolymers first.

2.2 BIOPOLYMERS

As will be seen, gel morphology is much the same independent of the neutral or polyelectrolyte character of the biopolymer. However, two main classes can be distinguished on consideration of the helical forms which, according to the prevailing views, are involved in the molecular structure of the physical junctions: **gelatin gels** and **polysaccharide gels**. The former consist of **triple helices** and the latter are currently said to be composed of **double helices**.

2.2.1 Gelatin gels

(a) The gel state

For several reasons that will be elaborated below, gelation of gelatin is regarded as the reversion towards the native collagen crystalline form. Accordingly, a short survey of the molecular structure of this protein is necessary.

The crystalline form of collagen is rope-like and consists of a triple helix (see Figure 2.1). The triple helix, which is right-handed, possesses a pitch of 8.6 nm while each individual chain adopts a left-handed helical form with three residues per turn and a pitch of 0.9 nm (Ramachandran and Kartha, 1955; Rich and Crick, 1955). This picture of the native collagen structure has been deduced from two characteristic reflections at 0.29 nm

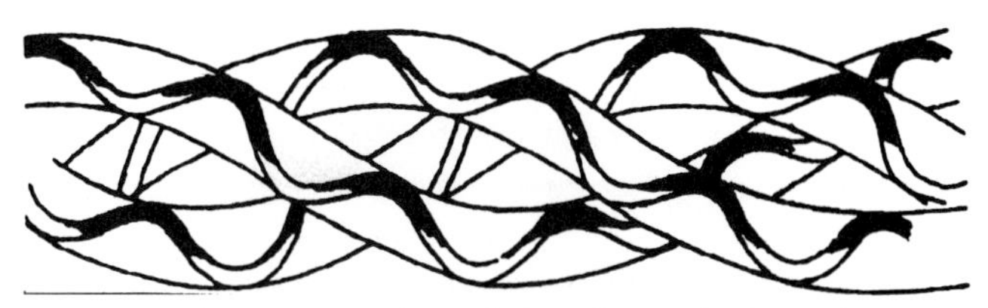

Figure 2.1 Schematic representation of the triple helix in native collagen. Each chain adopts a left-handed helical form while the super helix is right-handed. (Reproduced with permission from Clark and Ross-Murphy, 1987)

and 1.1 nm (meridional and equatorial respectively on oriented samples; see Figure 2.2a). It is worth noting that this structure was only elucidated about 30 years after its experimental observation. This emphasizes that elucidation of helical structure is not necessarily straightforward, especially when the diffraction patterns display a limited number of reflections.

It must be stressed that the equatorial reflection depends upon the so-called moisture, i.e. the degree of hydration. It can vary from about 1.1 nm to 1.45 nm which means that water is intercalated between the triple helices.

The stability of the triple helix is further enhanced by hydrogen bonds between chains through the C=O and N—H groups. However, it is now believed that water molecules can also be intercalated between these groups (Ramachandran and Chandrasekharan, 1968).

The length of the collagen molecule in the triple helix conformation has been shown by Boedtker and Doty (1955, 1956) to be around 300 nm.

Gelatin gels that have been concentrated by evaporation and stretched display reflections at 0.28, 0.78 and 1.15 nm plus a halo at 0.44 nm. Oriented gels (Figure 2.2b) exhibit the meridional reflection at 0.286 nm and an equatorial spacing at 1–1.6 nm depending upon the degree of swelling (Hermann *et al.*, 1930; Katz *et al.*, 1931). As a result it is concluded that the junctions responsible for the physical cross-links in gelatin–water gels have the same helical structure as collagen.

Recent X-ray investigations by Pezron *et al.* (1990) on the variation with the equatorial spacing as a function of moisture (g of water/g of dried gelatin) have shown it to be virtually identical with that seen for bovine achilles tendon (Sasaki, N. *et al.*, 1983).

Various experiments (see for instance Mrevlishvili and Sharimanov, 1978) as well as the discussion presented in Chapter 1 on the melting behaviour of gelatin–water gels suggest that water molecules may participate in the molecular structure of physical junctions. One way of assessing this assumption consists of studying by proton NMR proton mobility in the gel, as was done by Maquet *et al.* (1986). These investigators have observed below the freezing point of water two populations of protons, one of which is immobile (belonging to iced water) and the other one that is still mobile. They regard the latter population as being partly composed of protons of

(a)

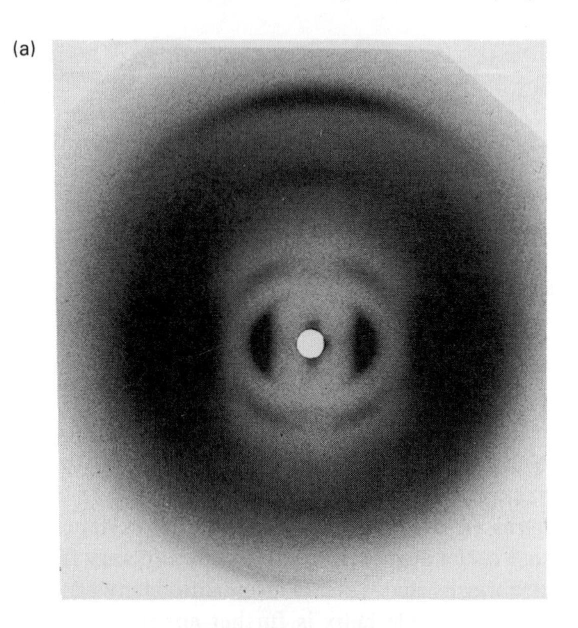

(b)

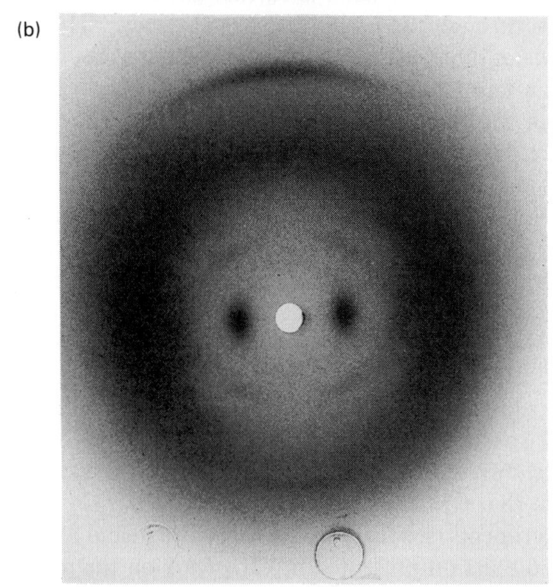

Figure 2.2 (a) X-ray diffraction pattern for an air-dried collagen fibre from rat-tail tendon. Low-angle diffraction arising from the long-range periodicity can be observed close to the direct beam. (b) Diffraction pattern of air-dried and stretched gelatin gel. The meridional spacings seen in (a) with collagen are also observed, although less well-defined. The location of the equatorial reflections is dependent upon the degree of hydration. Unlike collagen fibres, no low-angle diffraction can be detected (with permission from Pezron *et al.*, 1990).

the gelatin macromolecule and partly of protons of bound water. From these results they have computed the fraction of bound water, which turns out to be about 2.7 water molecules per amino acid residue, in agreement with the value derived by Mrevlishvili and Sharimanov (1978). Maquet *et al.* (1986) have determined, by pulsed NMR, the spin–lattice relaxation time T_1 and the spin–spin relaxation time T_2. Here they have studied the variation of T_1 and T_2 during the sol–gel transition. While T_1 is virtually unaffected, T_2 decreases with the gelation time. It is shown that T_2 is approximately proportional to the inverse of the amount of helicity χ_H, the latter parameter being measured by polarimetry (Figure 2.3). They accordingly conclude that a fraction of water molecules is in the gel junctions.

Here, it seems worth adding that such a reversion in moderately concentrated solutions poses some topological problems. As will be seen in Section 2.2.2, the formation of a double helix from totally disordered entangled random coils is now much questioned. As a matter of course, under these conditions where a chain has to drape around another one, physical entanglements and, still worse, the formation of a dcuble helix portion with a third chain should strongly inhibit or prevent this mechanism at a very early stage. Even the 'reel-in' mechanism currently invoked for the crystallization of polymers would be strongly impeded and of little use in draping the two chains around one another.

Evidently, reversion towards the triple helix is an even trickier process unless the helical structure has not been thoroughly broken up by heat,

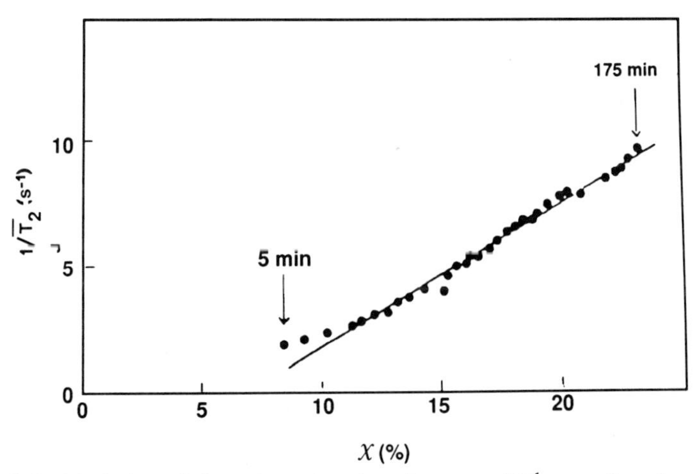

Figure 2.3 Variation of the spin–spin relaxation rate, T_2^{-1}, as a function of the amount of helices, χ, in a gelatin gel quenched at 24°C ($C_{pol} = 21\%$ w/w) (with permission from Maquet *et al.*, 1986, by permission of the publishers Butterworth-Heinemann).

leaving remnants of 'loose' triple helix, an idea originally put forward by Atkins (1986) for polysaccharide double helices. Recent light scattering experiments by Pezron *et al.* (1991) on moderately concentrated solutions at 50°C demonstrate the presence of 'aggregates', which suggests that the triple helix structure has not been completely destroyed. The feasibility of formation of multiple helices from disordered entangled chains in solution will be evoked throughout Section 2.2 since it is both the keystone and the Achilles' heel of the gelation mechanism of biopolymers which mostly prevails nowadays.

(b) The pregel state

Further information on the way reversion takes place can be gained from investigations in dilute solutions. According to Boedtker and Doty (1954), gelatin chains in dilute solutions exhibit a statistically random conformation in the sol. For $M_w = 9 \times 10^4$ they have obtained the z-average value of the single chain radius of gyration, $\langle R \rangle_z = 17$ nm. From the experimental value of $M_z/M_w \simeq 2.6$ (Williams *et al.*, 1954), they have derived the weight-average radius of gyration $\langle R \rangle_w \simeq 10.7$ nm. From this, the statistical segment b can be derived from (see Appendix 3):

$$R^2 = Mb/6\mu_L \tag{2.2}$$

in which μ_L is the linear mass which may be estimated from the work of Boedtker and Doty (1956) to be about 460 g/nm per mol $[(3.45 \times 10^5)/(2.5 \times 300)]$. From relation (2.2) one derives $b \simeq 3.5$ nm which is close to what has recently been found by low-angle neutron scattering ($b \simeq 4.0$ nm, Pezron *et al.*, 1991).

Gelation is therefore a transition of the type **random coil ⇒ helix** which can be followed by polarimetry. Whenever necessary, the amount of helicity, χ_H, can be calculated from the specific rotation of the collagen molecule, $[\alpha]_{collagen}$, and the random coil, $[\alpha]_{coil}$ with:

$$\chi_H = \frac{[\alpha] - [\alpha]_{coil}}{[\alpha]_{collagen} - [\alpha]_{coil}} \tag{2.3}$$

Studies by this technique in dilute solution have been carried out by several authors so as to gain further insight into the mechanism of the reversion process (Flory and Weaver, 1960; Harrington and von Hippel, 1961; Piez and Carrillo, 1964; Harrington and Rao, 1970). Basically, the rate of reversion, $d[\alpha]/dt$ has been studied as a function of gelatin concentration, C, so as to determine the order, n, of the process (not to be confused with the order of phase transition). This gives information on the number of chains involved in the 'reaction', i.e. n, according to the following equation:

$$\mathrm{d}[\alpha]/\mathrm{d}t = kC^{n-1} \tag{2.4}$$

Harrington and Rao (1970) have observed with α-chains of various origin that below $C = 0.1$ mg/cm^3, $n = 1$ whereas above this concentration $n > 1$.

Keeping in mind that polarimetry is only a reflection of the helix content of each individual chain, the value of $n = 1$ found at low concentrations suggests that the triple helix is not directly formed from the random coils, as this process should have given $n = 3$. To account for the kinetic data, two models have been put forward: (1) the **three-step model** by von Hippell and Harrington (1959) in which the first step consists of the nucleation of the poly(L-proline II)-type helix in the imino acid-rich portions, the second step the growth of this helix and the third step the specific association between the single chain helices; (2) the **intermediate state model** by Flory and Weaver (1960) where the chains first adopt their helical structure (intermediate state with transitory existence) and then associate to form the triple helix.

The three-step model requires the helical conformation of the single chain to be stabilized by some mechanism such as chain solvation. This was shown to be unlikely by Bensusan and Nielsen (1964) on the basis of infrared investigations. The latter experiments actually support Flory and Weaver's model.

Currently, however, the model of Flory and Weaver is no longer regarded as valid since there is strong evidence favouring intramolecular associations, thus explaining the first-order kinetics in dilute solutions. This was first suggested by Drake and Veis (1964) who pointed out that, for any segment within the random gelatin coil, the concentration was about 2×10^{-3} g/cm^3 so that interchain and intrachain contacts are about equally probable. In addition, the absence of interchain aggregation after reversion as demonstrated by light scattering experiments has given strong support to this model. The prevailing view at the moment is summarized in Figure 2.4.

According to Harrington and Rao (1970), in moderately concentrated solutions a significant proportion of reversion should still be intramolecular which, they claim, accounts for the non-observation of a third-order reaction ($n = 3$). Here again the topological problem raised by the triple helix re-formation should be taken into consideration. While this problem can be ignored in dilute solutions in which entanglements do not exist, it becomes more and more unavoidable as concentration is increased. As a result, the fact that $n = 3$ is never observed may simply arise from a misconception of the events occurring at the sol–gel transition.

It should be noted that the presence of intramolecular reversion and, correspondingly, ineffective junctions that add nothing to the strength of the network should have a direct bearing on mechanical properties.

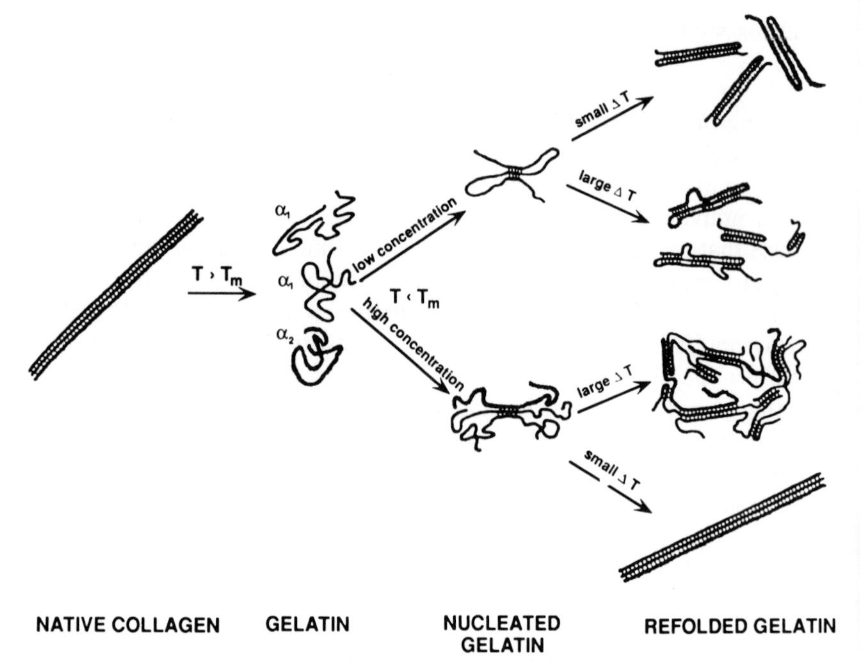

Figure 2.4 Schematization of the renaturation process occurring in gelatin solutions (dilute and concentrated) as suggested by Harrington and Rao, 1970 (with permission from Harrington and Rao, 1970).

(c) Gel morphology

Gelatin–water gel morphology has been found to be fibrillar (Mal'tseva *et al.*, 1972; Tomka *et al.*, 1975; Djabourov, 1986; Djabourov *et al.*, 1988*a*). A typical electron micrograph can be seen in Figure 2.5 of a sample prepared by Escaig's (1981) method (rapid freezing followed by sublimation). According to Djabourov (1986), the mesh size is about 0.5 μm and the fibre diameter varies from about 2 to 20 nm in a 0.5% gel ($C = 0.5 \times 10^{-2}$ g/cm^3). The mesh size is larger than the length of the native collagen molecule and, therefore, larger than the mean dimension of the gelatin chains ($\langle R_g \rangle_w \simeq 10.6$ nm according to Boedtker and Doty, 1954). Also, the fibre diameter can be larger than the triple helix cross-section. These results indicate that several chains participate in fibre formation.

Of further interest is the comparison of C_{gel} (in this case $C_{gel} \simeq 0.5 \times 10^{-2}$ g/cm^3) with the overlap concentration of the coil $C^* \simeq 2.8 \times 10^{-2}$ g/cm^3 as calculated from $\langle R_g \rangle_w = 10.7$ nm and $M_w = 9 \times 10^4$ (Boedtker and Doty, 1954). Evidently, gelation can occur well before entanglements can exist in the sol state. In fact, the comparison between C^* as derived from the conformation in the sol state and C_{gel} would only make

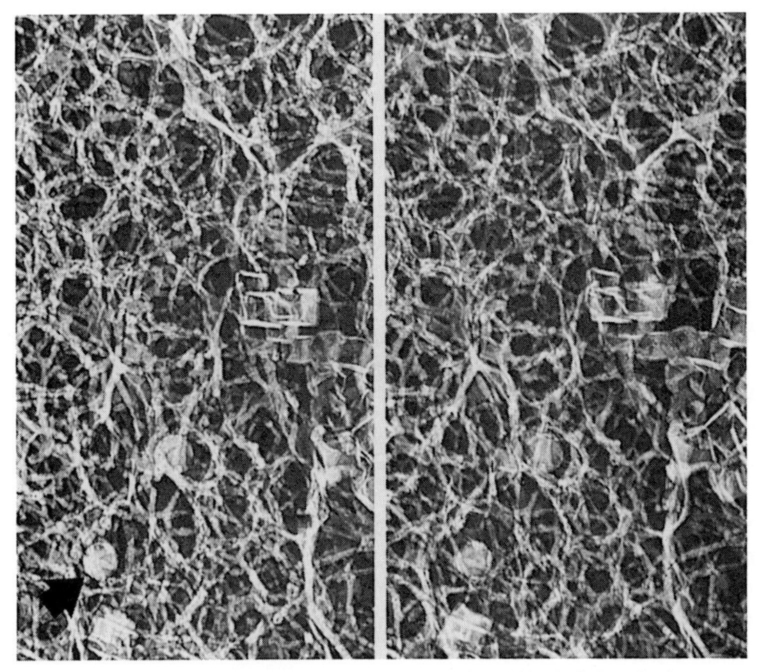

Figure 2.5 Electron micrograph (stereoview) of a 2% (w/w) gelatin–water gel obtained by the quick-freeze, depth-etch, rotary replication technique. Arrows indicate latex spheres of 90 nm diameter that were mixed with the solution prior to gelatin. The gel was matured at room temperature for one week before observation (from Favard *et al.*, 1989).

sense if the solution and the gel were to display a similar structure, as is the case for chemical gels (Munch *et al.*, 1977). Only then would it be relevant to draw a parallel between the screening length, ξ, and the mesh size of the gel.

Curiously enough, C_{gel} appears to be rather high for such a mesh size. Why this is so may find explanation in the fibrillar structure. Obviously, the gel will only exist if the system is above some 'overlap concentration' of the fibrils, C_f^*. The value C_f^* depends upon the size of the fibril, that is essentially upon its length, L, and upon its mass, μ_f. It can be expressed as:

$$C_f^* \simeq 2\mu_f/N_A L^3 \tag{2.5}$$

As a fibril may be made up of several chains without adding anything to its length but adding much to its mass, C_{gel} may be controlled by the growth of the lateral fibril. Equation (2.5) accordingly implies that the thinner the fibre the lower C_{gel} and vice versa.

2.2.2 Polysaccharides

(a) Gel morphology

(i) Carrageenan and agarose gels

Gel morphology of carrageenan gels and agarose gels has been studied by electron microscopy of samples prepared by different methods.

Snoeren *et al.* (1976) have investigated $\varkappa$- and ι-carrageenan gels by negative staining techniques and by replicas made after evaporation on freshly cleaved mica. They have observed an entanglement of thin threads which also display branching. They report an average thread diameter of 8 nm for the $\varkappa$-carrageenan gels and 20 nm for ι-carrageenan gels.

A similar morphology has been found by Rochas (1982*a*) and by Sugiyama *et al.* (1992) by means of cryoelectron microscopy which allows observation in vitrified ice (Figure 2.6). In agarose gels fibres of about 10 nm diameter are seen.

A more detailed study has been carried out on $\varkappa$-carrageenan gels by Hermansson (1989) who has also observed threads of about 13 nm diameter. In addition, Hermansson reports the observation of a 'fine network' just above the gelation temperature which she attributes to the

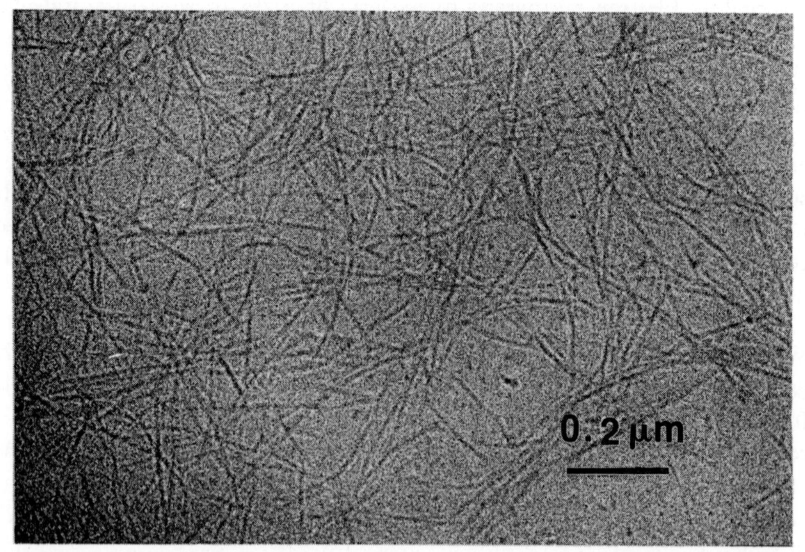

Figure 2.6 Cryo electron microscopy of $\varkappa$-carrageenan micro-gel observed in vitrified ice. The concentration in poly saccharide is 0.01×10^{-3} g/cm^3 in 5×10^{-2} under KCl in water and is allowed to stand for 24 h at room temperature prior to the cryo experiment. The diameter of the threads can be estimated to be about 6 to 10 nm (with permission from Sugiyama *et al.*, to be published).

coil ⇒ helix transition preceding gelation. Hermansson has shown that the salt concentration affects the morphology. Low salt concentration (0.01 M-KCl) produces only the 'fine network' morphology. At a salt concentration of 0.1 M-KCl the threads overwhelmingly dominate the structure. At higher salt content (0.2 M-KCl) there is a mixture of threads and 'fine network' structure.

Worth comparing is the morphology of these polysaccharide gels with gelatin gels. In both cases fibrils are observed, yet they are far more developed and regular in carrageenan gels than in gelatin gels. This difference should have a significant bearing upon both mechanical properties and performances.

(ii) Amylose

Amylose gels do not exhibit a fibrillar morphology as marked as that of agarose and carrageenans but are rather reminiscent of gelatin gels (Leloup, 1989) (Figure 2.7). The fibril diameter is about 50 nm, values virtually independent of concentration. Conversely, the mesh size is notably dependent upon concentration. Leloup insists, however, that measuring the average mesh size is difficult from the electron micrograph. According to Leloup the pore size can be measured by determining the accessibility of molecules of known hydrodynamic radius. Accessibility measurement relies upon the same principle as gel permeation chromatography (now named size exclusion chromatography). It consists of determining the diffusion of objects, such as lattices, of known size (usually of known hydrodynamic radius). The average size of the gel pores can be estimated, Leloup indicates, from the hydrodynamic size of the probe particles when 50% of the solvent within the gel is accessible to them. By assuming this, she obtains the following relation between the average pore size, R_p, and amylose concentration:

$$R_p(\text{nm}) = 0.23C^{-1.22} \tag{2.6}$$

This relation shows that the pore size decreases quite rapidly with concentration. This power law yields an exponent larger than that calculated for other systems such as entangled chains (Gaussian, excluded volume or rod-like) for which the screening length, ξ, varies as:

$$\xi \approx C^{-\nu/3\nu-1} \tag{2.7}$$

in which ν^{-1} is the fractal dimension. In the present case this would mean $\nu^{-1} \simeq 3.8$ which has no physical sense. Clearly, variation given in relation (2.6) conveys something deeper which remains to be understood. In addition, such a result possibly correlates with the unusual variation of the elastic modulus with concentration (see Section 3.3.2.c).

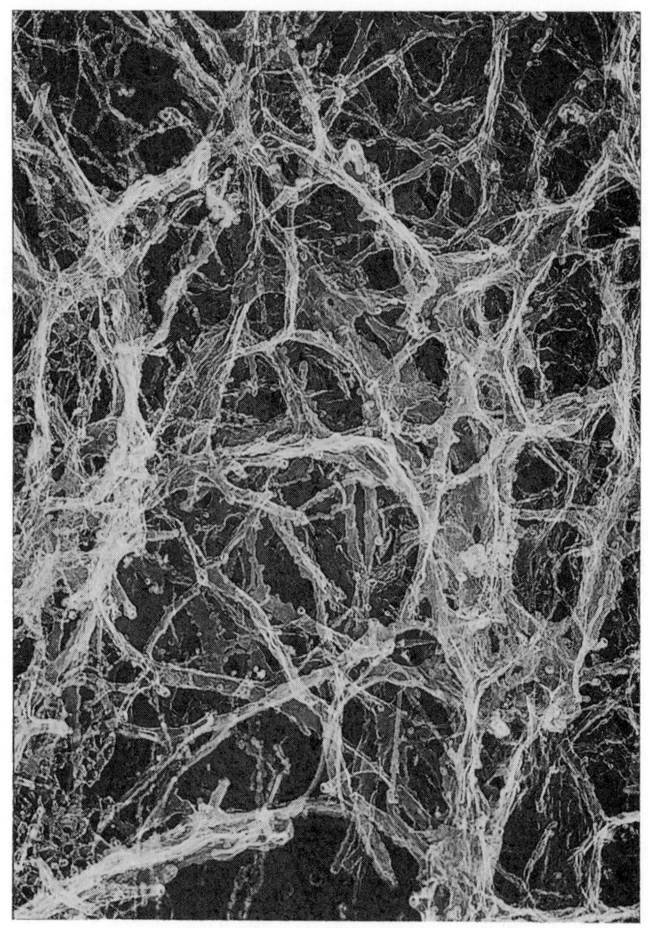

Figure 2.7 Pt-C replica of an amylose–water gel (C_{pol} = 7.8%) that had been frozen in and processed through the method developed by Escaig (1981). The diameter of the fibrils can be estimated to be about 50 nm (with permission from Leloup, 1989).

(iii) Cellulose derivatives

Papkov *et al.* (1971) also report the observation of fibrillar structures of cellulose diacetate prepared from aqueous 35% acetic acid solutions.

(b) The gel state

(i) Carrageenans

Investigations by X-ray diffraction on stretched ι- and κ-carrageenans (see Figure 2.8) led Anderson *et al.* (1969) to put forward the notion of a double

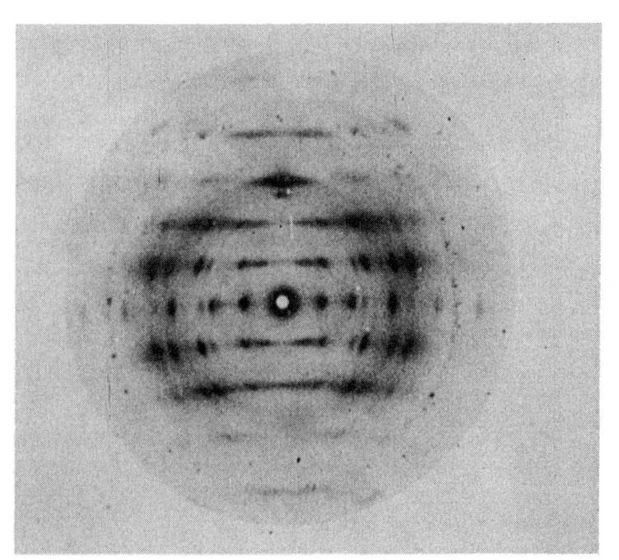

Figure 2.8 Fibre diffraction pattern of the Ca^{++} salt of ι-carrageenan. The fibres were tilted out of the plane normal to the X-ray beam 10° in order to record the 003 reflection (0.443 nm spacing). The relative humidity is maintained to 75% (from Arnott *et al.*, 1974*a*).

helix for polysaccharides. This structure can be viewed as two identical helices draping around a common axis which are translated by half the pitch value relative to each other (see Figure 2.9). In the present case each chain adopts a right-handed threefold helical structure with a pitch of 2.656 nm. In the diffraction diagram the layer line spacing amounts to 1.328 nm which represents half the pitch of the double helix. In fact, this is only strictly true for ι-carrageenans. The diffraction pattern for κ-carrageenans does not display any layer line at 1.37 nm. Still, Anderson *et al.* (1969) propose the same double helical structure for κ-carrageenan and account for the absent layer line by considering either parallel chains displaced from the iota arrangement by 'screwing' or two antiparallel chains.

Further studies on ι-carrageenans by Arnott *et al.* (1974*a*) led to the determination of the unit cell giving $a = b = 1.373$ nm and $c = 1.328$ nm the space group being $P3_212$. According to these authors most of the distances are large so that van der Waals forces and hydrogen bonding cannot play a major role in stabilizing the crystal. Ionic forces, that is cation–sulphate group interactions, are essential for crystal stability. Each cation is coordinated to two sulphate groups belonging to two different chains. This is reminiscent of the **'egg-box' model** proposed by Grant *et al.* (1973) for pectins. Water molecules are presumably occluded in the crystalline lattice by linking helices through hydrogen bonds. However, no experimental evidence is available to support the latter point.

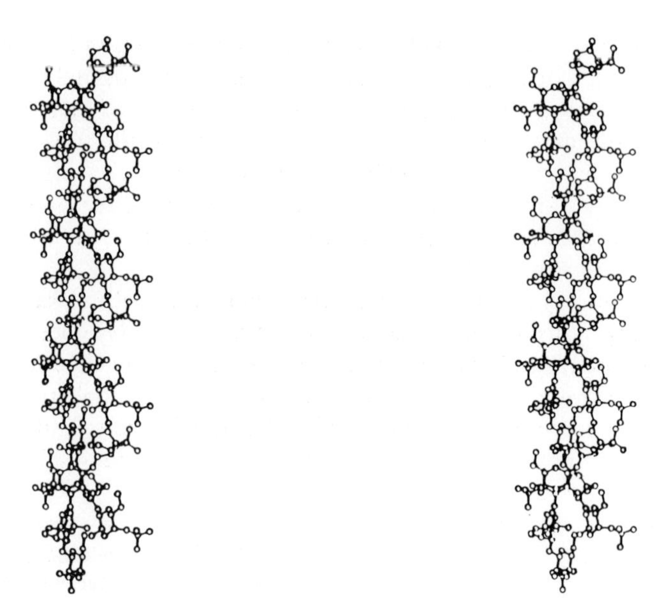

Figure 2.9 Stereoscopic drawing of the ι-carrageenan double helix (3_1) (with permission from Arnott *et al.*, 1974*a*).

On the basis of the double helix, Anderson *et al.* (1969) proposed the two-step gelation process which is sketched in Figure 2.10. At high temperature (Figure 2.10a) carrageenan chains possess a random structure (and are also independent of one another), then on lowering the temperature (Figure 2.10b) they start to form knots with double helices, the latter aggregating at lower temperature to form the physical junctions of

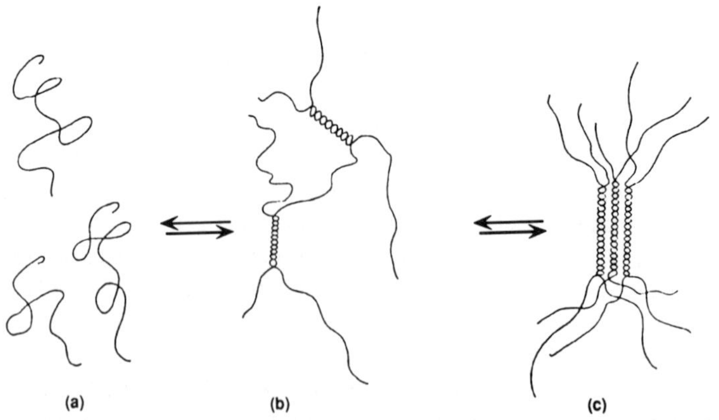

Figure 2.10 Schematic representation of the aggregation mechanism in polysaccharides according to Anderson *et al.* (a) Random coil; (b) coil–helix transition; (c) helix aggregation (from Anderson *et al.*, 1969).

the gel (Figure 2.10c). Actually, as has been seen in Section 2.2.2.a, the gel consists of well-defined fibres which contrasts with the partially disordered structure drawn in Figure 2.10(c). It should be emphasized that the gel morphology had not been determined at the time when Anderson *et al.* (1969) proposed their model. This model is now but of historical significance.

According to Anderson *et al.* this scheme accounts for the large hysteresis which exists between gel formation and gel melting temperatures (see Sections 1.3.1.b.i and 1.3.2.a).

Yet, as already discussed above, the topological problem raised by the intertwining of disordered random chains in a moderately concentrated solution is virtually insoluble considering that gelation occurs in a finite period of time. To overcome this difficulty Smidsrød and Grasdalen (1982) have instead proposed the formation of pairs of parallel helices. Unfortunately, this model cannot account for the X-ray diffraction data at all, at least as far as ι-carrageenan is concerned.

To summarize, although there is little doubt that the double helix occurs in carrageenan gels, its creation out of a truly disordered moderately concentrated solution is less than likely. An elegant explanation has been proposed by Atkins (1986). It consists of considering a loosely coupled double strand which could be flexible at high temperature but, when the temperature is lowered, the interaction between the chains could increase, locking them into a rigid double helix capable of forming gels (mechanism of **loosely bound ⇒ tightly bound** double strand). To the author's knowledge, no experiments have been set up to test this assumption although this should be feasible with neutron scattering by comparing, in the solstate, the actual linear mass with those calculated from single or double strands (see Section 2.3.1).

(ii) Agarose

In the early 70s, Arnott *et al.* (1974*b*) investigated agarose fibres drawn from the gel state. Unlike ι-carrageenans, agarose diffraction patterns display only a small number of reflections. Three layer lines can usually be observed giving a spacing of 0.95 nm, the third one giving a meridional reflection at 0.317 nm (Figure 2.11). These layer lines have been interpreted as even orders of a threefold left-handed helical structure of pitch 1.9 nm (Figure 2.12). The intertwining of two such helices (double helix) with an axial translation of 1.9/2 nm accounts for the absence of odd layer lines.

The current situation is, however, not so clear for agarose as it is for carrageenans, first of all, because the number of reflections is lower than for carrageenans and second, and perhaps more importantly, there is convincing experimental evidence for the probable occurrence of single helices in the solid state. Foord and Atkins (1989) have re-examined the question and found that under given conditions a diffraction pattern characteristic

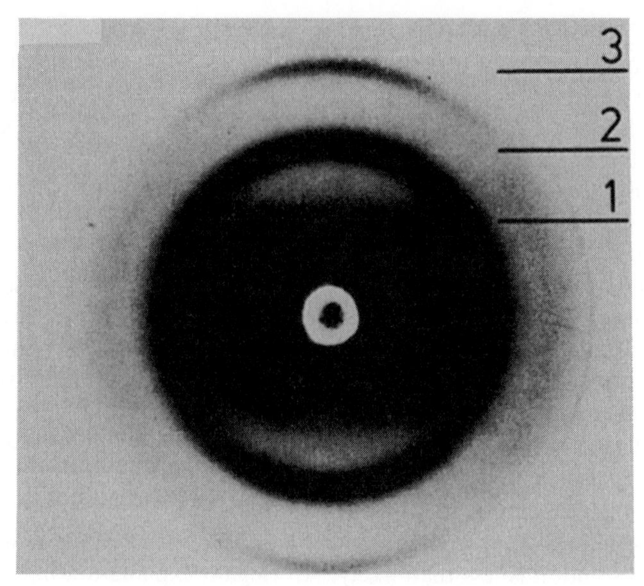

Figure 2.11 X-ray diffraction pattern of a partially oriented agarose–water gel produced by stretching (stretch axis vertical). The first three layer lines with spacing 0.95 nm are visible, as well as a meridional arc on the third layer line (with permission from Foord and Atkins, 1989).

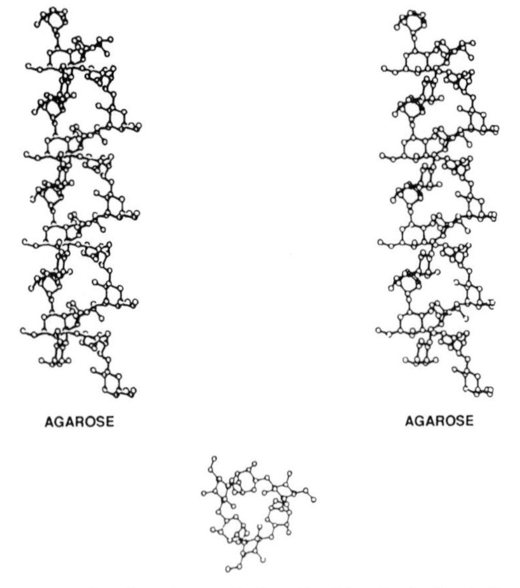

Figure 2.12 Stereoscopic drawing of the double 3_1 helical form proposed by Arnott *et al.* to account for the diffraction pattern of agarose gels (with permission from Arnott *et al.*, 1974*b*).

of highly crystalline material can be obtained giving additional and valuable information.

Foord and Atkins' samples were prepared by evaporating an aqueous solution of agarose near 100°C, then stretching at different temperatures and degrees of humidity. Typically, films stretched from 20 to 53°C display a diffraction pattern with layer line spacing of 0.95 nm similar to the one obtained with dried and stretched gel samples. Now, if the sample's drawing is achieved between 80 and 100°C instead, the diffraction pattern exhibits a high degree of orientation with a layer line spacing of 3.552 nm and a well-defined meridional reflection at a spacing of 0.888 nm on the fourth layer line. By preparing the samples in *N*,*N*-dimethylacetamide, similar patterns to those of the dried gel state are observed, but different patterns are also obtained revealing a series of new crystalline forms.

As emphasized by Foord and Atkins (1989), these new forms are produced from films that, at lower temperatures, yield patterns that are supposed to arise from double helices. However, only single helices with different pitches depending upon the stretching conditions fit the X-ray data of these new forms (Figure 2.13). They accordingly seriously question the occurrence of double helices in the gel state and suggest a threefold single helix instead, provided that allowance is made for the position of the meridional reflection (0.98 nm instead of 0.95 nm which lies within experimental uncertainty). Admittedly, unless stretching itself induces transitions from double to single helix in the solid state (solid–solid transition), the process of which remains to be imagined, Foord and Atkins' results seriously question the very existence of the double helix in agarose gels. It is worth underlining that, were it not for the X-ray data, the double-helix model for the gel state would find little if any support from other techniques and is of the kind where the model fits but may not be strictly indispensable. Of further note is the support given to the 3_1 single helix by recent modelling studies carried out by Jimenez-Barbero *et al.* (1989).

The gel structure has been shown by electron microscopy to be fibre-like. Through this technique the size of the gel fibres can be determined. They possess diameters of about 10 nm. However, one may still question the validity of such a measurement, as electron microscopy requires some processing, the effect of which on the original gel structure is not well known. (It is worth mentioning, however, that new techniques of sample preparation have been used over the past few years that enable one to have more confidence in what is observed.)

Attempts have been made by Djabourov *et al.* (1989) to measure the diameter of the elements (fibres) constituting the gel by means of small-angle X-ray scattering. If the element is rod-like in the range of investigated transfer momentum q ($q = 4\pi/\lambda \sin \theta/2$) and possesses a given cross-section characterized by a so-called transverse radius of gyration R_σ then the

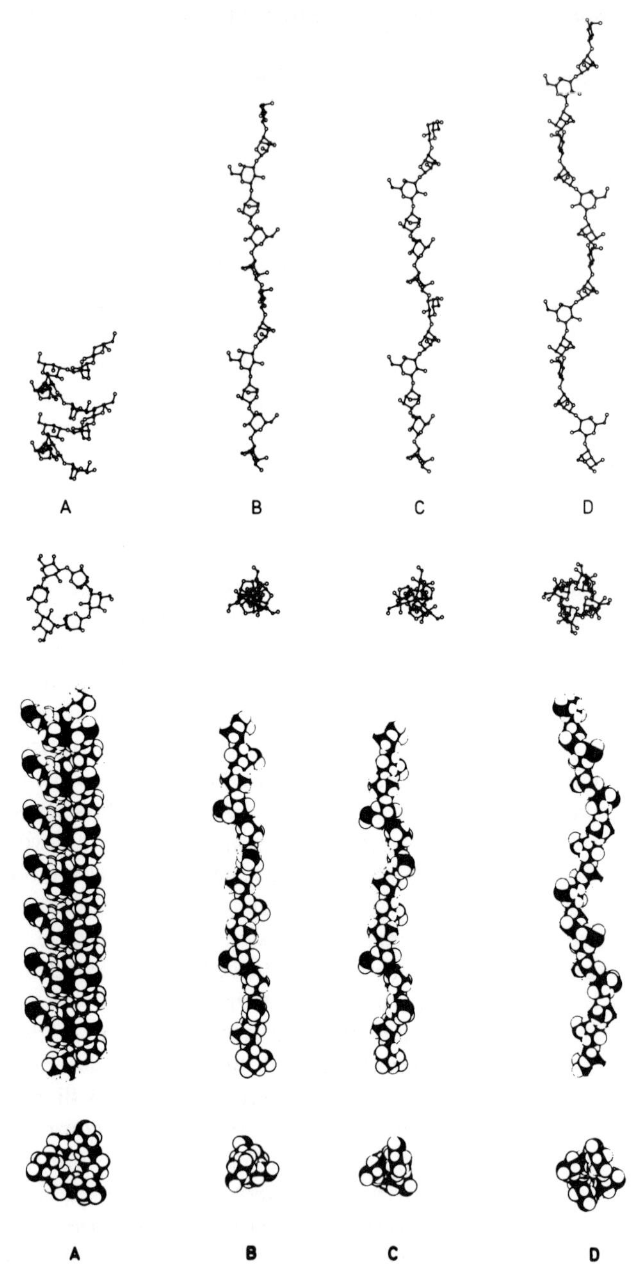

Figure 2.13 Alternative models proposed by Foord and Atkins to account for the different diffraction patterns observed with samples produced from agarose gels: A, B and C are 3_1 forms with different axial rise (0.317 nm, 0.973 nm and 0.938 nm, respectively). D is a 4_1 helix with axial rise 0.888 nm. (Upper models: ball and stick projections perpendicular and parallel to the helix axis; lower models: equivalent space-filling models.) (With permission from Foord and Atkins, 1989)

intensity scattered may be written:

$$I(q) \approx (\pi\mu_L/q) \exp - (q^2 R_\sigma^2/2) \qquad \text{for } qR_\sigma < 1 \tag{2.8}$$

in which μ_L is the linear mass.

By plotting log $qI(q)$ *vs* q^2 one should obtain a straight line, the slope of which gives R_σ^2. Figure 2.14 shows the results of Djabourov *et al.* (1989). As can be seen, if a straight line is drawn from the behaviour at small angles, significant deviation occurs at larger angles. Djabourov *et al.* (1989) have interpreted these data by considering two populations of fibre cross-section, as they consider that two straight lines can be obtained in Figure 2.14 leading to two different values of R_σ. This interpretation is, however, incorrect and in conflict with morphological evidence. This can be easily shown by writing the intensity expected with such a two-population system of proportions x and $1 - x$:

$$I(q) = xI_1(q) + (1 - x)I_2(q) + 2x(1 - x)I_{12}(q) \tag{2.9}$$

Ignoring the cross term $I_{12}(q)$ for the q-domain under consideration, relation (2.9) can be rewritten using Porod representation ($qI(q)$) for qR_{σ_1}

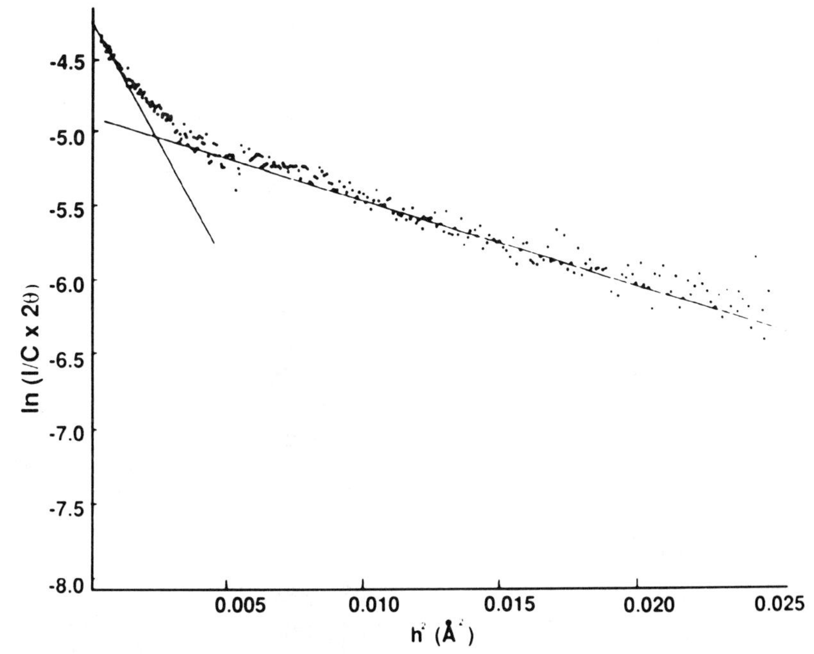

Figure 2.14 Plot of the logarithm of the intensity times the scattering angle *vs* q^2 (Porod-plot) for agarose gels. The two straight lines correspond to those drawn by Djabourov *et al.* (with permission from Djabourov *et al.*, 1989).

and $qR_{\sigma_2} < 1$:

$$qI(q) = x \exp - (q^2 R_{\sigma_1}^2/2) + (1 - x) \exp - (q^2 R_{\sigma_2}^2/2) \qquad (2.10)$$

which finally reads by approximating the exponential function $\exp - y \approx 1 - y$:

$$qI(q) \approx 1 - q^2 \langle R_\sigma^2 \rangle /2 \approx \exp - (q^2 \langle R_\sigma^2 \rangle /2) \qquad (2.11)$$

with $\langle R_\sigma^2 \rangle = xR_{\sigma_1}^2 + (1 - x)R_{\sigma_2}^2$. Therefore, the slope seen at small angles is not proportional to the transverse radius of gyration of one type of cross-section but is a combination of both.

Now, if $qR_{\sigma_1} > 1$ and $qR_{\sigma_2} < 1$ then $qI(q)$ reads:

$$qI(q) = xI_1(q) + (1 - x) \exp - (q^2 R_{\sigma_2}^2/2) \qquad (2.12)$$

in which $I_1(q)$ can no longer be approximated to a simple form but depends on the pair-correlation function. There is no argument for regarding the first term in eqn (2.12) as negligible. The only case where this could be conceived would be when Porod's limit is reached for one population in which case $I_1(q) \approx q^{-4}$ (Guinier, 1956). In the domain of transfer momentum presently explored this should not be the case. As a result, determining a cross-section radius from the second slope is most probably misleading.

Worth adding, however, is that the cross-section average radius derived by Djabourov *et al.* (1989) from the first slope, a procedure that is correct provided that the inter-rod cross-term is negligible, is $\langle R_c \rangle = 2.9 \pm 0.3$ nm which corresponds to an average fibre diameter of 8.2 nm if one considers a circular cross-section. This result agrees quite well with morphological evidence.

(iii) Amylose and amylopectin

Depending on its origin, native amylose occurs in two different major crystalline forms: A-type from cereals and B-type from tubers (see for instance French, 1984). An intermediate, C-type, is less frequently observed. The most recent analyses of crystalline amylose diffraction patterns together with molecular modelling suggest that the same double helices are involved in both A and B types (Imberty and Pérez, 1988; Imberty *et al.*, 1988*b*). These double-stranded structures are said to be composed of sixfold left-handed helices possessing a pitch of 2.138 nm (Figure 2.15). The only difference between A and B type lies in the crystalline lattice (Wu and Sarko 1978*a*,*b*; Imberty and Pérez, 1988; Imberty *et al.*, 1988). Polymorph A is organized into a monoclinic cell ($a = 2.14$ nm, $b = 1.172$ nm, $c = 1.069$ nm and $\gamma = 123°$) containing four water molecules only. Polymorph B consists of an hexagonal cell ($a = b = 1.85$ nm, $c = 1.04$ nm), each double helix possessing three first neighbours. This particular arrangement creates a hollow

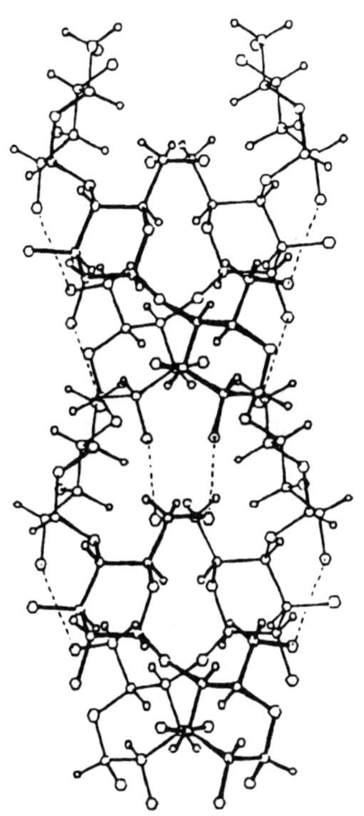

Figure 2.15 Drawing of the double helical form generated by the association of left-handed amylosic strands adopting a 6_1 helix. Dotted lines represent hydrogen bonds (with permission from Imberty *et al.*, 1988*a*).

which contains 36 water molecules. It is worth noting that short chains crystallize in the A-type crystalline form while longer chains crystallize in the B-type (see for instance Pfannemüller, 1987).

In the gel state, only the B-type crystalline structure has been detected so far (Katz, 1930; Miles *et al.*, 1985). Recent investigations by cross-polarization/magic angle spinning ^{13}C-NMR have led Gidley (1989) to the suggestion that amylose gels contain two distinctive structures: rigid B-type double helices on the one hand and mobile amorphous single chains on the other hand. Double helices are said to aggregate so as to form the physical junction. This model is most reminiscent of, if not identical with, that of Anderson *et al.* (1969) for carrageenans.

Since amylopectin differs only from amylose in possessing a branched structure, it is not surprising that the crystalline arrangement in the gels are also of the B type (Ring, 1985). From acid hydrolysis combined with GPC, Ring *et al.* (1987) have concluded that chain segments involved in the crystalline region of the gel are characterized by a weight-average degree of polymerization $DP_w \simeq 15$ whereas this value amounts to $DP_w \simeq 53$ in amylose. This accounts for the large discrepancies observed at a given concentration between the melting point of amylose gels and amylopectin gels.

(iv) Gellan

Early diffraction patterns obtained from gellan samples of low crystallinity (prepared from gels of the acetylated form) hinted at a threefold helical structure with a repeat of 2.8 nm (Carroll *et al.*, 1982, 1983). The diffraction pattern was greatly improved by Upstill *et al.* (1986) by using deacetylated material in the form of a lithium salt (Figure 2.16). The threefold helical symmetry is well demonstrated, as the meridional reflections only occur on layer lines of index $l = 3n$. From this diffraction pattern, Upstill *et al.* (1986) gave the correct unit cell ($a = b = 1.56$ nm, c (fibre axis) = 2.82 nm) but failed to determine the right type of helical form as they proposed a 3_2 double form which was later revealed to be erroneous. Chandrasekaran *et al.* (1988) re-examined the X-ray data together with

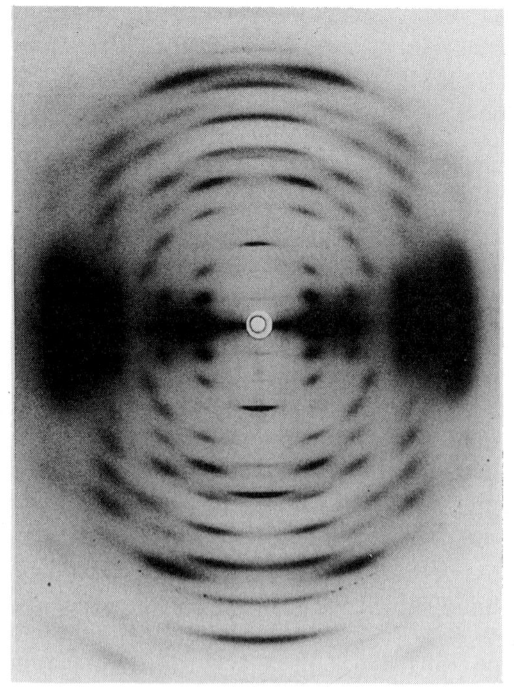

Figure 2.16 X-ray fibre diffraction from an oriented, polycrystalline sample of lithium salt of gellan (with permission from Chandrasekaran *et al.*, 1988).

computer modelling and found a parallel half-staggered 3_1 left-handed double helix with a pitch of 5.64 nm.

These investigators also add that, from their model, acetylation of the D-glucose A residue should not disrupt helix formation but affect aggregation with respect to the deacetylated material which is what is seen experimentally (see Chapter 1, Section 1.3.2.b).

(v) Cellulose derivatives

Apparently, no investigation of the molecular structure of cellulose derivatives in the gel state has been carried out so far. On the other hand studies on cellulose fibres nitrated *in situ* have been performed by Meader *et al.* (1978). Although not directly related to the gel state, the results of Meader *et al.* might be of relevance for this book and for future investigations.

The diffraction pattern of these nitrocellulose fibres (13.9% nitrogen content) exhibit a fivefold helical symmetry with a layer line spacing of 2.54 nm and a strong meridional reflection at 0.508 nm. Four possible helices possessing fivefold symmetry can be considered on the basis of these results: 5_1, 5_2, 5_3, 5_4. 5_1 and 5_4 helices have to be dismissed at once since they cannot occur for $1 \rightarrow 4$-linked β-D-glucose backbone (Sathyarayana and Rao, 1971). Conformational analysis led Meader *et al.* (1978) to propose a 5_2 helix. It would be of interest to discover whether this helical form is present in nitrocellulose gels.

(c) The pregel state

(i) Carrageenans

Although carrageenans gel have been extensively studied, there have been few investigations on the pregel state.

Systematic studies on pure $\varkappa$-carrageenan in dilute solutions have been carried out by Vreeman *et al.* (1980). They have determined both the intrinsic viscosity $[\eta]$ and the dependence of the radius of gyration ($\langle R \rangle_z$) on molecular weight for the sodium salt. As a reminder, this system gives gels with very low melting temperatures (see Chapter 1, Section 1.3.2.a). Consequently, although of interest, the results gathered at room temperature for this salt form cannot necessarily be extrapolated to all carrageenan systems, particularly those such as the potassium salt which display far higher gel melting temperature.

Concerning the intrinsic viscosity, they have found that the K_η and α parameters of the Mark–Houvink law ($[\eta] = K_\eta M^\alpha$) are dependent upon the ionic strength. Their results are summarized in Table 2.1.

Table 2.1 Mark–Houvink parameters ($[\eta] = K_\eta M^\alpha$) for $\varkappa$-carrageenans in solutions of different ionic strength (solutions in NaCl) (Vreeman *et al.*, 1980).

Ionic strength	$K_\eta \times 10^3$	α
0.028	7.78	0.90
0.188	8.84	0.86
0.218	20.9	0.78
∞	52.0	0.67

Values of α gathered in Table 2.1 show that the lower the ionic strength the more the chain is expanded. The limiting value of the exponent of the Mark–Houvink law for chains displaying excluded volume effect should not be larger than $\alpha = 0.8$. In fact, the intrinsic viscosity is given by (see, for instance, Weil and Des Cloizeaux, 1979):

$$[\eta] = \frac{R_H R_G^2}{M} \tag{2.13}$$

in which R_H is the hydrodynamic radius, R_G^2 the radius of gyration, and M the molecular weight. The different radii are expressed as follows:

$$R_H = N^{-2} \sum_i^N \sum_j^N 1/|r_{ij}| \quad \text{and} \quad R_G^2 = N^{-2} \sum_i^N \sum_j^N r_{ij}^2 \tag{2.14}$$

If both radii exhibit the same variation with molecular weight, then for a chain possessing excluded volume, $[\eta] \approx M^{0.8}$. However, as can be seen from relations (2.14), R_H is much more sensitive to the short-distance statistics while R_G mainly reflects the long-distance statistics. As a result the value of $\alpha = 0.8$ might not be observed if the short-distance statistics differ from the long-range statistics as was suggested by Weill and Des Cloizeaux (1979). The same reasoning holds for chains with a large persistence length, a. In fact, two regimes exist:

$$\text{for} \quad r_{ij} < 1 \quad r_{ij} = |i - j| \quad \text{and for} \quad r_{ij} > a \quad r_{ij}^2 = |i - j| \tag{2.15}$$

As a result, exponents higher than 0.8 can be experimentally observed even if the chain possesses global Gaussian statistics.

More detailed equations have been derived by Yamakawa and Fuji (1974). Be L/b the reduced contour length of a worm-like chain of statistical segment b (persistence length $a = b/2$) and diameter $2r_c$ and molecular weight M. Then, these authors define two regimes:

* for $L/b < 2.28$ $[\eta]$ reads:

$$[\eta] = \frac{\pi N_A L^3}{24Mb^3 Ln(L/2r_c)} \times \frac{f(L/b)}{1 + \sum_{i=1}^{i=4} A_i[Ln(2r_c/L)]^{-1}}$$

in which A_i are constants and $f(L/b)$ reads:

$$f(L/b) = (3b^4/2L^4)[\exp(-2L/b) - 1 + 2L/b - 2L^2/b^2 + (4/3)L^3/b^3]$$

* for $L/b \geqslant 2.28$ $[\eta]$ reads:

$$[\eta] = \frac{\Phi_\infty L^{3/2}}{Mb^{3/2}} \times \frac{1}{1 - \sum_{i=1}^{i=4} C_i(L/b)^{-i/2}}$$

wherein $\Phi_\infty = 2.87 \times 10^{23}$ and C_i are numerical terms that depend on $2r_c$.

This might be the case for the system where the ionic force is 0.118, since Vreeman *et al.* (1980) found, after correction for molecular weight, polydispersity:

$$R_G \text{ (in nm)} = 0.045\ M^{0.565} \tag{2.16}$$

As was concluded by Vreeman *et al.*, $\varkappa$-carrageenan possesses a coil shape in dilute solution, the persistence length of which decreases with the ionic force as is usual for polyelectrolytes. The coil shape theory was later borne out by investigations of dilute solutions with a combination of elastic and quasielastic light scattering (Drifford *et al.*, 1984).

In other systems, in particular potassium or rubidium salts, the pregel state has been shown to consist of helical dimers that aggregate at lower temperature (Rochas and Rinaudo, 1984).

(ii) Agarose

An extensive study of the pregel state has been achieved by Dormoy and Candau (1991) by means of electric birefringence coupled to electron microscopy in the range of concentration 5×10^{-6} g/cm^3 to 5×10^{-4} g/cm^3, concentrations at which macroscopic gels do not form. The decay of birefringence of the aggregates, $\Delta n(t)$ is not purely exponential but can be described by a stretched exponential:

$$\Delta n(t) = \Delta n_0 \exp(-6D_r t)^\beta \tag{2.17}$$

where D_r is the rotational diffusion coefficient and β is found experimentally to be 0.53. The mean rotational diffusion coefficient $\langle D_r \rangle$ then reads:

$$\langle D_r \rangle = D_r[\beta/\Gamma(1/\beta)] \tag{2.18}$$

where Γ is the gamma function.

This behaviour indicates a large distribution in aggregate size.

The mean diffusion coefficient $\langle D_r \rangle$ has been found to be proportional to the inverse of the concentration ($\langle D_r \rangle \approx C^{-1}$) which means that in this range of concentration the aggregate size increases with increasing concentration.

By means of electron microscopy, Dormoy and Candau (1991) have observed that the aggregates possess a rigid cylinder shape or, in other words, a fibre-like shape. Further, in contrast with the claim of Djabourov *et al.* (1989), Dormoy and Candau do not report on the existence of a bimodal distribution of cylinder cross-section. On the contrary, the fibre diameter shows little variation from one fibre to another and is independent of concentration whereas the fibre length, L, is (hence the stretched exponential). Estimations from platinium-shadowed samples yield a fibre diameter of about $2R = 5$ nm. Knowing R permits calculation of the rod length from the rotational diffusion coefficients from the relation derived by Broesma (1960):

$$D_r = \frac{3k_B T}{\pi \eta L^3} \left\{ \ln(L/R) - 1.57 + 7\left[\frac{1}{\ln(L/R)} - 0.28\right]^2 \right\} \qquad (2.19)$$

in which η is the solvent viscosity. The following variation with concentration is obtained:

$$\langle L \rangle = L_0 C^{0.37} \qquad (2.20)$$

in which L_0 is some initial length which is proportional to the agarose chain contour length.

Dormoy and Candau (1991) also examined the effect of temperature on aggregate structure. At 85°C, aggregation had not totally vanished, but the aggregates had undergone a considerable loss of organization, as their permanent electric dipole was close to zero and their optical anisotropy factor had drastically dropped. Interestingly enough, total disorganization only takes place at relatively high temperatures ($T \approx 130$°C).

The rigidity of the cylinder-like pregels accounts for the very low critical gelation concentration, as is apparent from equation (2.5) but also the fibre length increases with increasing concentration. However, it can be easily shown that the overlap concentration of these fibres does not show the same dependence with the initial chain molecular weight as would be if one had isolated rods. As a matter of fact, in the case of fibres (see below) C_f^* reads in the present case:

$$C_f^* \simeq L_0^{-1.14} \qquad (2.21)$$

whereas this concentration should vary as L_0^{-2} if one were dealing with isolated rods of length L_0. While the magnitude C_f^* has considerably decreased because of the continuous fibre's growth with increasing concentration, its variation with the initial molecular weight should be far less important than with rods of constant length.

Compare relation 2.5 which gives the C_f^* for a fibre of length L and molecular weight μ_f:

$$C_f^* \simeq 2\mu_f/N_A L^3$$

As the fibre's diameter is found to be a constant, μ_f depends only on L so that C_f^* reads:

$$C_f^* \simeq L^{-2}$$

At this concentration one also has:

$$L \simeq L_0 C_f^{*\alpha}$$

with $\alpha = 0.37$ in the case of agarose. Accordingly, it is easy to show that C_f^* will eventually read:

$$C_f^* \simeq L_0^{-2/(1+2a)}$$

if $\alpha = 0$ then one retrieves the case where the rods possess a constant length with increasing concentration. Now, for $\alpha = 0.37 C_f^*$ varies like $L^{-1.14}$.

(iii) Amylose and amylopectin

Several studies have been carried out on amylose solutions by elastic light scattering (Banks and Greenwood, 1975; Cowie, 1961; Burchard, 1963; Brandt and Dimpfl, 1970; Ring *et al.*, 1985). Dispersed amylose chains can be obtained at high dilution ($C \approx 3.4 \times 10^{-4}$ g/cm^3). However, aggregates form in these solutions upon ageing (significant effects are perceptible after 12 h according to Ring *et al.*, 1985). So far only the dispersed state has been extensively studied.

In neutral aqueous solutions amylose is a relatively stiff molecule as the persistence length is about $a \simeq 2.8$ nm (that is $b = 2a = 5.6$ nm) (Burchard, 1963; Ring *et al.*, 1985). The global statistics are most probably Gaussian as revealed by the variation of the radius of gyration as a function of molecular weight:

$$R_G \text{ (in nm)} \simeq 0.023\, M^{0.5} \tag{2.22}$$

The magnitude of R_G is sensitive to the ionic strength. Values from Banks and Greenwood (1975) differ markedly from relation (2.22) as they found a prefactor of 0.01 instead in a 0.33 M-KCl medium.

Investigations have also been carried out in dimethyl sulphoxide (DMSO) by Cowie (1961) and Fujii *et al.* (1973). They both found that the radius of gyration varies as $M^{0.5}$ yet their prefactors differ (0.0375 against 0.0308). From relation (2.2) the statistical length (or twice the persistence length) can be derived. One finds by using $\mu_L = 373$ g/nm × mole a statistical segment $b \simeq 8.4$ nm. This indicates that amylose in DMSO is a very stiff chain. For the Mark–Houvink viscosity law, Fujii *et al.* have obtained:

$$[\eta]\,(\mathrm{dl/g}) = 8.1 \times 10^{-6} M_w^{0.91} \quad (2.23)$$

while Cowie has established:

$$[\eta]\,(\mathrm{dl/g}) = 1.25 \times 10^{-5} M_w^{0.87} \quad (2.24)$$

Clearly, although slightly different, both variations displaying rather large exponents do confirm the amylose chain stiffness in DMSO.

Amylose is not reported to gel in DMSO while it does in water. This may be surprising since amylose chains are more rigid in the former solvent which might have favoured gelation. Evidently, something more is involved which remains to be understood.

(iv) Cellulose derivatives

The work by Newman *et al.* (1956) on nitrocellulose aggregates represents to date the most thorough investigation. These authors have studied the aggregate size by light scattering. From the relative constancy of the ratio $M_w/\langle R^2\rangle^{3/2}$ they came to the conclusion that the aggregates were most probably of spherical symmetry. It is to be noted, however, that the aggregates possess molecular weights larger than 5×10^6 (from 6.35 to 19.7×10^6). It therefore remains uncertain whether the measurement of the radius of gyration is valid, i.e. whether the condition $qR < 1$ is still fulfilled.

From osmotic pressure measurements these authors have determined the variation of the second virial coefficient:

$$\pi/C = \pi_0/C \quad (2.25)$$

where A_2 reads in the dilute solution treatment (Flory, 1953):

$$A_2 = (\bar{v}^2/V_1) \times (1 - \Theta/T) \times \Psi_1 \times F(X) \quad (2.26)$$

in which $\bar{v}$ and V_1 are the partial specific volume of polymer and the molar volume of the solvent, Ψ_1 the entropy parameter and $F(X)$ a function close to unity in the vicinity of the Θ-temperature.

From relation (2.26), Newman *et al.* (1956) obtained for the lower Θ-temperature $\Theta \approx 301$ to 310 K in ethanol. They accordingly conclude that aggregation takes place before the condition for liquid–liquid phase separation is attained which undoubtedly contradicts the mechanism imagined by Rees (1972) (see Section 1.3.1.b.iii).

Finally it should be noted that cellulose nitrate is a relatively stiff chain in solution. Holtzer *et al.* (1954) have deduced a statistical segment length $b = 3.8 \pm 0.3$ nm.

There seems to be a general trend among cellulose derivatives to display chain stiffness in solution. For instance, persistence lengths of $a = 7.5$ nm ($b = 15$ nm) are reported for hydroxypropylcellulose in dimethyl acetamide (Conio *et al.*, 1983).

2.2.3 Blends of biopolymers

One major concern in investigations carried out on blends of polysaccharides relates to the kind of molecular interaction involved. Cairns *et al.* (1987) consider four types of structures corresponding to four types of interaction.

(1) One polysaccharide forms the gel whereas the other one remains in solution.

(2) Interpenetrated networks – these may originate from a solid–liquid phase separation. One polysaccharide may crystallize first. This implies that the polysaccharides are compatible in the sense used for synthetic polymers.

(3) Phase-separated networks – wherein a liquid–liquid phase separation is involved at the early stage of gelation. This implies that the polysaccharides are incompatible in the sense used for synthetic polymers.

(4) Coupled networks.

Most probably, synergetic effects will probably only arise with cases 2 to 4.

Currently, there is no definite answer to this question. X-ray diffraction studies carried out by Cairns *et al.* (1987) suggest that no intermolecular binding, that is complex formation between parts of two different polysaccharides, occurs. They also conclude that galactomannan is randomly oriented within the gel, and moreover report an exception to this trend: xanthan + galactomannan.

The assertion that galactomannan is not incorporated into the network and is further oriented randomly within the gel is based on the absence of any reflections due to galactomannan in the gel when stretched and dried. Under these conditions Cairns *et al.* have only observed those reflections arising from $\varkappa$-carrageenan fibres.

Recently, ^{13}C-NMR studies by Rochas *et al.* (1990) have given new insight into the problem although contradicting the findings of Cairns *et al.* Studying galactomannan/$\varkappa$-carrageenan gels, the investigators found that the ratio galactose/mannose in their galactomannan sample was G/M = 22/78, and showed that there were three populations of galactomannan. One population is involved in the gelation phenomenon as about 40% of the signal due to galactomannan disappears below the sol–gel transition.

These chains are characterized by a ratio of G/M = 8.5/91.5 that contains a lower degree of branching than the overall sample. Another population (about 30%) displays reduced mobility with a ratio of G/M ≃ 31/69, and in a final population, 30% of the galactomannan chains exhibit the same mobility with a ratio of G/M = 31/69. Rochas *et al.* therefore conclude that galactomannan is undoubtedly involved in the gelation process, otherwise its mobility would not be affected. Nevertheless, from NMR there is no way to tell whether the galactomannan chains form a complex with $\varkappa$-carrageenan or whether they gel on their own, in which case they would give an interpenetrated network.

2.3 SYNTHETIC POLYMERS

In Chapter 1 synthetic polymers were classified, depending upon the way gels are produced, into two major groups, polymers of type I giving solvent-induced gels and polymers of type II giving crystallization-induced gels. From the existing investigations into the molecular structure it appears that polymers possessing bulky side groups belong in the former class while polymers without bulky side groups belong in the latter. As will be discussed below, it seems to be so because type I polymers possess the intrinsic tendency to take on helicoidal arrangements in the ordered state, which favours the formation of polymer–solvent compounds, whereas type II polymers usually do not.

Interestingly enough, solvent-induced gels (prepared from polymers of type I) are in certain cases chiefly non-crystalline. While the absence of crystallinity seems particularly well established for polystyrene gels, as hinted at by numerous studies, and suspected for others, it remains to be discovered whether this can be a more general principle. That the gels are said to be non-crystalline does not, however, imply a totally disordered state as is the case for chemical gels. These non-crystalline physical gels at least display nematic-like order, i.e. one-dimensional order, as opposed to crystalline gels which possess three-dimensional order.

Of final note is the disparity in the varieties of ordered states for stereoregular synthetic polymers and most biopolymers. As a rule, the latter, as they undergo molecular ordering, form only a gel, unlike the former which can either develop crystals (spherulites) or produce gels. The crystalline state of synthetic polymers has received continuous attention for the past thirty years and is accordingly well documented. As a consequence, it comes as no surprise that the crystalline state serves as a reference throughout this section to evaluate how far the molecular structure of a gel deviates from it.

2.3.1 Solvent-induced gels

(a) Isotactic polystyrene

The isotactic polystyrene crystalline unit cell has been determined by Natta (1955) and found to belong to the spatial group of symmetry $R\bar{3}c$ (see Figure 2.17). In the unit cell there is an equal number of right-handed and left-handed 3_1 helices of pitch 0.665 nm (6.65 Å) (trans-gauche, *tg*, arrangements). The 3_1 helix has been found to be energetically stable (Natta *et al.*, 1960).

From energy calculations, Sundararajan (1979) and Corradini *et al.* (1980) have shown that the 3_1 helix is probably not the only stable form and that a 12_1 helix (near *tt*, trans–trans conformation) displays equivalent stability. However, this helical form has never been observed in crystals, particularly in single crystals, so that the 3_1 form remains the reference helical form (both types of helix are drawn in Figure 2.17).

The 12_1 helix has nevertheless been regarded by some authors as being the constituent element of the nascent gel structure on the basis of X-ray diffraction results obtained on partially dried and stretched samples (Atkins *et al.*, 1980, 1984). This point of view has been more recently questioned by Guenet (1986) from neutron diffraction experiments on nascent gel samples. These different views will be discussed in Section 2.3.1.a.ii.

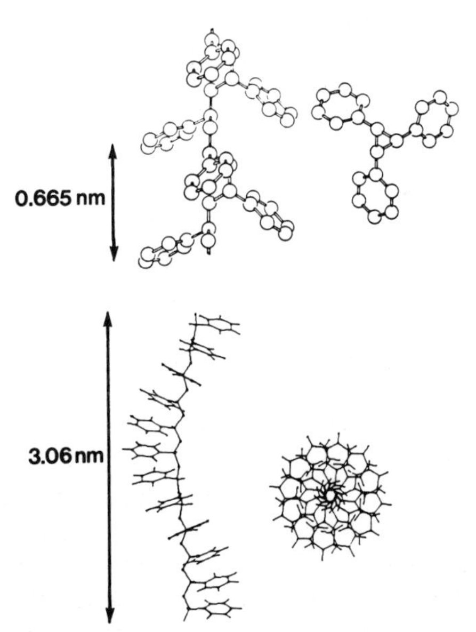

Figure 2.17 Schematic drawing of the 3_1 helical form (upper figure) and the 12_1 helical form (lower figure) of isotactic polystyrene seen perpendicular and parallel to the helix axis, respectively. Pitch in nanometres as indicated (with permission from Sundararajan *et al.*, 1982).

(i) Gel morphology

Earlier work by Lemstra and Challa (1975) had indicated that iPS–decalin gels possessed a fibrillar structure. Later, Atkins *et al.* (1984) and Guenet *et al.* (1985) confirmed these findings. Figure 2.18 shows a typical electron micrograph obtained by Guenet *et al.* (1985) of a gel prepared in 1-chlorododecane, annealed at 50°C, a process which transforms the original structure into the 3_1 form (see Section 1.2.1.a.i) and finally dried out. The fibres have a diameter of about 40 nm and the mesh size is of the order of 1 μm (for a concentration of 7% in polymer). Isolated crystals that have grown from the dilute phase can also be seen (dark spots). They are not well formed in that they do not possess the hexagonal shape. Here it must be stressed that these crystals have grown while the gel still had its original structure (nascent gel structure).

Another system was investigated by Guenet *et al.* (1985): iPS–diethyl malonate. As emphasized in Section 1.2.1.a.i, the nascent molecular structure of these iPS–diethyl malonate gels is highly unstable in that it transforms spontaneously into the 3_1 form. Three types of morphology have been observed: fibres, isolated aggregates of single crystals and shish-kebabs. Considering that the gel consists of fibre-like entities, Guenet *et al.*

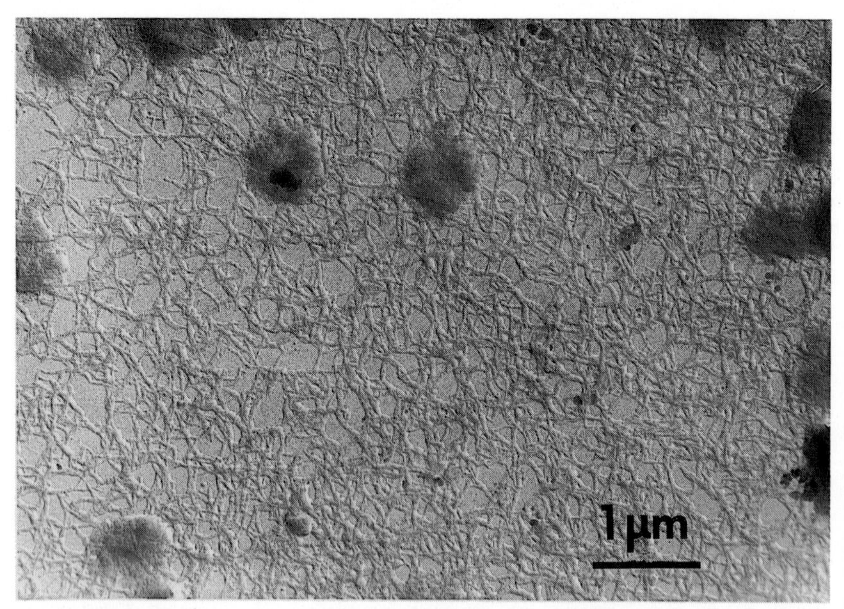

Figure 2.18 Electron micrograph of a thin iPS gel particle obtained from a 7% (w/w) gel in 1-chlorododecane (quenching temperature −5°C, aged one week at 20°C) after transformation of the nascent molecular structure into the 3_1 form as achieved by annealing at 72°C and toluene extraction (with permission from Guenet *et al.*, 1985).

(1985) proposed a three-step mechanism to account for all the observed morphological species: (1) aggregates of single crystals form from the dilute phase while the gel (the network) is still in its nascent form; (2) when the gel transforms into the 3_1 form it nucleates crystallization of the chains that are still in solution in the dilute phase, hence the formation of shish-kebabs; (3) once the dilute phase does not contain enough chains in solution, this process stops, hence the presence of bare fibres.

Similar observations of shish-kebabs were made by Atkins *et al.* (1984) on gels prepared from *trans*- or *cis*-decalin which had been partially dried and then annealed at various temperatures.

(ii) Gel molecular structure

Early work by Girolamo *et al.* (1976) had shown that the gel diffraction pattern does not exhibit any of the reflections known for the 3_1 form. Instead, a meridional reflection at 0.51 nm was observed on partially dried and stretched gels. If this reflection arises from a helical form, it may imply that the axial rise is 0.51 nm (see Appendix 2).

To account for this meridional reflection, Girolamo *et al.* (1976) suggested first that syndiotactic sequences were present and were actually responsible for the gelation phenomenon. In fact, these sequences are liable to adopt a planar zig-zag conformation, the axial rise of which could be about the right value provided the orientation of the phenyl rings showed a period of 1.02 nm. Head-to-head as well as tail-to-tail arrangements were also considered. However, defects of this type altogether account for less than 1% of the chain which throws doubts on this explanation. In following papers (Atkins *et al.*, 1980, 1981, 1984), the 12_1 helix suggested by Sundararajan (1979) and Corradini *et al.* (1980) was considered to account for the 0.51 nm reflection. However, it had to be supposed that the helix unit was composed of a dimer otherwise the axial rise would only be 0.25 nm. This means that the helix must possess sixfold symmetry, an assumption not backed up by conformational analysis which gives twelvefold symmetry (see Figure 2.17).

Atkins *et al.* (1980) also noticed that the experimental intensity could not be theoretically reproduced with this model, but they attached greater importance to the position of the reflection rather than to its intensity.

Aware of the packing problems caused by a twelvefold helix, Sundararajan and Tyrer (1982) and Sundararajan *et al.* (1982*b*) have proposed this helix to be solvated, which implies that any ordered structure made up of this helical form cannot be grown without the presence of solvent.

This point was partially questioned by Atkins *et al.* (1984). While this form may need solvent to grow, they said that it can exist without it, at least in *trans*-decalin. By investigating stretched and partially dried samples, these authors have been able to observe reflections that they attribute to the first and second layer lines. The first layer line gives a distance of about 3.06 nm which is in agreement with the pitch of a twelvefold helix. So there seems to be substantial evidence that in these samples a 12_1 helix is present. It must nevertheless be emphasized that there still remain many unresolved problems such as the inversion of the ratio between the first and the second layer line intensities when using *cis*- or *trans*-decalin.

In addition to these questions, the issue of interest still remains: was the 12_1 form already present in the nascent gel or has this form been obtained by the drying and stretching process? After all Foord and Atkins (1989) have clearly shown in the case of agarose that, depending upon both the stretching process and the stretching temperatures, different helical forms of different symmetry can be obtained.

The only reason for drying a gel sample before investigating it by X-ray diffraction is to attenuate the signal coming from the liquid which usually gives off a strong parasitic halo. However, many different studies point towards polymer–solvent compounds, a point not questioned by Atkins *et al.* (1984) and no-one knows to what extent the drying step modifies the gel nascent structure.

To overcome this difficulty, Guenet (1986) performed neutron diffraction experiments on nascent gels. The advantage of this method over X-ray diffraction rests upon the possibility of labelling differently either component of the system without altering its intrinsic properties and particularly its structure (see for instance Cotton *et al.*, 1974). Since deuterium possesses a scattering length about twice that of hydrogen ($b_D = 0.65$ against $b_H = -0.337$), the signal arising from the deuterated component is enhanced by a factor of about 4 with respect to that of the protonated components. Making use of this property, various gels were investigated: iPSD–*cis*-decalinH gels and iPSD–*cis*-decalinD gels. These gels were investigated without stretching.

The early experiments by Guenet (1986) showed two major results (Figure 2.19): (i) while for iPSD–*cis*-decalinD samples a strong reflection is observed at 0.53 ± 0.01 nm (Figure 2.19a), for iPSD–*cis*-decalinH samples this reflection is considerably attenuated (Figure 2.19b); (ii) the diffraction spectrum of liquid *cis*-decalin (deuterated) exhibits the same pattern (Figure 2.19c). Admittedly, the maximum at 0.53 nm is comparatively broad, but no narrow reflection at 0.51 nm can be detected. Since the 0.51 nm reflection is said to be the strongest one (Sundararajan and Tyrer, 1982; Sundararajan *et al.*, 1982b; Atkins *et al.*, 1984), it should have been unmistakably detected in the iPSD–*cis*-decalin gel sample, as are the main reflections of the 3_1 helix detected for samples of similar polymer concentration (Klein *et al.*, 1990*b*; see Figure 2.20).

These neutron diffraction experiments show that in the nascent state, there is no valid argument for a twelvefold helical form. However, the fact

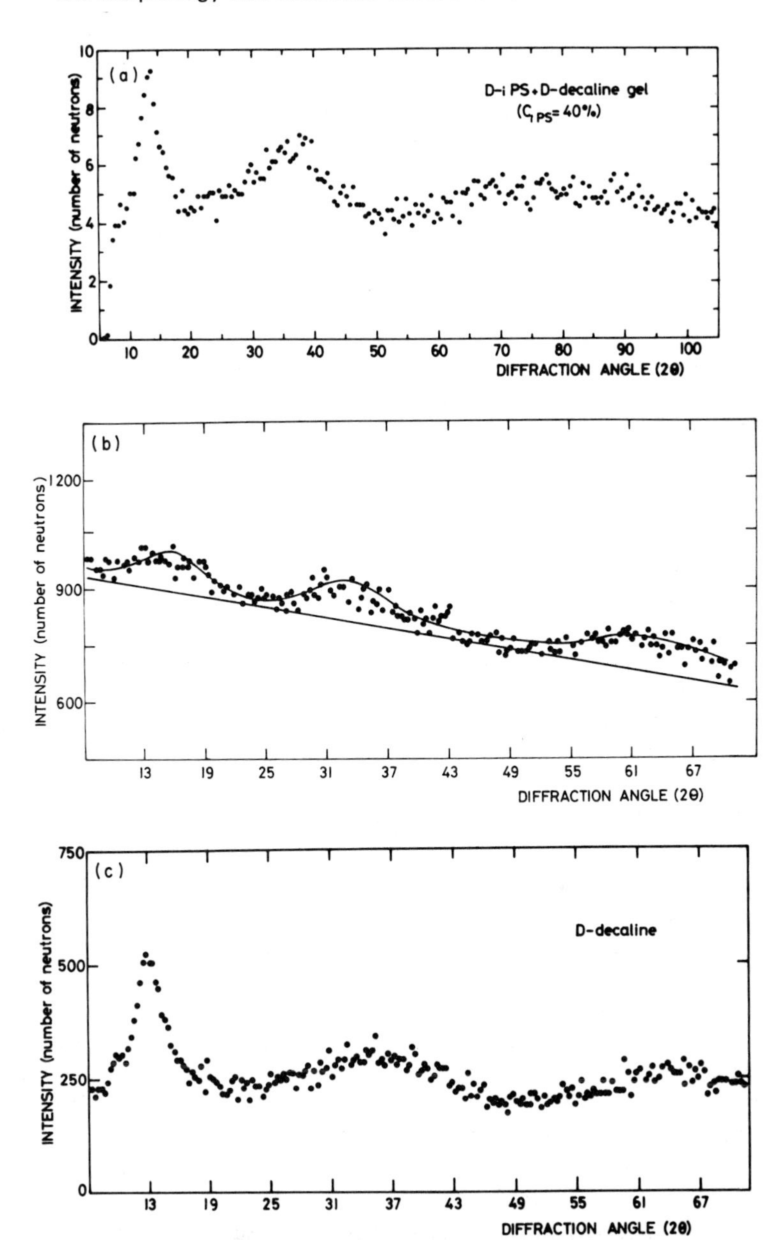

Figure 2.19 Neutron diffraction patterns, $\lambda = 0.1225$ nm. Upper curve: diffraction pattern of a gel containing perdeuterated isotactic polystyrene and perdeuterated *cis*-decalin. Gel obtained by a quench at 0°C, $C_{pol} = 40\%$ (w/w). Middle curve: diffraction pattern of a gel prepared with perdeuterated isotactic polystyrene and protonated *cis*-decalin. Gel obtained by a quench at 0°C, $C_{pol} = 20\%$ (w/w). Lower curve: diffraction pattern of *pure* perdeuterated *cis*-decalin at 20°C (therefore in the liquid state) (with permission from Guenet, 1986).

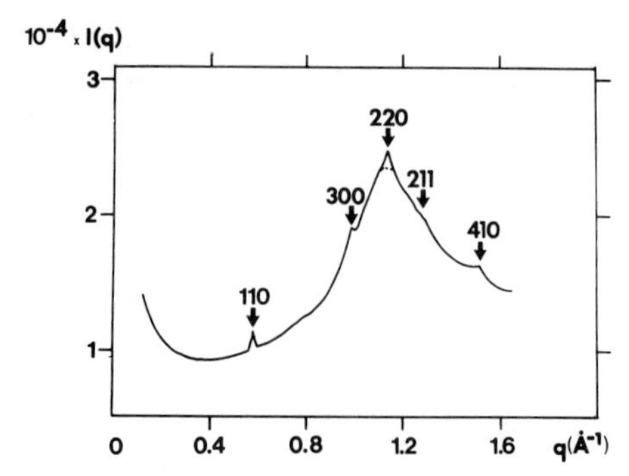

Figure 2.20 Neutron diffraction pattern, $I(q)$ *vs* q, for a 30%-solution of perdeuterated isotactic polystyrene in perdeuterated *cis*-decalin annealed at 30°C for 24 h. Under these conditions chain-folded crystals containing the usual 3_1 form are produced as indicated by the reflections of the crystallographic planes observed (indicated by arrows) (with permission from Klein *et al.*, 1990*b*).

that *cis*-decalin in the liquid state exhibits a maximum at 0.53 nm, i.e. within experimental uncertainty, a distance similar to that reported for the gels is puzzling. Guenet (1986) did not accept this result as being merely fortuitous and suggested re-examination of the interpretation of this distance on the basis of two arguments. (1) Inspection of the gel diffraction pattern points towards a liquid-like order and not a crystalline order at all (compare Figures 2.19a and 2.20). (2) When dealing with liquids, the first diffraction maximum is directly related to the distance between first neighbours and is given in the case of spherically averaged liquids by:

$$1.23\lambda = 2d\,\sin(\theta/2) \tag{2.27}$$

This relation, originally due to Ehrenfest (1915), can be derived quite simply from the Debye relation for the diffracted intensity:

$$I(q) \approx 1 + \Sigma\,\Sigma\,\sin qr_{ij}/qr_{ij}$$

in which the second term represents a summation of all the possible pairs. If there are a large number of pairs spaced by a distance r_0 (which is the case for first neighbours), then one can show that the first maximum of $I(q)$ will occur for $qr_0 = 7.72$ which eventually gives relation (2.27). Similarly, if one now considers an assembly of parallel rods, as is encountered in liquid crystals, then $I(q)$ reads:

$$I(q) \approx 1 + \Sigma\,\Sigma\,J_0(qr_{ij})$$

in which J_0 is the Bessel function of zero order. Using the same argument as above, one finally obtains:

$$1.115\lambda = 2d\,\sin\theta/2$$

However, as emphasized by Guinier (1956), these are approximated relations that allow one to give only an estimate of the true distance between first neighbours. The correct procedure remains the calculation of the Fourier transform of the diffracted intensity, whenever possible, so as to derive the radial distribution function $g(r)$, the first maximum of which gives the distance between first neighbours.

This entails the distance between *cis*-decalin molecules in the liquid state being about $d \simeq 0.65 \pm 0.015$ nm. Guenet noticed that this value happens to be very close to the pitch of the 3_1 helix (0.665 nm). In other words, the distance between two phenyl groups after one turn of the threefold helical form is in register with the distance between first-neighbour decalin molecules. Inspection of Figure 2.21, in which a 3_1 helix is represented, reveals that between two such phenyl groups there is a cavity large enough to house one solvent molecule of about the size of decalin.

On the basis of this circumstantial evidence, Guenet (1986) has proposed a new model (designated the **ladder-like model**) which was slightly modified later (the latest version is given in Figure 2.22; Klein *et al.*, 1991). In this model, which is to some extent reminiscent of the **egg-box model**, the chain adopts a near-3_1 helix so as to accommodate the decalin molecules. If the helix has no strict regularity, then one cannot expect to see its characteristic reflections. On the other hand, the diffraction pattern should be dominated

Figure 2.21 Representation by means of space-filling models of an isotactic polystyrene chain under a near-3_1 helical form. Arrows indicate the cavity created by phenyl rings.

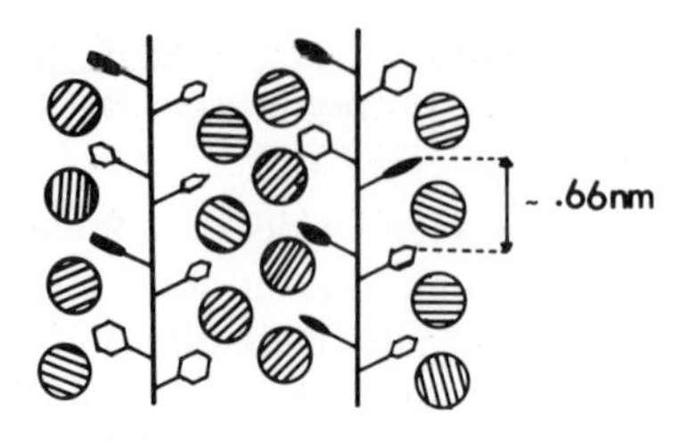

Figure 2.22 Schematic representation of the ladder-like model. Hatched spheres represent the solvent molecules. The phenyl rings are drawn with random orientation to emphasize the absence of well-defined order such as would be found in crystals.

by the diffraction of the decalin molecules that still possess an order close to their liquid order. As Guenet (1986) points out, the 0.51 nm reflection is sometimes observed at 0.48 nm, a result totally incompatible with a 12_1 helix (Sundararajan *et al.*, 1982). Also, the 0.51 nm reflection has mostly been observed with samples prepared from solvents of nearly identical size (140 to 156 cm^3/mol) which are therefore liable to have diffraction maxima in the vicinity of 0.5 nm, such as hexahydroindan or *trans*-decalin. In the latter case, the first maximum of the liquid diffraction pattern turns out to be at exactly 0.51 nm.

The ladder-like model, in which the chain takes on a near-3_1 helical form, has some decisive advantages. First, it offers the simplest solution for elucidating the helical structure since the 3_1 form is known to exist. Then it accounts for the polymer–solvent compounds of different types demonstrated in the temperature–concentration phase diagrams. Finally, it allows an understanding of why there exist stable nascent gel structures and metastable ones. A closer look at the molecular size of the solvents with which the nascent gel structure is metastable shows that these solvents are bulkier than decalin. According to Guenet (1986), solvents larger than the cavities are not so well accommodated, an effect that is supposed to worsen with bulkier solvents. As a result, the 'anhydrous' 3_1 form becomes more stable than the solvated one for purely entropic reasons, hence the **nascent structure** $\Rightarrow$ **3_1** transformation which occurs spontaneously below the metastable gel melting point in some solvents. This transformation is then nothing else than a desolvation process of the nascent structure according to Guenet.

This assumption receives support from the study of the degree of solvation carried out with the solvent crystallization method (see Section 1.2.1.a.ii, Figure 1.15). The average number of solvent molecules per monomeric unit, α, decreases with increasing solvent size. Gels that are known to have a metastable nascent structure are characterized by $\alpha < 1$.

Obviously, the above statement is only valid at constant polymer–solvent interaction in which case entropic effects play the main role. If the polymer–solvent interaction becomes poorer at constant solvent molecular size, as is the case from decalin to diethyl malonate, then the transformation also occurs but for energetic reasons.

The early version of the ladder-like model was later questioned by Pérez *et al.* (1988) on the basis of NMR experiments. These authors have investigated the line broadening of the solvent incorporated in the gel. Their results show that solvent mobility is virtually the same in the gel as in the liquid state and that solvent motion is isotropic. By lowering the temperature they have observed that the fraction of solvent in the gel displaying a line broadening similar to pure frozen solvent increases abruptly in *trans*-decalin at a temperature corresponding to the solvent solidification point, whereas in *cis*-decalin this fraction increases smoothly (Figure 2.23).

This result was used by Pérez *et al.* (1988) as an argument against the existence of a polymer–solvent compound as deduced from DSC measurements of the solvent crystallization behaviour (Guenet, 1986) (albeit the concept of a polymer–solvent compound did not rely solely on this experiment).

They have accordingly concluded that a polymer–solvent compound with the ladder-like structure was not correct and suggested that instead the solvent was mainly excluded from the gel junction. Small-angle neutron scattering experiments reported by Klein *et al.* (1990*a*) on 30% gels prepared in *cis*-decalin show that solvent is indeed intercalated between the polymer chains. In this experiment only the polymer is deuterium-labelled. Under these experimental conditions, if the solvent were absent from the gel junctions, then the gel could be regarded as a two-density system for which the

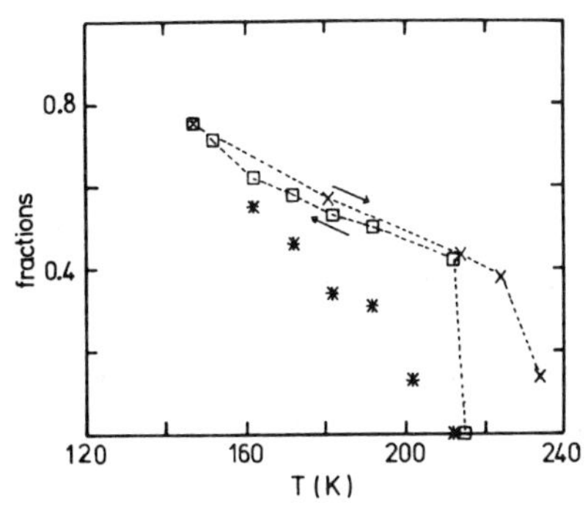

Figure 2.23 Fractions of solvent protons in gel samples having line widths as broad as in the pure frozen solvent in proton NMR experiments. * = iPS–*cis*-decalin cooling cycle; □ = iPS–*trans*-decalin, cooling cycle; x = iPS–*trans*-decalin, warming cycle. Uncertainties are ±0.06 (with permission from Perez *et al.*, 1988).

scattering intensity is given by the Debye–Bueche relation (Debye and Bueche, 1949):

$$I(q) \simeq a_d^3/(1+q^2a_d^2)^2 \tag{2.28}$$

in which a_d is the range of the density fluctuations. For $qa_d > 1$ relation (2.28) reduces to the well-known Porod's law (Porod, 1951):

$$I(q) \simeq 1/q^4 \tag{2.29}$$

Results given in Figure 2.24 do not support this model. On the contrary, the scattering pattern is in agreement with an intercalated model for which the following relation applies for parallel rods of infinite length and radius r_c, the arrangement of which is described by the radial distribution function $g(r)$ (Oster and Riley, 1952):

$$I(q) \approx \frac{\pi\mu_L}{q} \times \frac{J_1^2(qr_c)}{(qr_c)^2}\left\{1 - 2\pi n_A \int_0^\infty [1-g(r)]\, J_0(qr)\,dr\right\} \tag{2.30}$$

in which n_A is the number of rods per unit area, J_0 and J_1 the Bessel function of first kind and of zero and first order respectively.

If the rods are separated by a distance $d > r_c$, then there is a domain $(d^{-1} < q < r_c^{-1})$ where, by considering $g(r) \approx 0$ and developing the Bessel function for $qr > 1$, relation (2.30) reduces to:

$$I(q) \approx \left(\frac{\pi\mu_L}{q}\right)\exp(q^2r_c^2/4)\,[1 - 2\pi n_A q^{-0.5} f(r)] \tag{2.31}$$

in which $f(r)$ is an oscillating function.

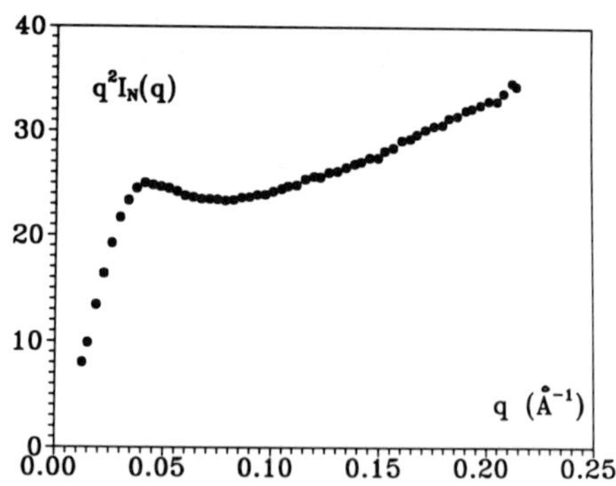

Figure 2.24 Determination of the order between chains: Kratky-plot ($q^2 I_N(q)$ *vs* q) for a 30% (w/w) iPS–*cis*-decalin gel in which only the polymer is perdeuterated. The same type of scattering pattern is seen with iPS–*trans*-decalin gels. The calibrated intensity, $I_N(q)$, is obtained by renormalization by means of a water spectrum and then by division by $K \times C(1-C) \times \rho_p$, in which K is a constant dependent upon the scattering amplitudes of the different species, C the polymer concentration (v/v) and ρ_p the polymer density. $I_N(q)$ can be expressed in cm^{-1} by multiplying it by the incoherent scattering cross-section of water (here for $\lambda = 1.2$ nm) (with permission from Klein *et al.*, 1990*a* and Guenet, unpublished results).

Relation (2.31) implies that $I(q)$ is proportional to q^{-1} in this range, which agrees qualitatively with what is found experimentally.

It is, however, possible that the modified version of the ladder-like model (Figure 2.22), which still accounts for the neutron scattering data, may now be in better agreement with the high solvent mobility found by NMR (Pérez *et al.*, 1988). Here a solvent molecule is in contact with other solvent molecules and thus is liable to exchange position quite freely and move isotropically, which was not feasible in the frame of the earlier version.

Finally, the ladder-like model has recently received additional support from experiments carried out by Nakaoki and Kobayashi (1991) by infrared spectroscopy combined with ^{13}C-NMR and X-ray diffraction.

(iii) Morphology and molecular structure near the gelation threshold

As has been stated in Section 1.2.1.a.i, the gelation threshold is located at a well-defined temperature, T_{gel}. The transition can also be evidenced by optical microscopy (Figure 2.25a). Below T_{gel}, a salt-and-pepper structure characteristic of the gel state can be observed whereas above, spherulites appear (Klein *et al.*, 1990*b*).

Investigating the morphology and molecular structure just above the gelation threshold again demonstrates the differences between the two decalin isomers. Whereas, in *trans*-decalin, species grown just above the gelation threshold are small spherulites consisting of normal crystals (3_1 form with the usual unit cell), in *cis*-decalin not only are the spherulites far larger but also the diffraction pattern is quite unusual (Klein *et al.*, 1990*b*). In *cis*-decalin the spherulite size can be as high as 180 μm (Figure 2.25b), values that have never been observed so far for this polymer. Yet, what is more surprising is that, despite the apparent long range order, these spherulites are not crystalline. In fact, diffraction experiments (neutron or X-ray, see Figure 2.26) reveal only three maxima at 2.5 nm, 1.7 nm and 1.1 nm which are relatively broad compared with diffraction peaks (for the sake of comparison see Figure 2.20).

Here, according to Klein *et al.* (1990*b*) the spherulitic state is reminiscent of smectic arrangements since spherulites unmistakably indicate the presence of lamellae and the diffraction pattern points towards a relatively poor organization within these spherulites (smectic F type).

Neutron diffraction experiments carried out on samples in which only the polymer is deuterated suggest that the peaks correspond to distances between polymer chains (Figure 2.26). Investigation of an all-labelled sample (both the solvent and polymer are deuterated, Figure 2.26) shows that these reflections vanish and, accordingly, do not arise from a helical structure but characterize distances between chains, giving strong support

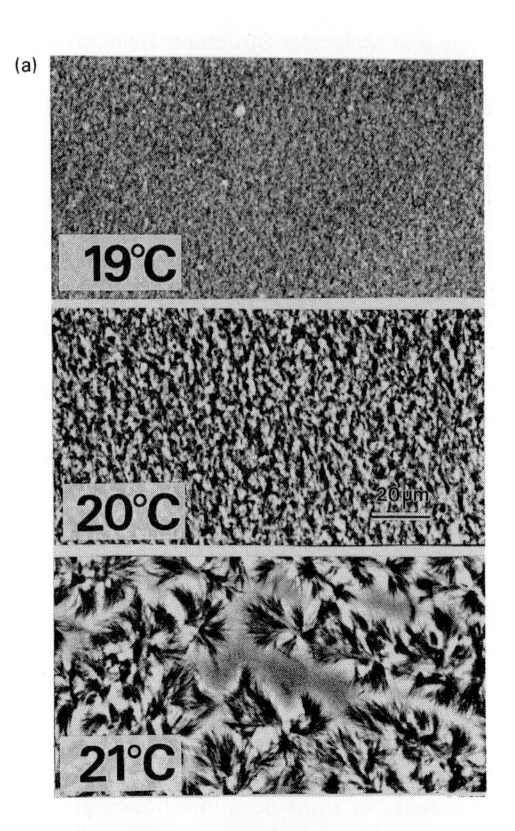

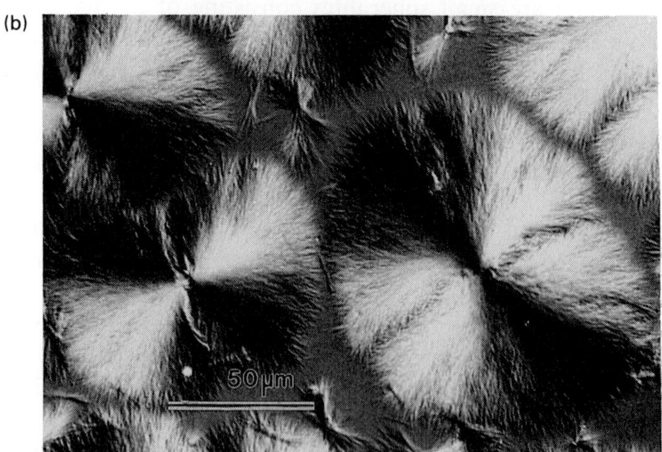

Figure 2.25 Optical micrographs showing: (a) modification of the texture of the structures grown from a 30% (w/w)-iPS–*cis*-decalin solution at the indicated temperatures. The transition from a gel texture (salt and pepper) to another state (spherulites) occurs around 20 ± 1°C. (b) Spherulitic structures (designated as carnation-like structures) obtained from a 30%-iPS–*cis*-decalin solution annealed 24 h at 27°C (nomarsky phase contrast) (with permission from Klein *et al.*, 1990*b*).

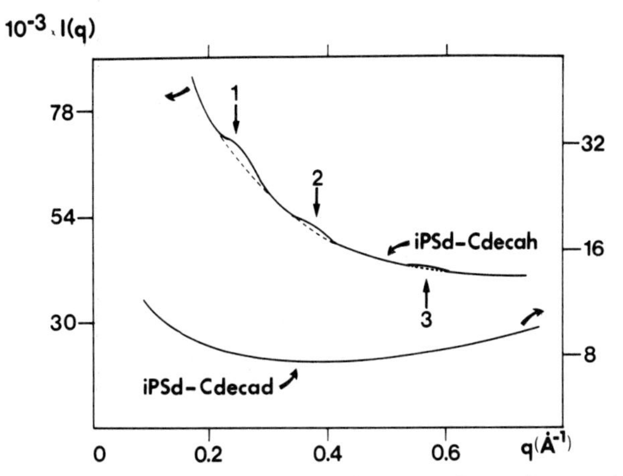

Figure 2.26 Neutron diffraction pattern, $I(q)$ *vs* q, from the structures grown from a 30%-iPS–*cis*-decalin annealed at 28°C for 48 h. Upper curve: deuterated isotactic polystyrene + protonated *cis*-decalin (iPSd + Cdecah); lower curve: deuterated isotactic polystyrene + deuterated *cis*-decalin (iPSd + Cdecad). Arrows highlight the position of the different diffraction maxima corresponding to the following distances: (1) 2.47 nm, (2) 1.7 nm and (3) 1.1 nm (with permission from Klein *et al.*, 1990*b*).

to the existence of a polymer–solvent compound. The diffracted intensity reads:

$$I(q) \approx a_p^2 S_p(q) + a_s^2 S_s(q) + 2a_p a_s S_{ps}(q) \tag{2.32}$$

in which a and $S(q)$ denote scattering amplitudes and scattering functions, the subscripts corresponding to polymer (p) and solvent (s). If the system is non-solvated, the cross-term in relation (2.32) is negligible and then $I(q)$ reduces to:

$$I(q) \approx a_p^2 S_p(q) + a_s^2 S_s(q) \tag{2.33}$$

Alternatively, in the case of intercalated solvent and applying the incompressibility hypothesis since we are dealing with distances larger than 1 nm, incompressibility hypothesis applies and then $I(q)$ reads:*

$$I(q) \approx (q_p - a_s)^2 S(q) \tag{2.34}$$

in which $S(q)$ is the scattering function of the system. If $a_s = a_p$, as is virtually the case when both the polymer and the solvent are deuterated, then the diffraction peaks vanish. Note that in the case of 'anhydrous' 3_1 helices,

---

* Even if the incompressibility hypothesis cannot be used, only relation 2.32 applies in the case of polymer–solvent compounds. Accordingly, the diffraction pattern will strongly depend upon the combination of labelled and non-labelled species.

relation (2.33) applies and diffraction peaks are observed despite the use of both deuterated polymer and solvent. This shows the considerable advantage of neutrons over X-rays when dealing with polymer–solvent compounds.

From these results, Klein *et al.* (1990*b*) have tentatively proposed a centred quadratic unit cell with $a = 2.5$ nm, $b = 1.7$ nm and $\alpha \simeq 100°$. These are obviously averaged values and one must keep in mind that the unit cells are probably decorrelated one with another so as to account for the few diffraction maxima.

Since these spherulites and their gel counterparts melt at nearly the same temperature, one can expect the chains to possess the same type of helicity in both systems. This offers an opportunity to test again the existence of the 12_1 helical form, as it cannot be argued, as it can for the gel, that the poor diffraction pattern results from a small amount of 12_1 helices. Here again, Klein *et al.* (1990*b*) report that the 0.51 nm reflection cannot be observed, either in the undried or the dried state. It then seems that the stretching process is involved in the formation of the type of helices that give the diffraction pattern recorded by Atkins *et al.* (1984).

(iv) Chain trajectory

Thus far, the trajectory (or conformation) of a chain embedded in a physical gel has only been determined in isotactic polystyrene by means of neutron scattering together with the isotopic labelling method (Guenet, 1987*a,b*; Klein *et al.*, 1990*a*). By mixing a small amount of deuterated chain with protonated ones, the single chain statistics can be determined. At the limit of infinite dilution of the labelled species ($C_D \to 0$), the scattered intensity, $I(q)$, then reads:

$$I(q) = K(a_D - a_H)^2 M_{wD} C_D P_D(q) \tag{2.36}$$

in which K is a constant depending upon the experimental set-up, $(a_D - a_H)$ the neutronic contrast, M_{wD}, C_D and $P_D(q)$ the weight-average molecular weight, the concentration and the form factor respectively of the deuterated chains. $P_D(q)$ is the Fourier transform of the monomer–monomer correlation function $\gamma(r)$ within one chain:

$$P_D(q) = \int_0^L \gamma(r)(\sin qr)/qr \; 4\pi r^2 \mathrm{d}r \tag{2.37}$$

Two regimes can be distinguished whether $qR < 1$ or $qR > 1$, where R is the chain radius of gyration. If $qR < 1$ (Guinier regime) $P_D(q)$ reduces to:

$$P_D(q) = 1 - (q^2R^2/3) \tag{2.38}$$

Alternatively, for $qR > 1$ (intermediate regime) $P_D(q)$ can be expressed in the general form:

$$P_D(q) \approx q^{-n} \tag{2.39}$$

in which n is an exponent which depends upon the short-range statistics and is sometimes equal to the fractal dimension of the object. On other occasions two different exponents can be found in the intermediate regime.

A similar study has already been performed on the crystalline state and it has been confirmed that a chain folds on itself, as has been demonstrated, among other results, by the q^{-2} terminal behaviour in the intermediate range (Guenet, 1980*b*, 1981).

Early neutron scattering investigations of iPS–*cis*-decalin gels ($C_{pol} = 15\%$) by Guenet (1987*a,b*) have shown the chain to possess a conformation intermediate between that in the amorphous state and that in the crystalline state. The variation in the chain radius of gyration, R_G, with molecular weight has been found to be:

$$R_G(\text{nm}) = 0.062 M_w^{0.5 \pm 0.03} \tag{2.40}$$

The exponent $\nu = 0.5$ suggests that the global behaviour remains Gaussian (or Brownian), but the chain dimensions are larger than in the amorphous state for which the prefactor has been found to be 0.028 for the unperturbed state at room temperature (Krigbaum *et al.*, 1958; Guenet *et al.*, 1979). This means that the statistical segment length ought to be about 4 times larger in the gel than in the amorphous state. Since $b_{\text{unperturbed}} \simeq 2$ nm, one expects b_{gel} to be about 8 nm. This is exactly the value found from the intermediate regime.

The intensity scattered in the intermediate regime (Figure 2.27) reveals a pattern typical of worm-like chains (see Appendix 3). In fact, for $q < q^*$ (in the present case $q^* \approx 0.5$ nm^{-1}), the intensity approaches a q^{-2} asymptote, whereas for $q > q^*$, the intensity reaches a q^{-1} behaviour. From the value of q^*, as determined from the intercept of the q^{-2} and the q^{-1} behaviour,

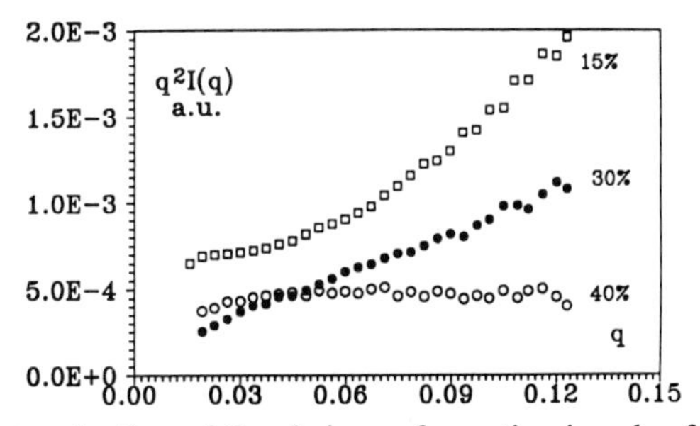

Figure 2.27 Determination of the chain conformation in gels of various polymer concentration: Kratky-plot ($q^2 I(q)$ in arbitrary units *vs* q in Å^{-1}) for deuterated chains (concentration around 1%) imbedded in a gel matrix prepared by a quench to 0°C. The indicated figures correspond to the total polymer concentration in the gel (data collected from Guenet, 1987*a*, 1987*b* and Klein *et al.*, 1990*a*).

(Guenet 1987*b*) deduced a statistical segment length of about $b \simeq 8$ nm. A fit of the experimental curve with Yoshisaki and Yamakawa's (1980) pseudo-analytical model (see Appendix 3) is also consistent with $b \simeq 8$ nm.

To summarize, in the gel state the chain is globally Gaussian as in the amorphous state but displays enhanced local rigidity as in the crystalline state. Correspondingly, chain folding is absent from the gel state. According to Guenet (1987*b*) the enhanced rigidity of the chain arises from helix stabilization by the solvent molecules, as has already been suggested for the ladder-like model (Guenet, 1986).

The worm-like statistics are typical of nematic polymers (see for instance, Arpin *et al.*, 1977). So, chain conformation and molecular structure are both reminiscent of nematic systems.

A study of the chain trajectory as a function of the overall polymer concentration C_{pol} has been carried out by Klein *et al.* (1990*a*) in *cis*-decalin gels. They have observed the radius of gyration to increase up to $C_{pol} = 30\%$ and then to decrease. Such behaviour is most unusual if compared with amorphous polymer solutions where R varies as $C_{pol}^{-1/8}$ (Daoud *et al.*, 1975). However, while the increase of R up to $C_{pol} = 30\%$ is not straightforwardly accounted for, the decrease beyond $C_{pol} = 30\%$ can again be simply explained by the existence of a polymer–solvent compound of the type described by the ladder-like model. $C_{pol} = 30\%$ turns out to be very close to the stoichiometric composition ($C_{\gamma} \simeq 28\%$) in iPS–*cis*-decalin (see Section 1.2.1.a.i). Klein *et al.* (1990*a*) reason that as long as there exists an excess of solvent with regard to the stoichiometric composition, one expects the helices to be fully stabilized. Once there appears to be a solvent deficit, helices are then partly stabilized, hence a decrease in rigidity and, correspondingly, in chain dimension.

The changes in the intermediate range happen to be more dramatic. The intensities reported in Figure 2.27 show that the q^{-1} behaviour observed for the 30% gel vanishes to the detriment of a q^{-2} behaviour for the 40% gel (Klein *et al.*, 1990*a*). It should be noted further that at $C_{pol} = 30\%$ the chain trajectory can no longer be modeled with worm-like statistics. Klein *et al.* have used Muroga's (1988) model in which long rods of length Λ and fraction f alternate with disordered portions made up of smaller rods of length λ (see Appendix 3). In addition, they found that the amount of large rods depends upon the labelled chain molecular weight. Typically, the curves can be fitted with $\Lambda \simeq 40$ nm and $\lambda \simeq 4$ nm with f varying from $f = 0.8$ for $M_w = 7.4 \times 10^4$ to $f = 0.4$ for $M_w = 2.5 \times 10^5$. These effects probably arise from the fact that the system lies near the stoichiometric composition, the latter being possibly slightly molecular weight dependent. Also, the presence of smaller rods hints at the appearance of 'amorphous' material, the amount of which becomes more and more important as C_{pol} is increased. On the other hand, Klein *et al.* report that the chain trajectory is not significantly altered by the gel preparation temperature from $-30°C$ to $14°C$.

The dependence of the chain rigidity on the solvent is well illustrated by experiments performed with *trans*-decalin and a mixture of *cis*- and *trans*-decalin. Guenet (1987*b*) has observed that the q^{-1} behaviour is absent from gels prepared in a mixture of *cis*- and *trans*-decalin. However, the radius of gyration is only about 15% smaller than in iPS–*cis*-decalin gel. Guenet therefore accounts for these results by considering a 'distorted' statistical segment arising from imperfect helix stabilization. This effect should not alter dramatically the chain dimension but it should increase the average cross-section of the conformation. The latter effect manifests itself through an additional term which, if sufficiently significant, can obliterate the q^{-1} behaviour (see Appendix 3, relation (A3.19)).

These results, if correctly interpreted, therefore imply that the chain conformation is very sensitive to the solvent type and, particularly, to the solvent conformation (either *cis*- or *trans*-decalin). That the solvent conformation has a dramatic effect has already been seen above the gelation threshold (see Section 2.3.1.a.iii). Recent neutron scattering experiments by Guenet *et al.* (1991) on *trans*-decalin gels have provided further illustration of the solvent-type effect: the rigidity is less enhanced in iPS–*trans*-decalin gels than in iPS–*cis*-decalin gels ($b \simeq 5$ nm against 8 nm). In addition, the decrease in rigidity (loss of the q^{-1} behaviour) is not seen at $C_{pol} = 40\%$ unlike in *cis*-decalin gels, which is anticipated from the solvent-stabilization principle as $C_\gamma \simeq 38$–40% in *trans*-decalin (Section 1.2.1.a.i). In fact, desolvation, and correspondingly the onset of helix destabilization, should only occur beyond $C_{pol} = 38$–40%.

Finally, it seems worth discussing results obtained at 25°C in iPS–*cis*-decalin systems, at which temperature giant spherulites grow (Section 2.3.1.a.iii). On the basis of the observation of two different domains in which $I_1(q) \approx q^{-1}$ and $I_2(q) \approx q^{-1}$ with $I_1(q)/I_2(q) \simeq 3$ (Figure 2.28),

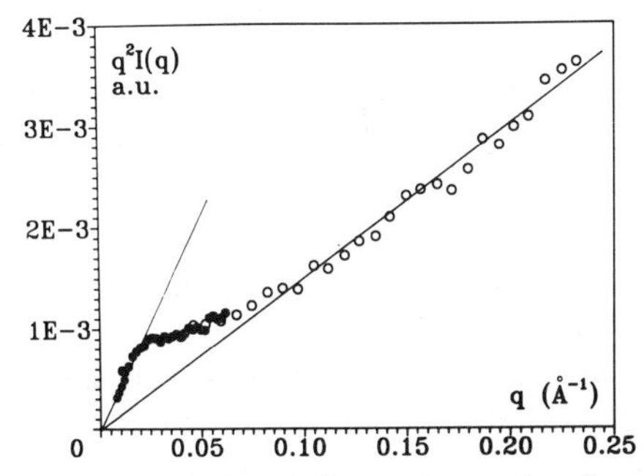

Figure 2.28 Determination of the chain conformation in carnation-like structures: Kratky-plot ($q^2 I_N(q)$ in reduced units, $I(q) = I_{exp}(q)/K = MC_D P_D(q)$) for deuterated chains ($C = 2 \times 10^{-2}$ g/cm^3) imbedded in a 30%-iPS–*cis*-decalin matrix annealed at 25°C (Guenet, unpublished).

Klein (1989) has suggested portrayal of the chain trajectory with a double hair-pin model. For such a model, an approximate expression of the form factor, in which the loops are neglected, reads:

$$P(q) = \pi/3qL\,[3 + 4J_0(qd) + 2J_0(2qd)] \tag{2.41}$$

in which L is the total contour length of the chain, d the distance between the branch and J_0 the Bessel function of first kind and zero order.

Two asymptotes may be observed in the intermediate range: in domain 1 where $qL > 1$ and $qd < 1$, relation (2.41) reduces, by developing the Bessel function, to

$$P_1(q) \approx 3\pi/qL(1 - q^2d^2/3) \approx (3\pi/qL)\exp(-q^2d^2/3) \tag{2.42}$$

In domain 2 where $qL \gg 1$ and $qd \gg 1$, relation (2.42) tends to the asymptotic value $P_2(q) \approx \pi/qL$ which yields in the end a ratio $P_1(q)/P_2(q) \simeq 3$.

These results show that chain-folding reappears just above the gelation threshold but with with many fewer folds than usually exist in the crystalline state. As far as chain trajectory and molecular structure are concerned, the state just above the gelation threshold differs from the gel state and the crystalline state.

(v) The pregel state

Thanks to deuterium labelling of a small number of chains, the chain conformation in the pregel state has been studied both in dilute and concentrated solutions by Klein *et al.* (1991). Their results have led to a thorough reconsideration of the gelation mechanism.

When a 15% iPS–*cis*-decalin gel is heated up to 66°C, i.e. about 10°C above its melting point, the **chain conformation** is found to be virtually **the same as in the gel state** (Figure 2.29). In both cases the chain trajectory can be modelled with a worm-like chain of statistical segment $b = 8$ nm.

Therefore, unlike crystalline systems for which melting produces amorphous material consisting of Gaussian chains possessing a short statistical segment length, gel melting more resembles a nematic–isotropic transition where chain rigidity is usually not altered.

To observe significant deviation from the original worm-like statistics, the sample must be heated up to 95°C (Figure 2.29). The q^{-1} behaviour then vanishes and $I(q)$ behaves like q^{-2}.

However, if the molten gel is kept at 66°C, the original worm-like conformation does not remain stable. Beyond 24 h, turbidity sets in which indicates crystallization of the sample and, similarly, the worm-like conformation vanishes, as shown in Figure 2.29. There the intensity varies like q^{-n} with $n > 2$ which is reminiscent of what has been observed for folded chains embedded in crystalline systems (Sadler and Keller, 1976) and can be

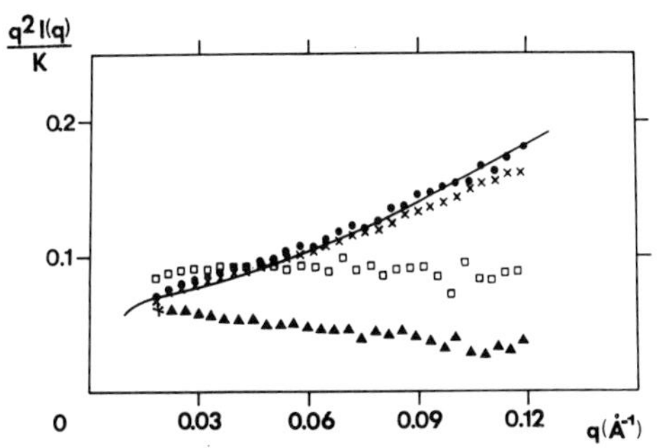

Figure 2.29 Determination of the chain conformation: Kratky plot ($q^2I(q)$ *vs* q in reduced units (see Figure 2.28 caption for further details)) for deuterated chains in a 15%-polymer matrix after going through the gel phase by a quench at 0°C and then heated first to 17°C (×), then to 66°C (●), and finally to 95°C (□). The solid triangles represent a sample that had been gelled at 0°C, heated up to 66°C and annealed 24 h at this temperature and finally annealed for a week at room temperature (with permission from Klein *et al.*, 1991, by permission of the publishers, Butterworth-Heinemann).

approximated to a slab (length L_c, width l_c and thickness δ) whose scattered intensity $I(q)$ reads:

$$I(q) \approx \frac{2\pi}{q^2 l_c L_c} \times \exp\left(-\frac{q^2\delta^2}{24}\right) \tag{2.43}$$

Relation (2.43) shows that apparent exponents larger than 2 can be observed.

Experiments carried out on dilute systems (here all the chains are labelled) confirm that the chain statistics remain worm-like at 66°C for a given period of time (Figure 2.30). Here again, the sample has to be heated up to

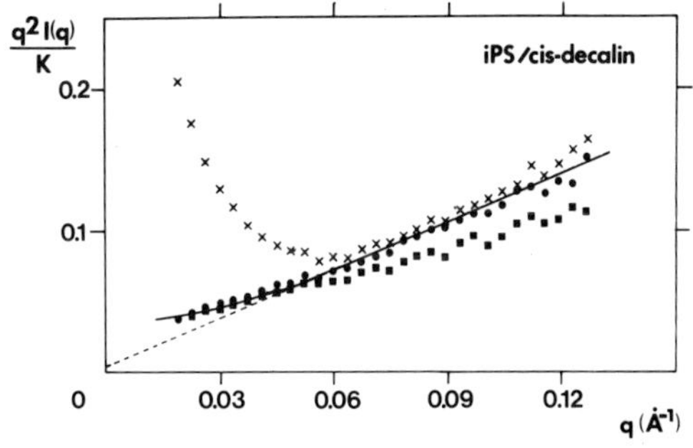

Figure 2.30 Determination of the chain conformation: Kratky plot ($q^2I(q)$ *vs* q in reduced units (see Figure 2.28 caption for further details)) for deuterated chains in dilute solutions ($C < C_{gel}$ and all the chains are deuterated) quenched at 0°C and then heated first to 17°C (×), then to 66°C (●), and finally to 90°C (□)). The solid line has been calculated by means of Yoshisaki and Yamakawa's relation for worm-like chains with $b = 10$ nm (Yoshisaki and Yamakawa, 1980) (with permission from Klein *et al.*, 1991, by permission of the publishers, Butterworth-Heinemann).

95°C to see significant departure from this bchaviour (here, at 95°C, $I(q) \simeq q^{-1.5}$).

From these results, Klein *et al.* (1991) have conjectured on the gelation mechanism. They have come to the conclusion that **the worm-like conformation is not the result of gelation** but, on the contrary, **gelation occurs on account of enhanced chain rigidity**. This chain rigidity is again believed to arise from helix stabilization by the solvent. This is again demonstrated when comparing results obtained in dilute iPS–*cis*-decalin pregels with those produced from *trans*-decalin: the q^{-1} behaviour appears at higher q value in *trans*-decalin which indicates shorter rigid segments than in *cis*-decalin.

These experiments throw new light on the gelation mechanism. Clearly, it is thought to be quite different from a crystallization process in which chains become rigid upon folding on to the crystal's growth faces. The gelation process requires rigid chains in the first place that simply align without folding at a temperature designated as the gelation temperature T_{gel}. It is probable that the parameter determining T_{gel} rests essentially upon the propensity of the chains to adopt a stable helical structure in a given solvent **without any folding**. Just above T_{gel}, folding reappears although helix solvation can still exist as in *cis*-decalin. In fact, gelation occurs when no folding can take place. This is what determines the gelation temperature in the first place, while gel melting depends only upon the short-range molecular structure. This statement is borne out by the fact that the gel possesses virtually the same melting temperature as the spherulites formed just above T_{gel}. It is then no wonder that there exists such a hysteresis between gel formation and gel melting.

(b) Atactic polystyrene

Atactic polystyrene is probably the paradigm of amorphous polymers as there does not exist the slightest trace of crystallinity in the bulk state. It has therefore been used as a model to test all sorts of theories for random polymers (solutions, bulk viscosity, etc.). Yet, some of its solutions can form physical gels which, based on thermodynamic considerations (see Section 1.2.1.a.ii), are thought to originate in the creation of order.

(i) Molecular structure

Solutions of aPS in carbon disulphide (CS_2) produce gels at relatively high temperature. Also, thermal analysis indicates a high content of organized structure. Guenet *et al.* (1989) have investigated this system by neutron diffraction. The use of neutrons rather than X-rays is better for two major reasons. (1) These organized structures are most probably solvated, as hinted at by the phase diagram, so that the system must compulsorily be studied in the wet state. Under these conditions, X-rays are strongly absorbed by the sulphur atoms of carbon disulphide whereas neutrons are not. (2) By using deuterated atactic polystyrene the diffraction power of the polymer can be markedly enhanced ($b_S = 0.214$ barns, $b_D = 0.662$ barns).

Figure 2.31 shows the results obtained on a solution ($C_{pol} = 40\%$, w/w) at 40°C and −60°C, the latter temperature being some 80°C below the gelation threshold. At −60°C a maximum has appeared which yields a characteristic distance of about 1.1 nm when calculated with Bragg's law. As emphasized by Guenet *et al.* (1989), provided that this maximum corresponds to the spacing between first neighbour chains, this result again suggests the presence of intercalated solvents, as close packing of either isotactic or syndiotactic polystyrene sequences should yield a distance of about 0.7–0.8 nm. Incidentally, the distance between chains in this case should be calculated with a relation derived for systems with cylindrical symmetry:

$$1.115\lambda = 2d \sin \theta/2 \tag{2.44}$$

which eventually yields $d \simeq 1.2$ nm instead.

Interestingly, another maximum giving a spacing of 0.34 nm also appears at −60°C (Guenet and Klein, 1990) which is still unexplained. It may arise from the helical form taken by the sequences involved in these ordered gel junctions.

It should be mentioned that Guenet and coworkers (Guenet *et al.*, 1989; Guenet and Klein, 1990) have ascertained that no maxima can be recorded in non-gelling systems such as aPS–methylene chloride.

Despite the existence of organized structures, the overall gel morphology must be rather loose as hinted at by forced Rayleigh scattering (FRS)

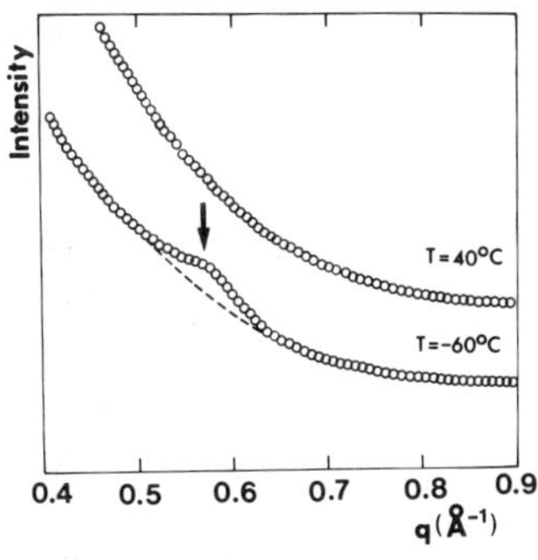

Figure 2.31 Diffracted intensity (in arbitrary units) as a function of q for an aPSD–CS_2 sample ($C_{pol} = 40\%$ w/w). Temperatures as indicated. The arrow highlights the position of the maximum which yields a distance $d = 1.095$ nm as calculated by means of Bragg's law (with permission from Guenet *et al.*, 1989).

experiments (Lee *et al.*, 1988). The technique consists of immersing some dye-labelled polystyrene chains among non-labelled ones (see, for instance, Léger *et al.*, 1981) and following the diffusion of the dye markers after photochromic excitation through a grating of spacing d. The signal output has the following form:

$$V(t) = [A \exp(-t/\tau) + B]^2 + C^2 \tag{2.45}$$

in which τ is the characteristic relaxation time, A the intensity factor, B the coherent background and C the incoherent one.

The diffusion coefficient D is related to the characteristic relaxation time τ and the lifetime of the excited state of the dye, τ_{dye}, through:

$$1/\tau = 4\pi^2/d^2D + 1/\tau_{dye} \tag{2.46}$$

From this investigation, Lee *et al.* (1988) have shown that there are two components of diffusion in the gel state, τ_1 for the slow mode and τ_2 for the rapid mode, separated by three orders of magnitude. The output signal then reads:

$$V(t) = [A_1 \exp(-t/\tau_1) + A_2 \exp(-t/\tau_2)]^2 + C^2 \tag{2.47}$$

They have also estimated that the chains participating in the physical junctions at $T = 250$ K amount to less than $X = 20\%$ for $C_{pol} \simeq 18$–25% depending on the polymer molecular weight. This estimate is derived from:

$$X = A_1/(A_1 + A_2) \tag{2.48}$$

It should be emphasized, however, that at these concentrations and temperature, the phase diagram (Figure 1.14) indicates that the system is composed of a polymer-rich phase (the physical junctions) and a polymer-poor phase, the concentration of which is still around $C_{pol} = 10\%$. To lower both the amount of polymer-poor phase and its polymer concentration, which will automatically result in an increase in the amount of polymer-rich phase, one should increase the overall polymer concentration and decrease the temperature. Accordingly, the results of Lee *et al.* (1988) do not provide the optimum amount of chains liable to be involved in the physical junctions but describe only a particular situation. In principle, the optimum amount should lie around the stoichiometric composition, i.e. $C_{pol} \approx 55\%$ (François *et al.*, 1986).

(ii) The pregel state

As has been briefly outlined in Chapter 1 (Section 1.2.1.a.ii), light scattering experiments have been carried out on concentrated solutions that display enhanced low-angle scattering (ELAS). Guenet *et al.* (1983) and later Gan *et al.* (1986) have attributed this scattering anomaly to the presence of the organized structures that may act as a junctions zone. According to Koberstein *et al.* (1985), ELAS can be expressed as the sum of two terms:

$$I(q) \sim (q^2 + \xi^{-2})^{-1} + I_{\mathrm{EX}} \qquad (2.49)$$

in which the first term represents the Lorentzian form of the scattered intensity in the absence of inhomogeneities and I_{EX} is the excess scattering, the latter being due, according to Gan *et al.*, to the physical junctions. It is further assumed that at larger angles, $I_{\mathrm{EX}} \approx 0$. This assumption is borne out by the determination of ξ from this domain (Gan *et al.*, 1986) which gives the expected variation in good solvents with polymer concentration (Daoud *et al.*, 1975):

$$\xi \sim C^{-0.78 \pm 0.3} \qquad (2.50)$$

On the other hand, I_{EX}, once properly extracted from the total intensity, is expressed through the Debye–Bueche relation (2.28). From this the range of density fluctuations, a_{d}, i.e. a distance related to the average spacing between physical junctions, can be measured. Gan *et al.* (1986) have determined the variation of a_{d} with polymer concentration C (in g/cm^3):

$$a_{\mathrm{d}} \text{ (in mm)} \simeq 9.7C^{-0.95} \qquad (2.51)$$

(c) Poly(methyl methacrylate)

Studies on the molecular structures have been mainly carried out on syndiotactic PMMA and on the stereocomplex. The crystalline structure of stereoregular PMMAs stands out among synthetic polymers. Isotactic PMMA is probably the only system thought to take on a double helical form whereas syndiotactic PMMA can only crystallize by solvent induction.

There appear to be no thorough studies on the molecular structure of isotactic PMMA aggregates and gels. The crystalline structure of bulk iPMMA is, however, worth presenting as it has a direct bearing on the crystalline structure of the stereocomplex. After the report by Tadokoro *et al.* (1970), on the basis of X-ray and infrared studies, of a 5_1 helix which showed no reasonable packing, the same group proposed to reanalyse their data using a double helix with periodocity $c = 2.08$ nm composed of 10_1 helices (Kusanagi *et al.*, 1976). All the side groups point outside the double-helical structure (Figure 2.32a). They also propose an orthorhombic unit cell with $a = 2.098$ nm and $b = 1.206$ nm.

Kusuyama *et al.* (1983) have shown that syndiotactic PMMA crystallizes by solvent induction only and accordingly forms polymer–solvent compounds (or intercalates). Desorption of the solvent leads to the disappearance of crystallinity. As a rule, the chain adopts a near-*trans–trans* (near-*tt*) conformation which eventually produces a helix with rather large radius ($r \approx 0.9$ nm) with 74 monomer units in four turns and a pitch of 3.54 nm.

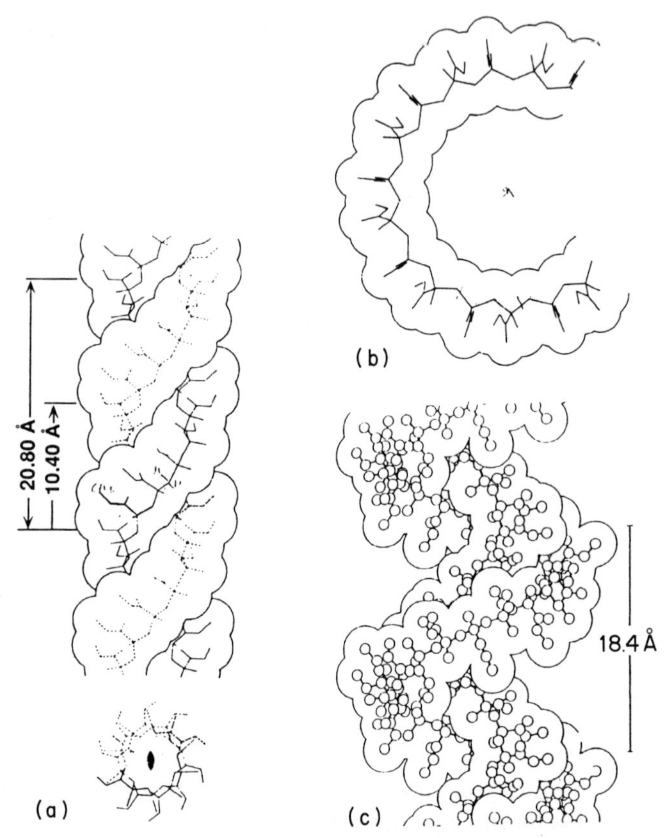

Figure 2.32 (a) Double strand helix of isotactic poly(methyl methacrylate) seen perpendicular (upper) and parallel (lower) to the helix axis (with permission from Kusanagi *et al.*, 1976). (b) Helical model for syndiotactic poly(methyl methacrylate) in polymer–solvent complexes seen parallel to the helix axis (with permission from Kusuyama *et al.*, 1983, by permission of the publishers, Butterworth-Heinemann). (c) Double helical form of the iPMMS–sPMMA stereocomplex. The syndiotactic chain wraps around the isotactic chains (with permission from Schomaker and Challa, 1989).

Solvent molecules occluded inside the helix are thought to be mobile while those located outside appear to be fixed (Figure 2.32b). In this helical form all the side groups stand outside the helix. A tentative unit cell is also given ($a = b = 4.35$ nm, $c = 3.54$ nm and $\gamma = 107^\circ$). One may, however, wonder at how such a crystal can be stable with such a helical form?

It is worth mentioning that Klein and Guenet (1989) have found, as with polystyrene, by the solvent crystallization method (see Section 1.2.1.a.ii), a direct correlation between the number α of solvent molecules 'bound' per monomer unit and the solvent molar volume V_m. They accordingly conclude that the 'bound' solvent molecules are housed by the cavities formed by the side groups. This molecular view of the polymer–solvent interaction may account for the fixed position of the solvent molecules outside the helix structure.

The first reasonable structure for the stereocomplex which was able to account for various properties such as template polymerization (see for instance Buter *et al.*, 1973) was given by Bosscher *et al.* (1982). It consisted of an iPMMA chain with a 30_4 helical form around which an sPMMA chain is wrapped with a 60_4 helix. More recently, Schomaker and Challa (1989) have proposed a somewhat modified version, a 9_1 double helix where both iPMMA and sPMMA have conformations close to *all-trans* (Figure 2.32c).

(i) The pregel state of sPMMA

Aggregation in solutions below the critical gel concentration occurs through a change of conformational state. Near-*tt* arrangements become more and more predominant. This change can be followed by infrared spectroscopy where the band intensity at 860 cm^{-1} characteristic of the near-*tt* arrangements grows at the expense of the band at 843 cm^{-1} which corresponds to other types of sequence (Dybal *et al.*, 1983, 1986). Dybal and Spěvaček (1988) have shown that this transformation can be studied in dilute solutions of toluene and CCl_4. This transformation does not occur instantaneously and it takes over 1 h to reach an equilibrium state (ratio of the integrated band intensities $(I_{860}/(I_{860}+I_{843}) \simeq 0.65)$.

Dybal and Spěvaček consider that the near-*tt* sequences produce a helix conformation with a large number of monomer units per turn similar to the conformation proposed by Kusuyama *et al.* (1983). Of further interest is the fact that in the non-aggregated state, the amount of near-*tt* sequences is very low in toluene but markedly higher in acetonitrile and benzene. One may wonder whether this situation could not be accounted for by taking into consideration Klein and Guenet's (1989) cavity effect, as benzene and acetonitrile possess smaller molar volume than toluene. They could accordingly enter the cavity more easily and stabilize the helical form leading eventually to fairly rigid sequences.

Interestingly enough, Schneider *et al.* (1987) report that isotactic PMMA displays a short-range conformation in solution virtually identical with the one in the amorphous state but markedly different from the one in the crystalline state. In the framework of Klein and Guenet's (1989) approach, these observations may not be surprising as the cavity of iPMMA is far less spacious than that of sPMMA. As a result, only very small solvent molecules, smaller than those suitable for sPMMA, should fulfil this space requirement and promote helix stabilization. Klein and Guenet have suggested the use of acetonitrile, as it possesses the smallest molar volume. Indeed, gelation occurs instantaneously in this solvent at low temperature (Klein and

Guenet, 1989) whereas it does not with the other solvents used so far (Könnecke and Rehage, 1981, 1983).

(ii) The gel state of sPMMA

As has been observed with dilute aggregates, gelation occurs through a conformational change from conformers, mainly *tg*, that transform into near-*tt* conformers as demonstrated by infrared spectroscopy for gels prepared in *o*-xylene (Berghmans *et al.*, 1987). At a given temperature (below the onset of gelation), this transformation takes place almost instantaneously, which led Berghmans *et al.* to propose a mechanism of gelation different from a crystallization mechanism, quite similar to the two-step process imagined by Sledaček *et al.* (1984): chain rigidification occurs first followed by chain aggregation. Interestingly enough, this mechanism is quite similar to the one proposed for isotactic polystyrene (see Sections 2.3.1.a.iv and 2.3.1.a.v). There exists, however, a difference in that iPS chain rigidity remains for a rather long period above gel melting which is apparently not the case, as indicated by the infrared spectroscopic results of Berghmans *et al.*

(iii) The pregel state of the stereocomplex

Vorenkamp and Challa (1981) have investigated the stereocomplex structure in dilute solutions by light scattering. They have observed that at the early stage of complexation in dimethyl formamide (DMF), the apparent molecular weight of the aggregates increases whereas their radii of gyration do not vary significantly (Figure 2.33). Alternatively, gel permeation chromatography (GPC) spectra show a shift of the stereocomplex aggregates towards shorter times of elution compared with the iPMMA and sPMMA parent molecules. Vorenkamp and Challa accordingly conclude that the aggregates have a size larger than the original PMMA parents, a conclusion in apparent contradiction with the light scattering results. They propose to overcome this paradox by considering a kind of fringed-micelle model wherein the organized parts are mostly in the centre whereas the disorganized ones are in the outer region (see inset, Figure 2.33). According to Vorenkamp and Challa, GPC is sensitive to the hydrodynamic volume, which they regard as being directly related to the external dimension of the aggregate, whereas they argue that R_G is given by $(\Sigma m_i r_i^2/\Sigma m_i)^{1/2}$, in which m_i is the weight of segment i located r_i from the centre of mass, so the central core ought to contribute more, hence the discrepancy between the two techniques. This explanation may not satisfactorily account for the experimental facts. First, R_G is known to be mainly sensitive to the largest distances within a particle and thus reflects mainly the largest dimension.

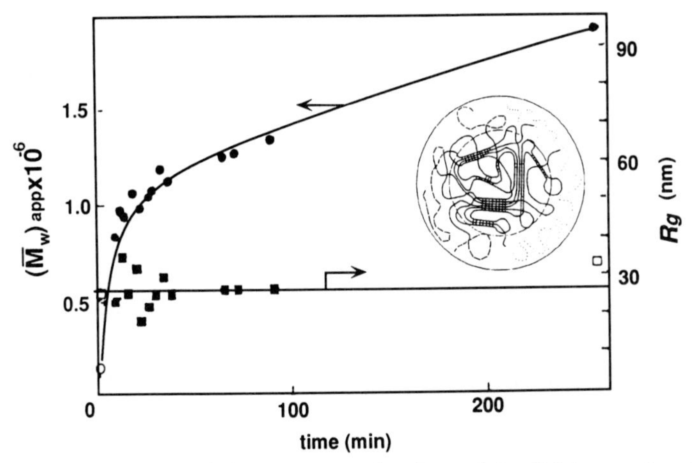

Figure 2.33 Apparent weight average molecular weight $(M_w)_{app}$ and radius of gyration (R_g) plotted against time after mixing isotactic and syndiotactic PMMAs with a ratio 1/2 at 0.86×10^{-2} g/cm^3 at 25°C in DMF. Inset: molecular model proposed by Vorenkamp and Challa (with permission from Vorenkamp and Challa, 1981, by permission of the publishers, Butterworth-Heinemann).

The density of the central core ought to be very high to cancel out this intrinsic property. Second, the hydrodynamic volume, $[\eta]\,M$ reads (see relation (2.13)):

$$[\eta]\,M \simeq R_G^2 R_H \tag{2.52}$$

As usually $R_H < R_G$, one does not expect to observe an increase in the hydrodynamic volume when R_G remains constant.

One may then wonder whether the value found for the radius of gyration by light scattering represents the true one or simply an apparent one.

That the segment density in the aggregates is larger than in a coil finds additional support in several reports (Belnikevitch *et al.*, 1983; Katime *et al.*, 1986). Katime and Quintana (1988*b*) have attempted to establish a relation between the radius of gyration and the molecular weight of the aggregates. To achieve this goal, they have collected fractions of aggregates from GPC columns and analysed them by light scattering. They have finally deduced the following behaviour:

$$R_{G_{agg}}\ (\text{nm}) = 0.054 M_w^{0.44} \tag{2.53}$$

These authors notice that relation (2.53) is close to that found for aPMMA ($R_G = 0.026M^{0.5}$) and that the exponent $\nu = 0.44$ may indicate a tendency towards a spherical shape (a sphere should lead to $\nu = 0.33$).

However, one ought to keep in mind that Katime and Quintana's analysis strictly holds if, and only if, the shape of the aggregate remains unchanged with increasing molecular weight. This condition may not be satisfied. Suppose for the sake of argument that the low-molecular-weight samples give fibre-like structures (flexible rods) while the high-molecular-weight samples consist of aggregates of these structures. This situation will probably yield also a variation of R_G *vs* M_w with an exponent lower than 0.5.

Many experiments point towards rigid rod-like structures for sPMMA. Could it not be the same for the stereocomplex, as the double helix may promote rigidity especially in DMF, in which Spěvaček and Schneider (1987) report degree p of associated units of 90% (which incidentally is not reflected in Vorenkamp and Challa's model)? In fact, Mekenitskaya *et al.* (1980) have concluded likewise from flow birefringence experiments in DMF that stereocomplex particles display a rod-like shape as opposed to a spherical one. They further surmise that the aggregates in the stereocomplex solutions consist of three to four stereocomplex particles.

Further support of the rod-like structure in the gelation of PMMA is found in the work by Fazel *et al.* (1992*b*), who studied the short-range structure (10–1 nm) of the aggregates by neutron scattering. They have used two different sPMMA samples with different tacticities (sPMMA1, with $M_w = 4.8 \times 10^4$, $M_w/M_n = 1.2$, $s \simeq 0.66$, $h \simeq 0.3$, $i \simeq 0.04$ and sPMMA2 with $M_w = 1.06 \times 10^5$, $M_w/M_n = 2.86$, $s = 0.66$, $i = 0.1$, $h = 0.24$). sPPMA1 does not gel as expected from the investigation of Dybal *et al.* (1986) which suggest the absence of sPMMA self-aggregation for samples of syndiotacticity lower than $s \simeq 0.66$. Therefore, aggregation of sPMMA2 may come as a surprise. In fact, according to Fazel *et al.* this sample is most probably a mixture of highly syndiotactic PMMA and isotactic PMMA. As a result, these authors conclude that aggregation arises both from SPMM self-association and from stereocomplex formation.

The aggregate molecular structure depends markedly upon the solvent type. Aggregates prepared in bromobenzene differ from those formed in *o*-xylene or in *p*-xylene. It also slightly depends on whether the sample is used as-received or dissolved in chloroform and then recovered prior to use.

In bromobenzene Fazel *et al.* estimate that sPMMA2 aggregates are composed of cylinder-like objects, as the intensity can be shown to obey the following relation (Figure 2.34):

$$I(q) \approx \mu_L[(\pi f(qr)/q) + (2\pi^2 n_L/q^2)] \tag{2.54}$$

in which the first term:

$$I(q) \approx \mu_L \pi q^{-1} f(qr) \tag{2.55}$$

is the rod scattering corrected for the effect of the cylinder cross-section $f(qr)$ (see Appendix 3), while the second term, in which n_L is the number of contacts per rod unit length, corresponds to intercylinder correlations (Luzzati and Benoit, 1961).

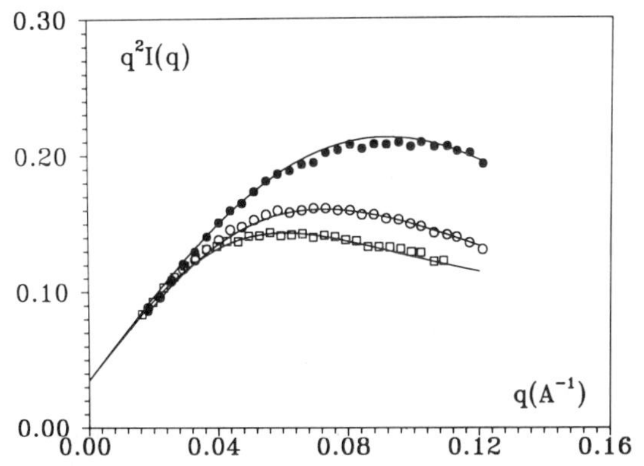

Figure 2.34 Kratky-plot ($q^2I(q)$ in arbitrary units *vs* q) for the determination of the short range gel and pregel structure. PMMA in deuterated bromobenzene. All curves have been reduced to the same scale and fitted with relation (2.54) in which $f(qr)$ is given by relation (A3.25). In the case of the aggregates and of the swollen gel cross-section heterogeneity has been considered by

$$q^2I(q) = C_p\left(\sum_i \pi q w_i \mu_{L_i} \frac{4J_1^2(qr_i)}{(qr_i)^2} + \text{const}\right)$$

in which w_i is the weight fraction corresponding to radius r_i with linear mass μ_{L_i}. For aggregates (□), two different radii are enough to fit the data: $r_1 = 1.5$ nm ($w_1 = 0.84$) and $r_2 = 3.4$ nm ($w_2 = 0.16$), for the nascent gel (•) one radius is enough: $r = 1.5$ nm and for the swollen gel (○) two radii are necessary: $r_1 = 1.5$ nm ($w_1 = 0.88$) and $r_2 = 3.0$ nm ($w_2 = 0.12$).

By means of this theoretical relation Fazel *et al.* have derived a linear mass of $\mu_L \simeq 1000$ g/nm per mol, a value that would be consistent with an all-extended conformation for which μ_L amounts approximately to $\mu_L \simeq 500$ g/nm per mol. However, the corresponding radius derived from $f(qr)$ by considering a full cylinder gives a value of $r_C \simeq 1.5$ nm. This is irreconcilable with the experimental value for μ_L and such a radius suggests that the rod consists of more than two chains. Alternatively, the results are not correctly fitted with the helical form proposed by Kusuyama for SPMMA (see Fig. 2.32b) which should give $\mu_L = 2090$ g/nm × mol. The double helix for LPMMA1sPMMA stereocomplex would yield $\mu_L \simeq 1500$ g/nm × mol with a radius of about 1.1. nm (Schomaker and Challa, 1989). Although these figures are closer to the experimental ones, the amount of stereocomplex liable to form is not high enough and cannot thus account for all the scattering pattern. To account for the paradox of a large cross-section radius together with a low linear mass, Fazel *et al.* consider, as Kusuyama *et al.* already had (Kusuyama *et al.*, 1983), the possible existence of a polymer–solvent compound which would modify the scattering amplitude of the structure. The scattering amplitude would then be diminished, provided that the solvent (bromobenzene) did not possess the same molar volume as in the liquid state. They also stress the fact that this may also mean that the cylinder's radius will ultimately be an apparent one, as the cylinder cross-section then consists of parts with different scattering powers (see Appendix 3). Therefore, the cross-section radius might well be underestimated or overestimated in the present circumstances.

In *o*- or *p*-xylene the scattering pattern is quite different The q^{-1} behaviour is totally absent and a q^{-n} behaviour with $n > 2$ is found instead. This, Fazel *et al.* argue, might point to the existence of fibre-like structures or in other words of rod-like structures that possess a cross-section far larger than that observed in bromobenzene. Why this is so remains unknown. It may arise from solvent quality (bromobenzene is probably a better solvent for PMMA than is xylene) or from the propensity of either solvent to promote the formation of a polymer–solvent complex.

(iv) The gel state of the stereocomplex

Together with the investigation of the pregel state, reported above, Fazel *et al.* (1991) have determined by small-angle neutron scattering the gel molecular structure. The gel scattering pattern resembles closely that of the aggregates. As with the aggregates, it differs from one solvent to another. It also depends, in the case of bromobenzene, upon whether the gel has been allowed to swell or not.

For a 17% (0.17 g/cm^3) gel in bromobenzene, rod-like behaviour is seen which can also be fitted to relation (2.54) (Figure 2.34). This shows that the short-range structure is little sensitive to the polymer concentration. The network is thus portrayed by jointed rods, a model reminiscent of that proposed for polysaccharides. The junction zones may arise from the ordering of double helices of the stereocomplex but may most probably be disordered.

The value of the cross-section radius depends markedly upon whether gels as-prepared or gels allowed to swell in an excess of preparation solvent are examined. In the former case the scattered intensity can be reproduced with relation (A3.19) by considering one population of rods with a cross-section radius of 1.5 ± 0.1 nm (see Figure 2.34). In the latter case, relation (A3.19) can still be used but two populations have to be considered with radii of 1.5 ± 0.1 nm, as previously, and 3 ± 0.1 nm in the ratio 88%/12% respectively. Again, the linear mass deduced from the experimental data, $\mu_L \simeq 680$ g/nm per mol, appears to be rather low for such values of the cross-section radius as has been already discussed in the previous section. The same effect (presence of two types of cross-sectional radii) is observed for 3% gels, which concentration is just above the critical gelation concentration C gel.

Fazel *et al.* (1991) attribute the increase in cross-section radius of a part of the rod-like elements to the chains that were not incorporated in the first place into the network after gelation had set (free chains). Thanks to gel swelling, whereby the mesh size of the network becomes larger, the free chains are given the chance to move more easily and to incorporate eventually by 'crystallizing' on to some of the existing rods. In the case of the 3% gels the mesh size is probably large enough to allow the phenomenon, that takes place on swelling the 17% gel, to occur readily here.

The schematic picture of jointed rods agrees quite satisfactorily with the mechanical properties, particularly the variation of the gel modulus with concentration (see Section 3.2.1.c).

Here it is worth dwelling upon the notion of gel junctions consisting of crystallites. Clearly, the rod-like elements evidenced in gels prepared from bromobenzene solutions are too small to be regarded as three-dimensional crystallites but should rather be considered as one-dimensional objects. These gels may prove to be of considerable interest for evaluating the effect of dimensionality upon thermal properties and particularly for discovering whether gel melting is accompanied by a thermal event or not (see the similar case of PVC in Section 1.2.2.b.ii).

In *o*-xylene the scattering pattern plotted by means of a Kratky plot displays a prominent upturn at small angles (Figure 2.35), as has already been reported for dilute aggregates. As with aggregates in the same solvent, Fazel *et al.* (1991) surmise that this pattern arises from a fibre cross-section far larger than in bromobenzene, but hasten to add that other models such as those employing fractal concepts, i.e. where the elements are no longer strictly rod-like, ought also to be taken into consideration. A tentative fit

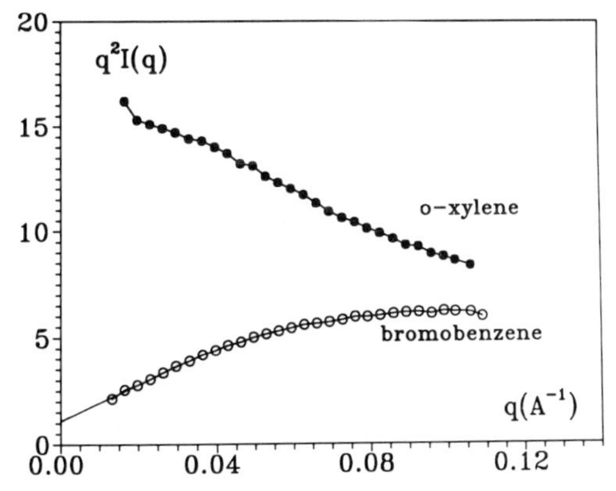

Figure 2.35 Kratky-plot ($q^2I(q)$ *vs* q) for a PMMA–deuterated bromobenzene gel (○) and a PMMA–deuterated *ortho*-xylene gel. In both cases $C = 0.17$ g/cm^3.

with rod-like objects of various cross-sectional radius gives: $r_1 = 1.5$ nm, $w_1 = 0.7$; $r_2 = 4.0$ nm, $w_2 = 0.2$; $r_3 = 10.0$ nm, $w_3 = 0.1$. If the approach is correct then this would suggest that some basic structure of cross-sectional radius 1.5 nm can be observed both in bromobenzene and orthoxylene. However orthoxylene would favour the growth of additional, larger structure. Still, it remains most important that there exists a definite effect of solvent type, a phenomenon expected in the light of the results obtained by Kusuyama *et al.* (1982, 1983) on solvent-induced crystalline sPMMA.

(v) Morphology of the gel state of the stereocomplex

From electron microscopic investigations, Quintana *et al.* (1989) have reported a fibrillar morphology, similar to that observed for amylose, for gels prepared in tetrahydrofuran (THF). Interestingly enough, this morphology is obtained after subsequent ageing (in the present case well over a week) and, therefore, cannot be presumed to arise from spinodal decomposition despite the network structure. This result emphasizes that gel morphology is not necessarily a decisive criterion for concluding on the nature of the gelation mechanism.

(d) Poly(4-methyl pentene-1)

Poly(4-methyl pentene-1) exists in several crystalline forms depending upon the preparation method. From the bulk the so-called **modification I** is obtained which consists of 7_2 helices with a pitch of $c = 1.38$ nm arranged in a tetragonal unit cell with $a = b = 1.866$ nm (Frank *et al.*, 1959).

Aharoni *et al.* (1981) have investigated gels prepared in cyclohexane and *cis*-decalin at high concentrations (50%, v/v) by means of X-ray diffraction. Basically, they have found that above 83°C for cyclohexane gels and 100°C for *cis*-decalin gels the diffraction pattern is similar to that obtained on amorphous medium (Figure 2.36), a result much reminiscent of that found later by Guenet (1986) for isotactic polystyrene (see Section 2.3.1.a.ii). On cooling gels prepared in cyclohexane a diffraction pattern typical of a new crystalline form, now named **modification IV** (in their 1981 paper Aharoni *et al.* designated this form modification II not being aware of the existence of form II, of differing unit cell parameter, discovered earlier by Nakajima *et al.* (1969)). It is not clear whether this form reflects the gel structure at lower temperature or corresponds to the growth of crystals as the gel becomes turbid when modification IV (II in Aharoni's paper) is seen by X-rays. Diffraction by the dried gel also reveals a pattern markedly different from the one recorded for a bulk-crystallized sample.

Alternatively, gels prepared in *cis*-decalin do not exhibit the modification IV pattern at low temperatures but modification I.

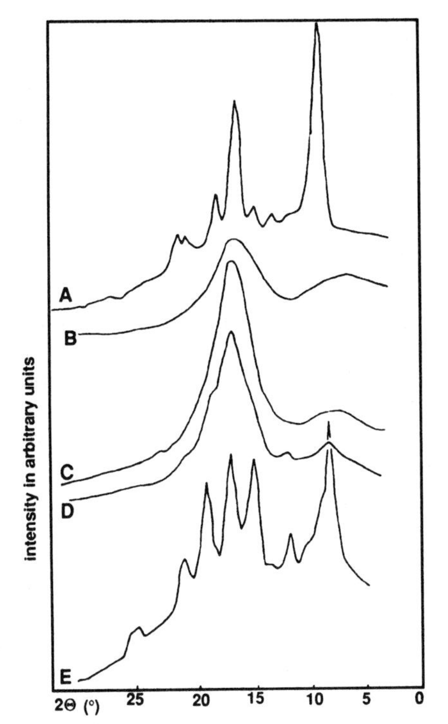

Figure 2.36 Diffraction patterns of poly(4-methyl pentene-1) in cyclohexane at different temperatures: A, pure polymer annealed; B, gel at 83°C; C, gel between 83°C and 38°C; D, gel below 38°C; E polymer dried from the gel showing a new structure (with permission from Aharoni *et al.*, 1981).

Charlet and Delmas (1982) later showed that an additional modification, named **modification V**, can be obtained from gels in cyclopentane by dissolution at 85°C and then cooling to room temperature. These authors have tentatively indexed the reflections with a hexagonal unit cell where $a = b = 2.217$ nm and $c = 0.669$ nm. No indication is given as to the helical structure of the chains in modification V nor in modification IV. As solvent inclusion is suggested (see Section 1.2.1.d), it would be of considerable interest to discover whether the solvent stabilizes the 7_2 helix or the rôle of the solvent goes as far as to monitor the helical form in which case more modifications may be to come.

One should note further that *cis*-decalin which possesses a size larger than cyclohexane or cyclopentane ($V_{\text{mcis-decalin}} \simeq 156$ cm^3/mol, $V_{\text{mcyclohexane}} \simeq 108$ cm^3/mol and $V_{\text{mcyclopentane}} \simeq 93.4$ cm^3/mol) allows formation of modification I only, i.e. without solvent occluded. This may again relate to the polymer cavity–solvent size effect discussed by Klein and Guenet (1989).

(e) Poly(γ-benzyl-L-glutamate)

PBLG follows the general rule of polypeptides in that it adopts an α-helix which, unlike helical forms discussed so far, is an irrational helix with a pitch of about 0.506 nm, and the ratio of the pitch to the axial rise is $P/h = 3.36$. This helical form is stabilized by the hydrogen bonds that establish between C=O and N—H chemical groups. There exist several ways whereby α-chains organize together. These are termed A, B and C. B and C consist of rather well-ordered phases while A, also designated by its inventors as the complex phase (Luzzati *et al.*, 1961), is poorly organized. The structure of A is not totally elucidated. McKinnon and Tobolsky (1968) have suggested the unit cell to be possibly hexagonal and of poor long-range order (see Table 2.2). As to form B, the same authors have deduced that helices are packed on an oblique two-dimensional net with a unit mesh of $a = 1.588$ nm, $b = 1.3$ nm and $\gamma = 113.7^\circ$ (see Table 2.2).

Surprisingly, the gel crystal structure has not received much attention. Sasaki *et al.* (1983) report that in benzyl alcohol the A form can be obtained when the gel is formed between 48°C and 60°C while the B modification grows when gelation is achieved by quenching and heating repeatedly from 61–62°C to below 48°C. They indicate that a single quench from 61°C to below 48°C produces forms A and B simultaneously. Sasaki *et al.* mention the possible crystal solvation (crystallosolvates) but do not provide any experimental support.

Gel morphology is basically the same in benzyl alcohol (Hikata *et al.*, 1977; Sasaki *et al.*, 1982) or in another solvent (Tohyama and Miller, 1981). In all cases the micrographs reveal a network with a mesh size of the order of 1 μm which consists of ramified fibres of diameter ranging from 10 nm to over 100 nm (Figure 2.37).

Table 2.2 Main reflections for form A and form B of poly(γ-benzyl-L-glutamate) (after [a] Parry and Elliott (1967) and [b] McKinnon and Tobolsky (1968)).

d (nm) in form A[a]	*d* (nm) in form B[b]	Index (form B)
1.03 ± 0.1	1.45	100
0.96	1.19	010, $\bar{1}10$
0.518	0.779	110, $\bar{2}10$
0.506	0.727	200
0.15	0.651	$\bar{1}20$
	0.594	020, $\bar{2}20$
	0.489	120, $\bar{3}20$
	0.43	$\bar{1}30$, $\bar{2}30$
	0.397	030, $\bar{3}30$

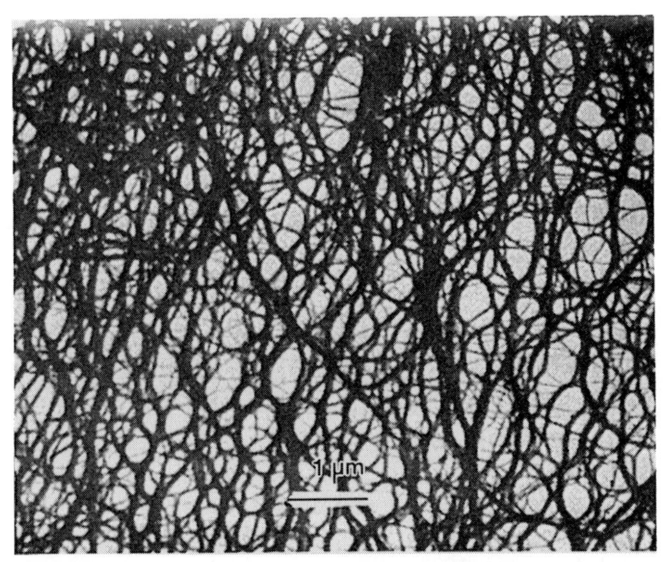

Figure 2.37 Electron micrograph of a poly(γ-benzyl-L-glutamate) gel (with permission from Tohyama and Miller, 1981, reprinted by permission from *Nature* **289**, 813, copyright (c) 1981 MacMillan Magazines Ltd).

The role of the solvent in the process of gelation has partially been unravelled by investigations in dilute solutions. One of the most thorough studies has been carried out by Doty and coworkers in several solvents (Doty *et al.*, 1956). These authors have shown that the viscosity–concentration relationship is dramatically different in DMF, chloroform–DMF mixtures on the one hand and dichloroacetic acid on the other hand. In the former case the relationship reads:

$$[\eta] \approx M^{1.7} \tag{2.56}$$

while in the latter case:

$$[\eta] \approx M^{0.87} \tag{2.57}$$

Relation (2.56) is analysed here with a power law, i.e. in terms of relation (2.13), but was fitted with Simha's (1940) equation for ellipsoïds of revolution of axial ratio a/b in Doty *et al.*'s paper:

$$\eta_{sp}/\Phi = (a/b)^2(\{15\,[\ln(2a/b) - 1.5]\}^{-1} + \{5\,[\ln(2a/b) - 0.5]\}^{-1}) + (14/15)$$

in which Φ is the polymer volume fraction.

Viscosity–concentration relations (2.56) and (2.57) indicate that PBLG behaves essentially as a relatively rigid rod in DMF and in a chloroform–DMF mixture while it possesses a more flexible conformation in dichloracetic acid. According to Doty *et al.* this finds explanation in the fact that formation of hydrogen bonds required between C=O and N—H groups for stabilizing the α-helix is prevented in dichloroacetic acid whereas it is promoted in the other solvents studied.

No further role is assigned to the solvent although one may wonder whether solvent molecules located between the benzyl groups could not enhance the hydrogen-bonding stabilization effect (cavity effect as in polystyrene).

2.3.2 Crystallization-induced gels

(a) Polyethylene

Polyethylene crystallizes in an orthorhombic unit cell containing two chains with $a = 0.74$ nm, $b = 0.493$ nm and $c = 0.253$ nm. The chains take on a planar zig-zag conformation.

Smith and Lemstra (1980*a*) have shown that the X-ray diffraction pattern obtained for wet gels prepared in decalin displays two sharp, if weak, reflections which they have identified as the 110 and 200 planes of the polyethylene orthorhombic cell. Drying and stretching of the gel reveals further reflections of the usual crystalline form (Smith *et al.*, 1981).

Morphology has been investigated by Barham *et al.* (1980) and Smith *et al.* (1981). By means of scanning electron microscopy, Smith *et al.* have clearly demonstrated the effect of stirring at high temperature above gel melting (see Section 1.3.1) on the gel morphology. A polyethylene solution in decalin held at 160°C then quenched at room temperature produces, once the solvent has been extracted, interconnected lamellar crystals (Figure 2.38A). Conversely, if the solution is stirred at 160°C prior to quenching a fibrous network is obtained which closely resembles those observed with many physical gels of polymers and biopolymers (Figure 2.38B).

Interestingly enough, while the gel melts below 140°C the system must be heated above 160°C to erase all memory effects. Failure to do so results in shish-kebabs, even in the absence of stirring, which definitely indicates that fibrous remnants of the gel are still present.

As has been discussed so far, such a morphology usually appears as a result of chain rigidification. This is similar with polyethylene, as stirring assuredly promotes chain extension. Yet, the curious thing is why rigidity is kept at such temperatures? In other words what is the mechanism whereby chain extension is stabilized? This so far remains an open question. Discovering to what extent the molecular and the crystalline structure at

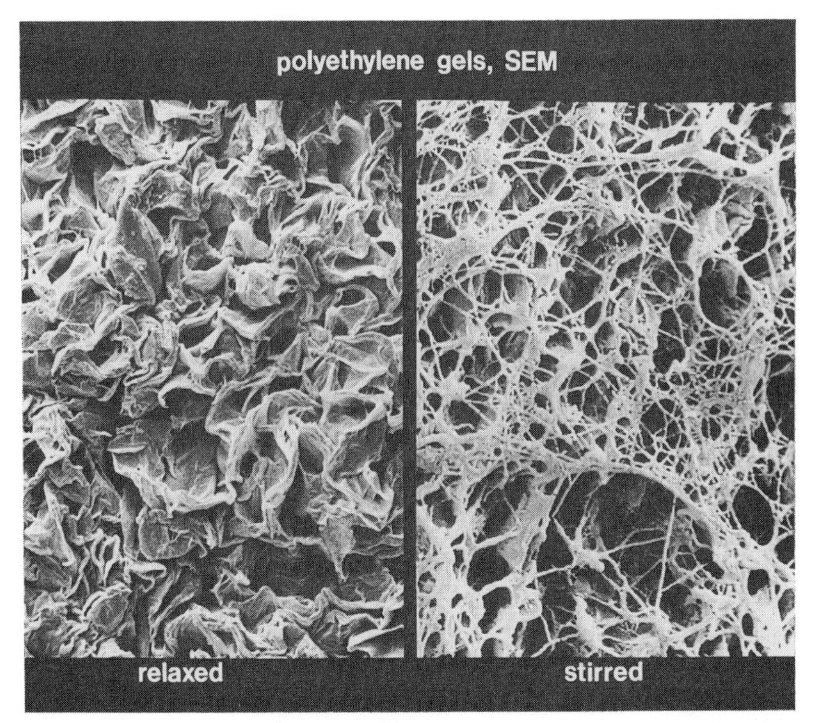

Figure 2.38 Scanning electron micrographs: (relaxed) = poly(ethylene) solution stored for 45 min at 160°C before quenching to room temperature; (stirred) = polymer solution stirred prior to gelation (with permission from Smith *et al.*, 1981).

high temperature depart from the ones established at room temperature might cast some light on the underlying phenomena.

(b) Poly(vinyl chloride)

The majority of studies on the gelation behaviour of poly(vinyl chloride) (PVC) has been carried out on the so-called atactic version. Unlike atactic polystyrene atactic PVC shows a small amount of crystallinity which undoubtedly arises from the syndiotactic or near-syndiotactic sequences (for further reading see Juijn *et al.*, 1973).

Single crystals from highly syndiotactic, but of low molecular weight, PVC have been prepared by Wilkes *et al.* (1973) and studied by X-ray diffraction. Under these conditions, PVC crystallizes in an orthorhombic unit cell with $a = 1.024$ nm, $b = 0.524$ nm and $c = 0.508$ nm, wherein the chains take on a planar zig-zag conformation. The main hk0 reflections

Table 2.3 hk0 reflections as obtained by Wilkes *et al.* (1973) from syndiotactic PVC single crystals.

d (nm)	hkl
0.524	010
0.512	200
0.466	110
0.366	210
0.286	310
0.262	020
0.256	400
0.254	120
0.233	220
0.23	410
0.208	320
0.183	420

together with the corresponding distances given by Wilkes *et al.* are presented in Table 2.3.

(i) Gel morphology

PVC gelation can be achieved in 'good' solvents and in 'bad' solvents. In the latter case liquid–liquid phase separation can be involved in the gelation process, whereas it is not in the former case. This offers an opportunity of finding out whether fibre-like morphology, observed in so many systems, originates only in spinodal decomposition.

Yang and Geil (1983) have determined PVC gel morphology in good solvents such as bromobenzene and nitrobenzene by means of electron microscopy and by preparing their samples by the freeze-etching technique. A typical micrograph obtained for gels formed in bromobenzene is reproduced in Figure 2.39(a). The morphology reveals a network with a mesh size of about 0.1 μm which consists essentially of fibre-like objects the diameters of which are roughly 10–20 nm.

The same or nearly the same features are seen with nitrobenzene and dioxane.

Gel morphology in bad solvents has been determined by Mutin and Guenet (1989) by optical microscopy. In benzyl alcohol and in cyclohexanol–hexanol, the gel morphology also consists of a fibrous network but the mesh size now reaches 5–10 μm (Figure 2.39b).

Interestingly enough, PVC gels prepared in benzyl alcohol exhibit a peculiar optical effect. When irradiated with visible white light they transmit the blue light only while scattering the other radiations, hence their yellowish visual aspect. An explanation

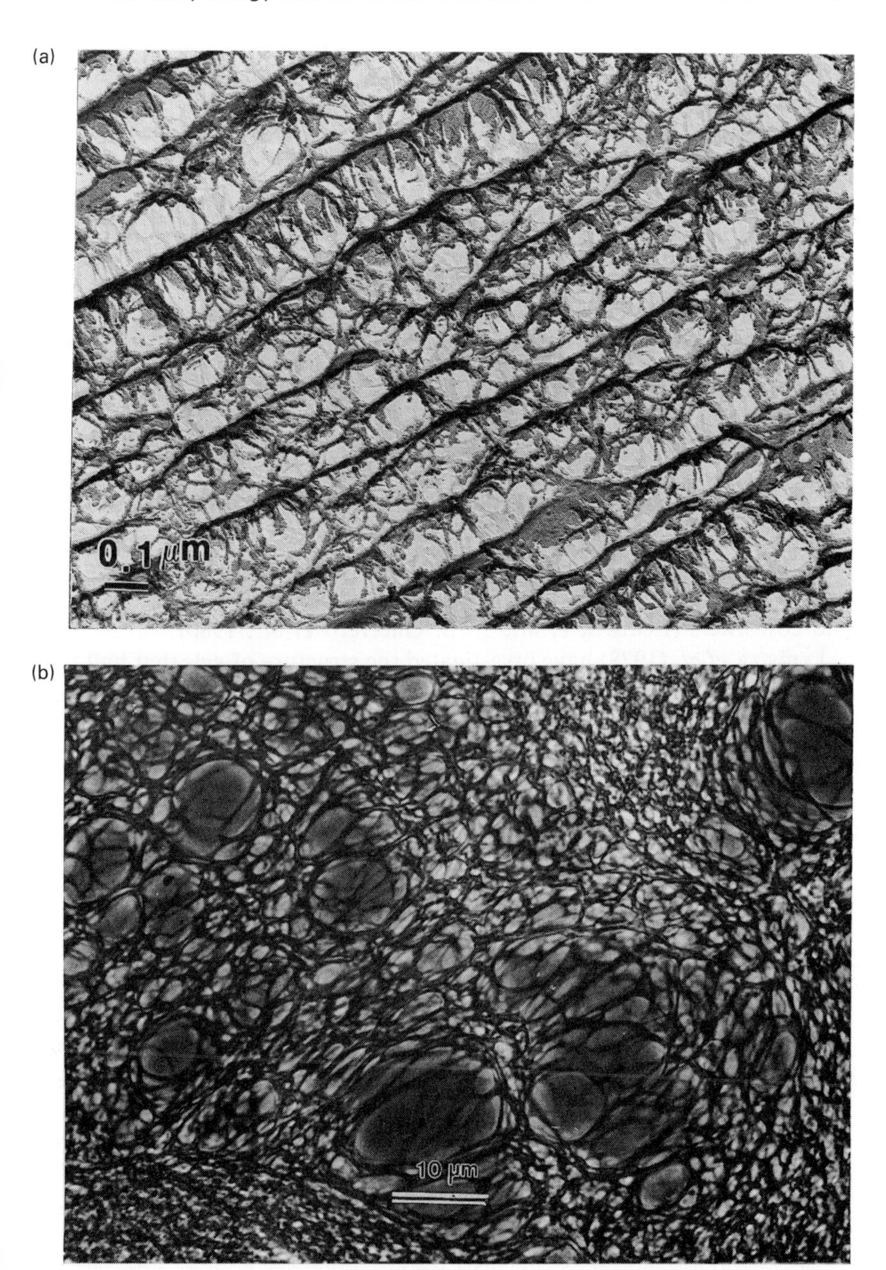

Figure 2.39 (a) Electron micrograph of freeze-etched (at -50°C for 30 min) of a PVC–bromobenzene gel ($C = 5 \times 10^{-2}$ g/cm^3) (with permission from Yang and Geil, 1983, reprinted from *J. Macromal. Sci. B* **(22)3**, 463, by courtesy of M. Dekker Inc.). (b) Optical micrograph (phase contrast) of a PVC–(cyclohexanone/hexanol 60:40) gel ($C = 0.125$ g/cm^3) obtained by a quench to room temperature between glass slides (with permission from Mutin and Guenet, 1989).

of this spectacular phenomenon might possibly lie in the match of refractive indices that exists between PVC and benzyl alcohol: it is sufficient that the difference between refractive indices of the gel structures and of benzyl alcohol be rigorously equal to zero for blue light slightly different from zero for the other radiations for this phenomenon to take place. In fact, blue light will then be transmitted while the large structures of the gel will produce a strong scattering of the other radiations which gives the sample the complementary colour to blue, i.e. yellow (red + green).

Admittedly, the fibrous morphology of PVC gels does not necessarily proceed from spinodal decomposition. The mechanism whereby fibre-like structures are formed must probably be understood through the elucidation of the molecular structure.

(ii) Gel molecular structure

Early investigations by X-ray diffraction showed that plasticized PVC contains crystallites (Stein and Tobolsky, 1948; Alfrey *et al.*, 1949). Takahashi *et al.* (1972) reported that gelation of PVC was probably due to crystallization, a statement that was later given strong support by the work of Keller and coworkers (Lemstra *et al.*, 1978; Guerrero *et al.*, 1980).

Lemstra *et al.* (1978) have investigated the structure of gels that had been stretched and dried, a method already employed for isotactic polystyrene gels. This method undoubtedly provides information about the gel state. However, one must always bear in mind that the preparation procedure of the sample might induce some alteration to the nascent structure.

A typical diffraction pattern displays a sharp meridional reflection at 0.52 nm and two broader equatorial ones at 0.54 and 0.47 nm. In addition, a broad circular halo is seen with its centre at 0.364 nm and intensification at 40° to the meridian. The 0.54 and 0.47 nm reflections are interpreted as arising from the 200 and 110 planes respectively of a crystalline unit cell composed of syndiotactic sequences (Figure 2.40).

These features are noticeably enhanced with a PVC sample of higher syndiotacticity, the latter being obtained by synthesis at low temperature (−30°C).

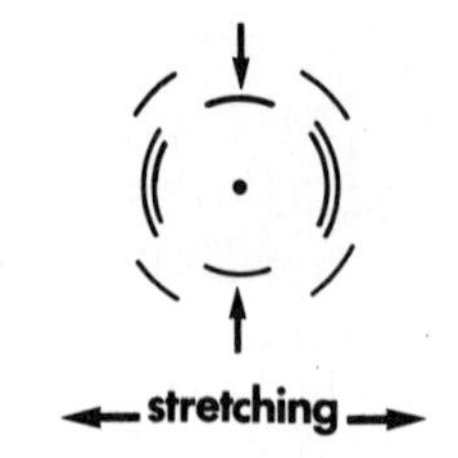

Figure 2.40 Schematic representation of an X-ray diffraction pattern of a PVC gel stretched and partially dried. Arrows indicate the meridional reflection at 0.51 nm; stretching direction as indicated (drawn after Guerrero and Keller, 1981).

The observation of a meridional reflection at 0.52 nm, which incidentally is much reminiscent of previous findings on isotactic polystyrene, was not accounted for in the paper of Lemstra *et al.* (1978). Later, Guerrero *et al.* (1980) came to the conclusion by further X-ray investigation combined with circular dichroïsm infrared studies that the sample contained two populations of crystals of the same crystalline unit cell. One population, named B, corresponds to the syndiotactic structures with *c*-axis orientation, i.e. the **fringed-micellar network**. The other population, named A, corresponds to **lamellar crystals**, essentially independent of the network, which, during the deformation process, adopt an *a*-axis orientation.

The appearance of the meridional reflection at 0.52 nm depends significantly on sample processing. It does not show up in melt-crystallized samples but only appears with gel samples. Guerrero *et al.* clearly mention that, although it is displayed by the majority of the gel samples, there are exceptions, the solvent used being the chief determining factor. Finally, there exists variability, in the same solvent, with regard to the absence or presence of this reflection, depending upon the stretching conditions.

Furthermore, irrespective of the PVC sample used, of high or low syndiotactic content, this reflection always exhibits the same sharpness.

As discussed in the previous chapter, Guerrero and Keller (1981) convinced themselves of the soundness of their interpretation through the observation of a high-melting endotherm which they assigned to the melting of crystal A. Since then it has been proved that this high-melting endotherm corresponds in fact to gel melting, i.e. to the so-called B crystals. Conversely, Guerrero and Keller attributed the low-melting endotherm erroneously to gel melting.

So far no other evidence for the existence of lamellar crystals has been gathered and one may wonder whether another explanation would not be worth considering. As has been mentioned in Chapter 1, Mutin and Guenet (1989) have suggested the possible existence of solvated structures, reminiscent of polymer–solvent compounds made up of solvent molecules and less stereoregular sequences. The 0.52 nm reflection might then arise from the same type of diffraction, as discussed for isotactic polystyrene. More evidence will be provided below for the involvement of solvent in some physical junctions.

Dorrestijn *et al.* (1981) have studied, by small-angle X-ray scattering, the gel structure on a larger scale as well as its evolution with ageing. Their investigation was carried out with two solvents that are common plasticizers of PVC: di(2-ethylhexyl)phthalate (DOP) and reomol (trimellitic acid ester of a mixture of branched alcohols).

The variation of scattered intensity with angle displays, once desmeared, a maximum which indicates the presence of domains, most probably crystallites (Figure 2.41). Interestingly, the correlation distance characterized by this maximum does not markedly vary with ageing time and amounts to about $d = 30 \pm 3$ nm if calculated with Bragg's law, i.e. $d = 2\pi/q$. It is worth pointing out that this distance may be calculated in two other ways. If this maximum were regarded as arising from a correlation hole effect, the correlation distance would be estimated from $d \simeq q^{-1}$, i.e. $d \simeq 5$ nm. Conversely, if one considered the diffraction pattern to be of a liquid type, then, $d \simeq 2.46\pi/q$, i.e. $d \simeq 37$ nm.

Dorrestijn *et al.* (1981) estimate the fraction of this crystalline material, w_1, by deriving the invariant Q from the slit-smeared experimental intensity $I(q)$:

$$Q = \int_0^\infty I(q)q \, dq = \text{const} \times w_1 \tag{2.58}$$

This approach shows that the overall 'crystallinity' is augmented by a factor of approximately 1.7. They accordingly come to the conclusion that ageing essentially promotes the growth of the existing crystallites, which accounts for both the virtual invariance of the correlation distance d and the increase in crystallinity. As to the network, they visualize it as a more or less cubic lattice (see inset of Figure 2.41), a model similar to the chemical gel structure except for the size of the cross-linkers. This model bears little

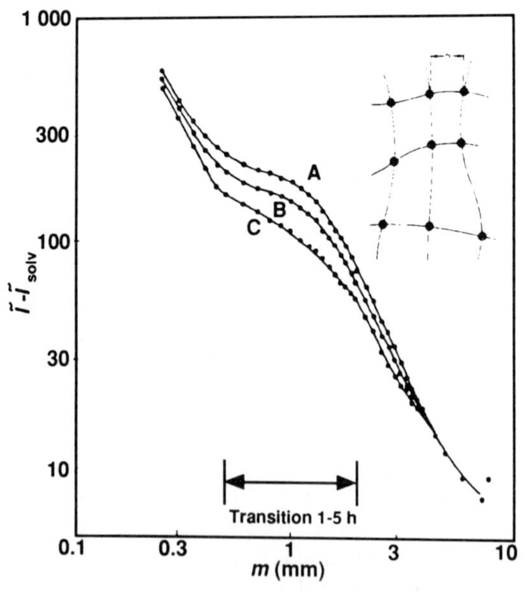

Figure 2.41 X-ray scattering pattern for PVC–dioctyl phthalate gels corrected for solvent scattering. Ageing temperature 28.5°C; ageing time A = 10 h, B = 1 h and C = 0.1 h. Inset represents the paracrystalline structure (with permission from Dorrestijn *et al.*, 1981, by permission of the publishers, Butterworth-Heinemann).

resemblance, if any, to the morphology of interconnected fibre-like objects reported by Yang and Geil (1983).

(iii) The pregel state

While much information has been gained over the past few years by studying the pregel state, new questions have also arisen that remain incompletely answered. New data have been obtained thanks to techniques not used, or little used, so far for PVC: light scattering, static and quasielastic, electric birefringence, neutron scattering, etc.

Mutin and Guenet (1986) have come across unusual and unexpected results during investigation, by **static light scattering**, of pregels prepared from ester solvents. The intensity scattered, $I(q)$, by 0.5% solutions in diethyl malonate or ethyl heptanoate displays a maximum. In the usual Zimm plot ($I^{-1}(q)$ *vs* q^2), such behaviour yields **hook-shaped patterns** (Figure 2.42).

Usually, molecular aggregation produces Zimm plots displaying a downward curvature. Kratochvil *et al.* (1967) report this type of intensity pattern for PVC aggregated systems. However, their results have been obtained either in other solvents or in esters (amyl acetate and butyl acetate) for which, as observed by Mutin and Guenet, no hook-shaped Zimm plots are obtained. Had Kratochvil and coworkers used other ester solvents they would have most probably came across the phenomenon 20 years earlier.

Evidently, the interpretation of Mutin and Guenet's findings is not that straightforward. The physical origin of the phenomenon involved is far from being totally elucidated. In their paper Mutin and Guenet have contemplated two possible reasonable explanations: formation of a physical copolymer or an effect due to a correlation hole.

The physical copolymer does not arise from chemical alterations but from the chain aggregation process which is thought to create two types of ordered sequence possessing different refractive indices. Clearly, one type of sequence is said to be composed of syndiotactic crystals whereas the second type is thought to be made up of solvated structures reminiscent of polymer–solvent compounds. Provided that the solvent in these structures does not possess the same molar volume as in the liquid state, then the refractive indices of either sequence may differ significantly. If the experimental conditions are such that $qR_G < 1$, as is mostly the case with light scattering, then the scattered intensity reduces to:

$$I(q) \approx M_{W_{app}}(1 - q^2 R^2_{G_{app}}/3) \qquad (2.59)$$

in which the subscript app stands for apparent. The apparent mean square

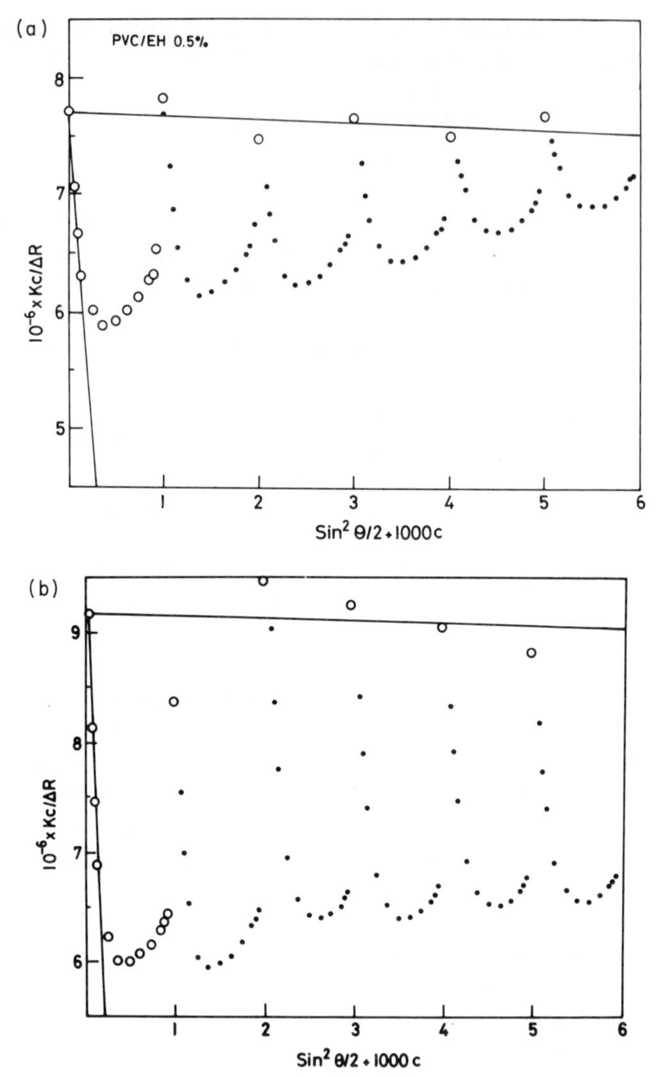

Figure 2.42 Zimm-plots for solutions of PVC (starting concentration 0.5×10^{-2} g/cm^3) in ethyl heptanoate (upper figure) and diethyl malonate (lower figure) (with permission from Mutin and Guenet, 1986, by permission of the publishers, Butterworth-Heinemann).

radius of gyration for multiblock copolymers reads:

$$R^2_{G_{app}} = x\nu_a^2 R_a^2 + (1-x)\nu_b^2 R_b^2 + 2\nu_a \nu_b x(1-x)\langle l^2 \rangle \qquad (2.60)$$

in which x and $1-x$ are the proportion of each sequence, R their radii of gyration, ν their refractive index increments and $\langle l^2 \rangle$ the mean square distance between their centres of mass. As the third term in relation (2.60) can be negative, this ultimately implies that $R^2_{G_{app}}$ can take negative values in the case of multiblock copolymers provided that two conditions are simultaneously fulfilled: (1) the refractive index increments of either sequence must be of opposite sign and (2) the number of sequences must be sufficiently low so as to have $\langle l^2 \rangle \neq 0$. If such is the case, then the scattering pattern may display the shape it has with PVC–ester aggregates.

However, one must bear in mind that the sequence consisting of solvated organized structures, should it exist, must possess a refractive index lower than that of the solvent ($n_D \simeq 1.41$ for the diesters used). As the refractive index of PVC is about $n_D \simeq 1.54$, these structures must inevitably contain voids. In other words the solvent molecule occupies a larger volume than it does in the liquid state and is characterized, accordingly, by a higher molar volume.

The case of a scattering pattern of correlation hole type also implies the existence of two types of sequence possessing different refractive indices. Yet, the refractive index increments with respect to the surrounding solvent need not be of opposite signs. It is, however, required that one sequence be embedded in the other, so that the difference between the refractive index of each sequence with respect to the average refractive index of the particle be of opposite signs.

If the aggregates are nothing other than microgels exhibiting the same structure as the gel, then this second model may also be valid.

While both models imply the presence of two types of domain with different refractive indices, the second one does not entail the existence of organized solvated structures. The model proposed by Dorrestijn *et al.* may be quite suitable for interpreting the experimental scattering curves, as the crystallite undoubtedly possess a refractive index that differs from the amorphous continuum.

Obviously, from such scattering curves no aggregate real dimension can be extracted. **Inelastic light scattering** proves to be helpful for obtaining a reasonable estimate. From this technique the temporal autocorrelation function, $C(t)$, can be obtained, which reads:

$$C(t) = \exp(-\langle \Gamma \rangle t)(1 + v\langle \Gamma \rangle^2 t^2 + \cdots) \tag{2.61}$$

in which $\langle \Gamma \rangle$ is the average decay rate and v the variance ($v = \langle \Gamma^2 \rangle - \langle \Gamma \rangle^2$). From relation (2.61) the particle diffusion coefficient can be determined through:

$$D(q) = \frac{\langle \Gamma \rangle}{2q} \tag{2.62}$$

When the diffusion coefficient derived from the autocorrelation function is plotted as a function of the square of the scattering vector, an upturn at low q-values is seen. This effect is thought to have the same origin as the anomalous static scattering curve. In spite of this, Candau *et al.* (1987*a*) have argued that an approximate value of the hydrodynamic radius R_H can be derived from a linear extrapolation to $q = 0$ of the values of $D(q)$ determined at large q-values (once the system is diluted to a suitable concentration to be rid of the virial effect for the highest concentrations). R_H and $D_{q=0}$ are then related through the Stokes–Einstein relation:

$$D(q)_{q=0} = k_B T / 6\pi\eta_0 R_H \tag{2.63}$$

in which η_0 is the solvent viscosity. The values obtained in diethyl malonate lie in the range 30–160 nm depending upon the starting concentration. Despite the questionable procedure used to determine these values, it turns out that they are in close agreement with those, more reliable results, gained from electric birefringence, as this technique is not sensitive to refractive index inhomogeneity of the particles.

Interestingly enough, when the aggregate hydrodynamic radius is plotted as a function of polymer concentration, divergence occurs near the critical gel concentration which is about 2.2% in the present case (see Figure 2.43).

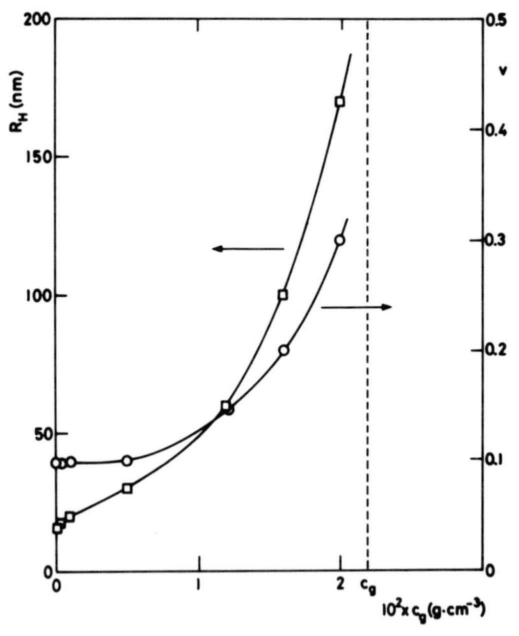

Figure 2.43 Hydrodynamic radius, R_H (■) and variance (○) as a function of PVC concentration. C_g stands for the critical gel concentration as determined visually. The variance is directly related to the polydispersity of the aggregates (with permission from Mutin *et al.*, 1988, by permission of the publishers, Butterworth-Heinemann).

It happens that this value is close to the calculated value of C^* for the PVC chains under study in the unperturbed state. Offhand, one would have probably found this coincidence obvious, as gelation is thought to occur only when chains overlap. However, one realizes that what matters is the aggregate overlap concentration. Since this concentration is quite high, it means that the molecular weight of the aggregates must be very high while their dimensions do not vary in the same proportion. As will be seen in what follows, the fibre-like structure proposed by Abied *et al.* (1990) may meet this requirement. At any rate, it seems that the virtual equality between the overlap concentration C^* of the non-aggregated chains and the critical gel concentration is purely fortuitous.

Mutin *et al.* (1988) have further observed that the position q_{max} of the maximum varies with the initial polymer concentration. Typically, it is found that, in diethyl malonate, for $C_{pol} = 0.5 \times 10^{-2}$ g/cm^3 $q_{max} \simeq 0.018$ nm^{-1} ($1/q_{max} \simeq 55$ nm) and $q_{max} \simeq 0.01$ nm^{-1} ($1/q_{max} \simeq 100$ nm) for $C_{pol} = 1.6 \times 10$ g/cm^3. This result apparently speaks against the correlation hole effect, as increasing the concentration should not lead to an increase in distance between crystallites but should lead to the reverse situation (the distance being proportional to $1/q_{max}$).

Mutin *et al.* overcome this apparent contradiction by invoking the finite size of the microgel. According to these authors, the intensity may be expressed as:

$$I(q) = P(q)S(q) \tag{2.64}$$

in which $P(q)$ is the microgel shape factor and $S(q)$ the structure factor. Only $S(q)$ contains a maximum while $P(q)$ is supposed to be a monotonically decreasing function.

They justify the use of this approach with an experiment carried out at two different temperatures on aggregates prepared from solutions of 1.6×10^{-2} g/cm^3. Figure 2.44 illustrates the change in intensity pattern during heating from 23°C to 62°C. At 62°C the scattered intensity is virtually similar to the one obtained for 0.5×10^{-2} g/cm^3 solutions at 23°C. From inelastic light scattering and electric birefringence, as discussed above, they have obtained an estimate of the hydrodynamic radius through the diffusion coefficient D at large angles. By assuming $S(q)$ to be invariant with temperature and microgel size they have derived the curve expected at 23°C from the curve at 62°C through:

$$I(q)_{23^\circ C} = I(q)_{62^\circ C} \times P(q)_{23^\circ C}/P(q)_{62^\circ C} \tag{2.65}$$

in which the $P(q)$s are calculated from the shape factor of a sphere with the appropriate radius. In fact, in the range of scattering vectors investigated $P(q)$ reduces to $P(q) \simeq 1 - q^2R_G^2/3$, so that any shape would do. As

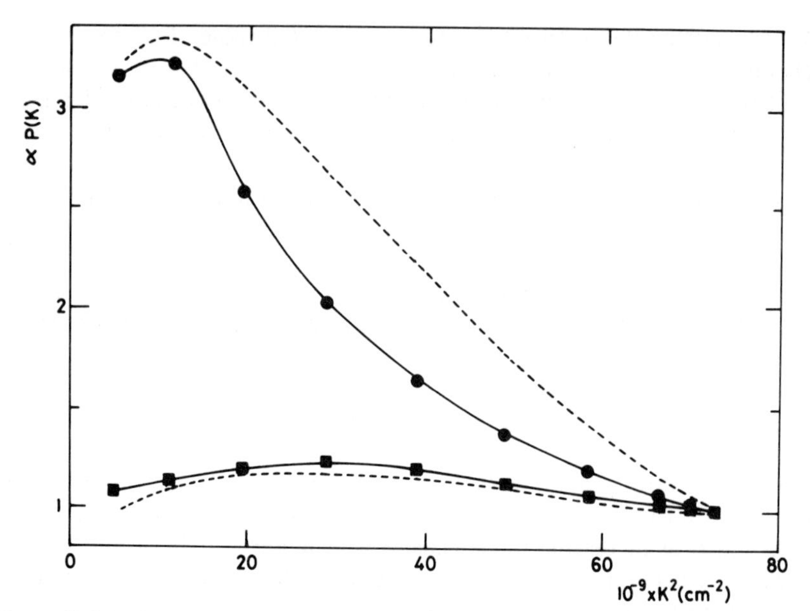

Figure 2.44 Scattered intensity (in arbitrary units) for PVC aggregates formed from dilute solution in diethyl malonate ($C = 1.6 \times 10^{-2}$ g/cm^3) diluted to 1×10^{-3} g/cm^3 prior to measurements and after 24 h annealing at: ■, $T = 23°$C; •, $T = 62°$C. The dotted line is calculated from relation (2.65) (with permission from Mutin *et al.*, 1988, by permission of the publishers, Butterworth-Heinemann).

can be seen in Figure 2.44, the agreement with the actual pattern is not that good, but the calculations do show that the maximum shifts toward smaller angles.

This approach, however, implies that the maximum cannot be seen with a gel which by definition possesses infinite dimension leading to $P(q) \simeq 0$ in the investigated range. This is certainly not what occurs in systems showing the correlation hole effect.

Conversely, from the physical copolymer approach one expects to observe the shift of the maximum toward smaller angles with an increase in microgel size.

Whatever the correct interpretation may be, the results of Mutin *et al.* (1988) indicate that there exists a type of physical junction that vanishes at relatively low temperature. They have designated this type of link as **weak** as opposed to the **strong links** that disappear at higher temperature. They have identified the strong links with the syndiotactic crystallites. They also suggest that the weak links are probably responsible for the ageing phenomena of PVC gels and suggested two alternatives for their origin: they arise from either the ordering of the less stereoregular sequences, an idea originally put forward by Juijn *et al.* (1973), or ordering of these sequences with solvent molecules.

Further information has also been obtained from electric birefringence (Kerr effect) experiments by Candau *et al.* (1987*a*,*b*). For low concentrations ($C_{pol} \leqslant 0.5 \times 10^{-3}$ g/cm^3), the birefringence decay $\Delta n(t)$ after removal of the electric field can be fitted with a single exponential:

$$\Delta n(t) = \Delta n_0 \exp(-6D_R t) \tag{2.66}$$

in which D_R is the rotational coefficient and Δn_0 the steady-state birefringence. From this relation they deduce, by assuming a spherical shape, a radius $R \simeq 40$ nm through the relation:

$$D_R = \frac{k_B T}{8\pi\eta_0 R^3} \tag{2.67}$$

in which η_0 is the solvent viscosity. The value of R agrees fairly well with that determined from the largest angles of the variation of D_0, as obtained from inelastic light scattering, with q^2 ($R \simeq 30$ nm).

By inversion of the electric field, they have shown further that the PVC particles possess a permanent dipole and, accordingly, regions of high ordering. After field inversion the birefringence reaches a value Δn_{min} at $t_{min} = 0.2747D^{-1}$. Δn_{min} is related to the permanent dipole moment, μ, by:

$$\frac{\mu^2}{(\alpha_1 - \alpha_2)k_B T} = \frac{P}{Q} = \frac{1 - \Delta n_{min}/\Delta n_0}{0.1547 + \Delta n_{min}/\Delta n_0} \tag{2.68}$$

in which α_1 and α_2 stand for the electrical polarizabilities of the particle parallel and perpendicular to the revolution axis respectively. If $\mu = 0$ (no permanent dipole), then no change occurs upon field inversion, i.e. $\Delta n_{min} = \Delta n_0$. Figure 2.45 shows that this is not the case, as $P/Q = 1.82 \pm 0.1$. It is worth emphasizing that the existence of well-ordered domains, which are most probably crystallites, receives here strong support from results obtained by a non-destructive technique.

At higher polymer concentrations ($C_{pol} = 1.2 \times 10^{-2}$ g/cm^3 and 1.6×10^{-2} g/cm^3) the birefringence decay cannot be fitted with a single exponential any longer which indicates aggregates of larger polydispersity. Meanwhile the average size increases with increasing concentration, as has also been observed with inelastic light scattering. Heating these solutions at 60°C and above gives values close to the one found for the 0.5×10^{-3} g/cm^3 solutions, a result in agreement with those derived from static light scattering.

The short-range structure, from about 10 nm down to 1 nm, has been investigated by Abied *et al.* (1990) as a function of concentration, solvent type and PVC tacticity by means of neutron scattering. Three deuterated solvents were used: diethyl oxalate, bromobenzene and nitrobenzene. In

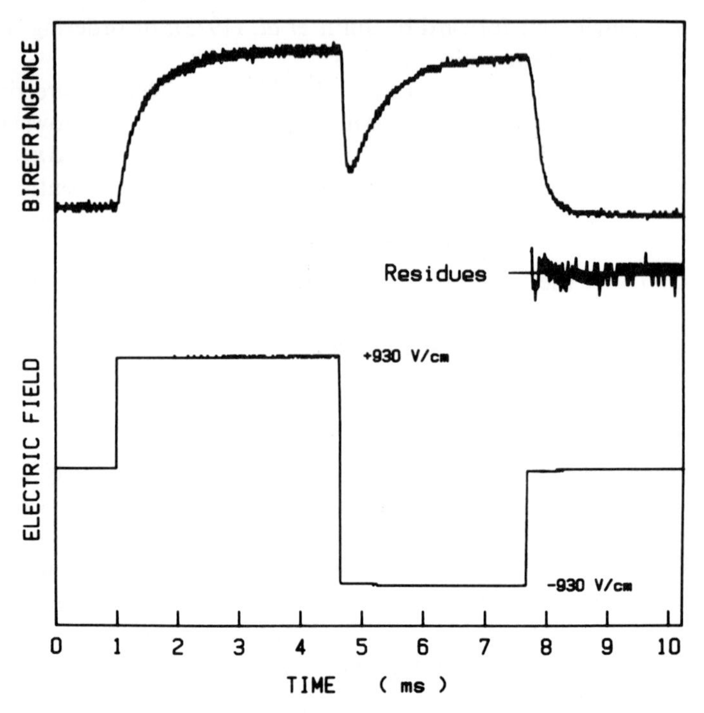

Figure 2.45 Time dependence of the electric birefringence signal under the action of a reversing electric pulse; PVC solution in diethyl malonate for $C = 10^{-3}$ g/cm^3 at 25°C obtained by dilution of an original concentration of 0.5×10^{-2} g/cm^3. The birefringence decay observed above 8 ms can be fitted perfectly with equation (2.66). Residues (i.e. the difference between the theoretical and experimental curves) are plotted below the birefringence curve (with permission from Candau *et al.*, 1987*b*).

diethyl oxalate the scattering pattern depends upon polymer concentration. For $C_{pol} = 1\%$ (w/v), $I(q)$ reaches first a q^{-4} asymptotic behaviour as seen from the plateau in a $q^4 I(q)$ plot, and then a behaviour close to q^{-2} appears beyond $q \simeq 0.9$ nm^{-1} (0.09 Å^{-1}) (Figure 2.46). As can be seen, there is little effect of ageing upon the short-range structure which agrees with the findings of Candau *et al.* (1987*a,b*). Abied *et al.* have interpreted these results as flexible cylinder-like (fibre-like) objects rather than spheres on the basis of viscosity measurements and the morphological findings of Yang and Geil (1983) described above. For long, rigid cylinders of mass M, length L and cross-section radius r ($L \gg r$), the intensity reads (Porod, 1948):

$$I(q) = KC_H(a_H - a_D)^2 \times \frac{M\pi}{qL} \times \frac{4J_1^2(qr)}{(qr)^2} \tag{2.69}$$

in which K is a constant depending upon the neutron scattering set-up, C_H the polymer concentration, $(a_H - a_D)$ the contrast between protonated and

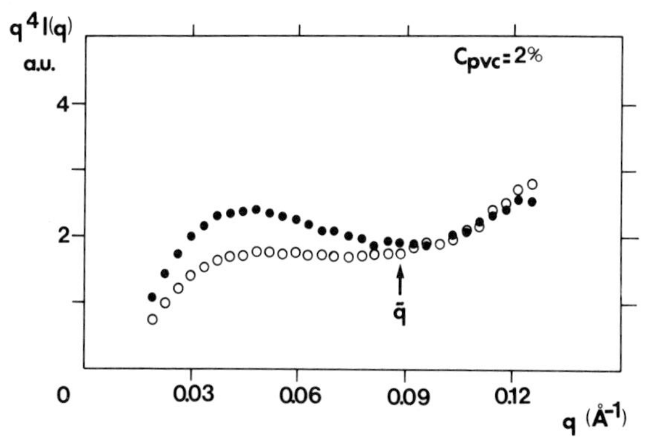

Figure 2.46 Determination of the PVC aggregates molecular structure by neutron scattering: $q^4 I(q)$ *vs* q in arbitrary units to 2%-PVC–diethyl oxalate systems (protonated PVC in deuterated solvent). ○ = nascent solution; • = the same solution aged one month at room temperature (with permission from Abied *et al.*, 1990).

deuterated species and J_1 the Bessel function of first kind. Equating $M = N_A L \pi r^2 d_p$ where d_p is the particle density, relation (2.69) reduces to:

$$I(q) = KC_H(a_H - a_D)^2 \times \frac{4\pi^2 N_A d_p J_1^2(qr)}{q^3} \tag{2.70}$$

For $qr > 1$ the Bessel function can be developed which eventually yields, by taking the density in g/cm^3 and the radius in Å:

$$I(q) = KC_H(a_H - a_D)^2 \times \frac{2.4\pi d_p}{q^4 r} \tag{2.71}$$

Using relation (2.71) they have determined a cross-section radius of about 35 nm. They further add that the smoothness of the scattering curve in the q^{-4} domain indicates that the aggregates must be at least polydispersed in cross-section, as monodisperse cylinders should have given an oscillating intensity due to the oscillatory character of the Bessel function.

The q^{-2} behaviour observed beyond $q \simeq 0.9\ \text{nm}^{-1}$ they interpret by regarding the cylinder as no longer strictly compact in this range of distances (i.e. distances smaller than about 1.1 nm). The lack of strict compactness may arise from solvent intercalation between chains. In the case of diethyl oxalate, Abied *et al.* (1990) suggest the existence of a polymer–solvent compound. Their model, which is illustrated in Figure 2.47(a), implies that the real density is lower than that of bulk PVC, so that the cross-section radius is most certainly overestimated.

The role of the intercalated solvent is seen as promoting linking between chains. Abied *et al.* remark, as was already suggested by Tabb and Koenig (1975) and Monteiro and Mano (1984), that the ester function, being

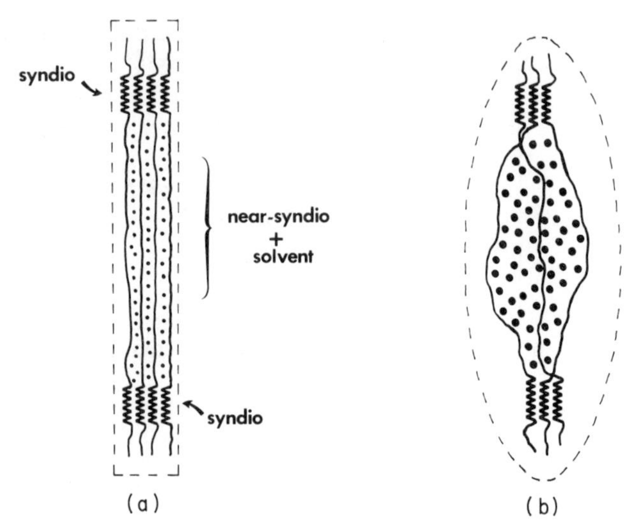

Figure 2.47 Molecular models for the aggregates formed in dilute solutions (i.e. $C < C_{gel}$) as deduced from the neutron scattering experiments on PVC–diethyl oxalate systems (a) and PVC–bromobenzene systems (b). The crystallites constituted of syndiotactic sequences are indicated by 'syndio'; the solvent molecules are represented by dots. The dotted lines outline the global shape of aggregates (with permission from Abied *et al.*, 1990).

negatively polarized, will probably interact with an H atom located on a H–C–Cl, which is positively polarized. As two ester functions are available they can interact with two different chains, thus establishing a physical link.

The question of a two-population system is also raised by these authors. In particular they examine the case where isolated chains may still be present in the solution, as suggested by Hengstenberg and Schuch (1964). If X is the proportion of aggregates and $1 - X$ the proportion of free chains then the intensity should read:

$$I(q) \approx Xq^{-4} + (1 - X)q^{-n} \tag{2.72}$$

In the case of Gaussian free chains the exponent n is equal to 2 so that in a $q^4 I(q)$ plot one should not observe any q^{-4} plateau but a continuously increasing curve. Incidentally, the model sketched by Dorrestijn *et al.* (1981) should also display a behaviour described by equation (2.72), apart from some cross-terms which rapidly become unimportant. Evidently, this model does not apply to the present case.

At higher polymer concentration ($C_{pol} = 2\%$, w/v) the ageing effect manifests itself quite distinctly. Nascent aggregates possess virtually the same structure as those formed at lower concentration apart from a cross-section radius slightly larger ($r \simeq 50$ nm but still probably overestimated).

However, the scattering pattern of the aged aggregates exhibits a maximum instead of a q^{-4} plateau (Figure 2.46). According to Abied *et al.* (1990), this arises from the appearance of terms of higher order in the expression of the intensity. In fact, relation (2.71) is only an approximation which neglects the terms higher than q^{-4}. For a cylinder of length L and radius r, the intensity reads, by developing the Bessel function to higher orders:

$$P_\infty(q) \simeq \frac{1}{q^4 L r^3}\left(1 + \frac{3}{8q^2 r^2} + \cdots\right) \tag{2.73}$$

The q^{-4} approximation holds as long as the cross-section radius is large enough but also, according to Abied *et al.*, as long as the cylinders do not aggregate. They show in the latter case that q^{-6} terms cannot be ignored. They accordingly account for the experimental scattering pattern by considering superaggregation of the flexible cylinders which occurs on ageing. This superaggregation between cylinders they believe occurs through the formation of weak links. Results obtained on heating the aged aggregates bear out this view, since the scattering pattern of nascent aggregates is retrieved above 50°C.

The effect of solvent type is shown by aggregates prepared in bromobenzene and nitrobenzene. As can be seen in Figure 2.48, they give different scattering patterns which also differ from that obtained in diethyl oxalate.

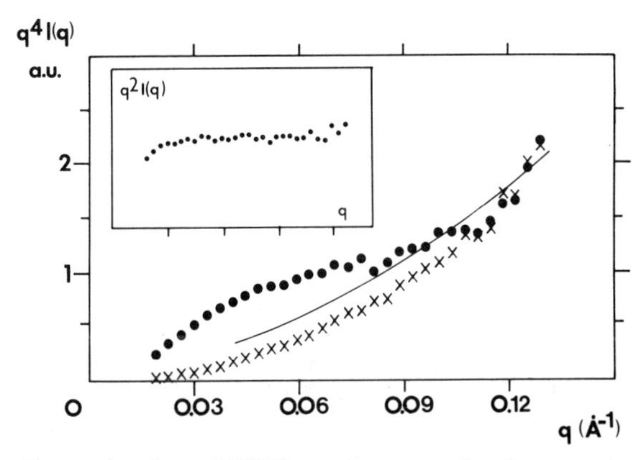

Figure 2.48 Determination of PVC aggregates molecular structure by neutron scattering: $q^4 I(q)$ *vs* q in arbitrary units for 1.5%-solutions in bromobenzene (•) and nitrobenzene (x) (protonated PVC in deuterated solvents). The solid line, calculated with $A + Bq^2$, is fitted to the results of PVC–bromobenzene for $q > 0.09$. This indicates that such a behaviour is not correct for PVC–bromobenzene systems but can apply to PVC–nitrobenzene aggregates. The inset shows that the latter these aggregates display a q^{-2} behaviour but no q^{-4} behaviour (with permission from Abied *et al.*, 1990).

In the case of bromobenzene, the q^{-4} plateau is barely reached and no such plateau can be observed in nitrobenzene. On the contrary, in the latter solvent a q^{-2} regime is seen. Abied *et al.* account for these results with a model that takes into account the solvent quality. In bromobenzene they consider the same type of model as with diethyl oxalate yet more swollen which in the end gives an ellipsoid-like structure (see Figure 2.47b). In nitrobenzene they consider the possibility of a still more swollen structure which they approximate with a hollow sphere of radius R the form factor of which is:

$$P(q) = \frac{\sin^2 qR}{q^2R^2} \tag{2.74}$$

For $qR > 1$, $P(q)$ oscillates around $1/2q^2R^2$, the oscillations being smoothed out, as with the Bessel function, by polydispersity. They also add that isolated chains are most certainly present in the solution.

They emphasize that these models are consistent with macroscopic properties such as swelling. In fact, swelling increases in the order diethyl oxalate, bromobenzene and nitrobenzene.

The effect of tacticity has been investigated in nitrobenzene with a PVC sample synthesized at lower temperature which is thus of higher syndiotacticity ($s = 0.33$, $i = 0.18$ and $h = 0.49$) and, accordingly, of higher 'crystallinity'. The scattering pattern obtained on 1.5% solutions is very similar to the one observed in diethyl oxalate. Therefore the solvent (diethyl oxalate) may play the same role as an increase in tacticity, i.e. promote the formation of additional physical links. This gives support to the concept of 'solvent difunctionality' developed by Abied *et al.* for the formation of a polymer–solvent compound.

Recent experiments by Najeh *et al.* (1992) show that the scattering pattern observed in light scattering differs depending on whether a diester or a monoester is used (Figure 2.49), the latter being not capable of linking two chains together while the former is. A maximum and then a marked decrease in intensity are observed with monoesters. The position of the maximum depends upon the solvent quality, i.e. the greater the swelling power the more the maximum is shifted towards the small angles. This maximum undoubtedly reflects some correlation between organized domains, the spacing of which expands with solvent quality. Conversely, aggregates formed in diesters always produce the same or nearly the same scattering pattern, i.e. a pronounced downturn at small angles followed by a plateau (the decrease at larger angles is not always discernible within experimental uncertainties).

Finally, as far as the short-range structure is concerned, Abied *et al.* report that aggregates and gels are very similar, as, once rescaled to the same concentration, the gel and the aggregate scattering patterns are virtually identical. This again suggests that the basic element in the gel resembles

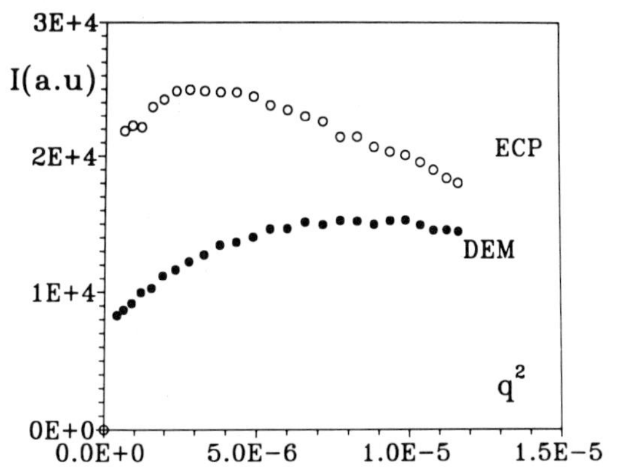

Figure 2.49 Light scattered (in arbitrary units) by PVC aggregates formed in dilute solutions ($C = 0.5 \times 10^{-2}$ g/cm^3) of diethyl malonate (• = DEM) and ethyl caprate (○ = ECP) by a quench to room temperature and aged 3 days minimum at this temperature (Najeh, Munch and Guenet, unpublished results).

a flexible cylinder, in agreement with the fibrillar nature of the gels reported by Yang and Geil (1983).

One may wonder why PVC in solution tends to form fibre-like structures instead of chain-folded crystals. The former are possibly more stable than the latter by virtue of minimum constraint taking place on the disordered chains. As crystallizable sequences located on a chain are supposedly widely spaced apart, a chain-folded crystal would be characterized by a highly disordered surface with constrained amorphous chain portions, hence a high surface free energy and correspondingly a poorly stable structure (see Appendix 1). Conversely, in a fibre-like structure the amorphous portions are far less constrained, as this structure allows for more flexibility. Similarly, a model such as the one proposed by Dorrestijn *et al.* (1981) where lamillar crystals are connected by amorphous chain portions is probably less stable than a fibril because the crystallites may not be capable of reaching the required critical size, especially as the surface free energy at the crystal interface may be unusually high. However, even if they would grow larger than the critical size, then the problem of density defect at the crystal-disordered domain interface would appear. This question has been raised by Guttmann *et al.* (1979) who have shown that the density may rise to an unphysical value (twice the polymer density) in the absence of chain folding.

(c) Poly(vinyl alcohol)

In spite of its atactic character, poly(vinyl alcohol) (PVA), like PVC, can display a non-negligible amount of crystallinity. This crystallinity in the case of the so-called atactic PVA arises mainly from the syndiotactic sequences. The crystalline unit cell of syndiotactic PVA, determined by Bunn (1948), is monoclinic with $a = 0.781$ nm, $c = 0.551$ nm, $b = 0.251$ nm and $\beta = 91^\circ 42'$ and contains two chains that take on a planar zig-zag conformation. Here the chain axis is taken to be parallel to the b-axis unlike the usual conventions.

PVA gelation has usually been studied in aqueous solutions, except for an investigation carried out with ethylene glycol by Stoks *et al.* (1988).

Takahashi and Hiramitsu (1974) have investigated PVA gels in water by means of X-ray diffraction. Despite the low concentration of their gel sample (2.5×10^{-2} g/cm^3) they have observed a halo at $d = 0.78$ nm, a value which turns out to be very close to the (100) spacing. Ogasawara *et al.* (1975) have observed on a 10% gel three diffraction rings at $d = 0.782$ nm, $d = 0.451$ nm and $d = 0.378$ nm which they assign to the (100), (101) and (200) planes of the unit cell of the bulk-crystallized PVA.

Stoks *et al.* (1988) report that a sample obtained from a gel prepared at 100°C ($C_{\text{pol}} = 8.3\%$ by weight) and then stretched while ethylene glycol is extracted exhibits a diffraction pattern consistent with crystals whose (101) plane is parallel and perpendicular to the stretching direction. For a gel prepared through a rapid quench to room temperature, only the ring characteristic of the (101) plane can be observed. For gels formed at 100°C and above, the diffraction pattern is accounted for by Stoks *et al.* by considering the existence of platelet crystals embedded in the gel, the latter being of fibre-like nature. On stretching, the (110) plane of the fibre will align parallel to the stretching direction, while the (101) plane of the platelets will orientate perpendicularly. This is exactly the same scheme as was propounded a few years earlier by Guerrero *et al.* (1980) for PVC gels.

A scattering study by Wu *et al.* (1990) has been carried out for samples encompassing the overlap concentration of a chain. Although in a Zimm representation the scattering curves are not strictly linear over the whole q-range, these authors have derived a so-called correlation length, ξ', from:

$$I^{-1}(q) \approx 1 + q^2 \xi'^2 \tag{2.75}$$

They regard ξ' as a measure of the heterogeneity of the system. Their results obtained as a function of temperature are summarized in Figure 2.50. While for 2%, 6% and 16%, the correlation length decreases monotonically with T, there is definitely a maximum for the 10% sample, which turns out to be very close to the calculated overlapping concentration (9% for this molecular weight). So far there seems to be no straightforward explanation for this result.

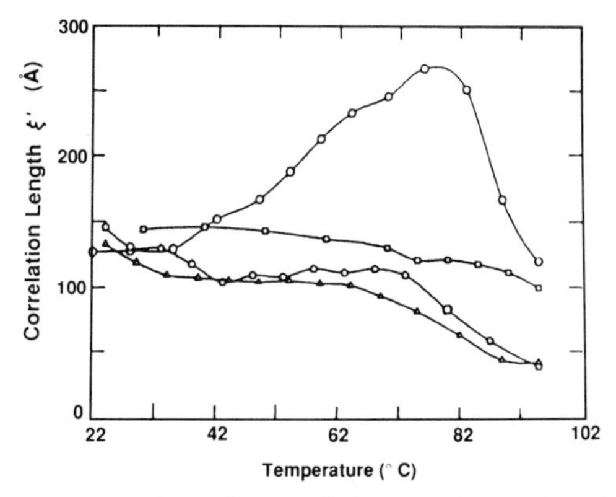

Figure 2.50 Temperature dependence of the correlation length ξ' measured by neutron scattering on aqueous PVA solutions at four different polymer concentrations (data collected on heating cycle): □ = 16%, ○ = 10%, △ = 6%, ⬠ = 2% (with permission from Wu *et al.*, 1990).

(d) Copolymers

Sufficient structural studies have not been carried out on copolymers to devote a section to each system, especially as they do not differ markedly.

Takahashi (1973) has determined that gels from **ethylene–vinyl acetate random copolymer** are produced through the crystallization of the ethylene sequences. His investigation was carried out on a 12% (w/v) gel containing about 71% of ethylene units. The diffraction pattern displays two intense rings at $d = 0.41$ nm and $d = 0.368$ nm which Takahashi assign to the 110 and 200 reflections of polyethylene crystals in the rhombohedric unit cell. Other much weaker rings can also be observed and are identified as the 020, 011, 211 and (320, 410) planes.

Berghmans *et al.* (1979) have investigated the crystalline structure of gels prepared from another random copolymer: **poly(ethylene terephthalate–co-isophthalate)**. Here the investigation was carried out on stretched samples from which the solvent had been removed. The diffraction pattern is similar to that of poly(ethylene phthalate) bulk-crystallized at low temperature. Again, the stretched sample shows two types of orientation and Berghmans *et al.* assign both of the gel junctions to *c*-axis orientation and *a*-axis orientation.

He *et al.* (1988) have shown that the gelation of a **silicic multiblock copolymer** in *trans*-decalin (see Section 1.2.2.d) occurs through the crystallization of the COSP-crystalline sequence with the same unit cell as in the bulk-crystallized state (see Table 2.4). The study was performed on a 50% gel which should possess the same crystalline structure as gels prepared at

Table 2.4 hk0 reflections observed on COSP in a 50% solution in *trans*-decalin and a multiblock copolymer gel made up of 50% of this sequence at the same concentration. *I* stands for the relative intensities with the following meaning for the letters: s = strong, m = medium and w = weak.

d_{COSP} (nm)	*I*	d_{COPO50} (nm)	*I*
0.6	s	0.6	s
0.571	w	?	?
0.518	m	0.513	w
0.438	s	0.438	s
0.375	w	?	?

much lower concentration since the experimental temperature–concentration phase diagram ascertains the absence in this solvent of any solid–solid phase transformation.

These authors report that gel morphology, as seen by optical microscopy is very similar to that of the phase-separated gels investigated so far. The mesh size extends from about 0.5 to 5 μm in 1-phenyldodecane and bromobenzene (He *et al.*, 1989). They, however, underline the fact that the texture in *trans*-decalin, which is a good solvent of the uncrystallizable sequence, appears salt-and-pepper-like. According to these investigators this texture possibly arises from a smaller mesh size.

3 Mechanical properties and rheology

'Circumstantial evidence ... is a very tricky thing. It may seem to point very straight to one thing, but if you shift your own point of view a little, you may find it pointing in an equally uncompromising manner to something entirely different'.

A. Conan Doyle in The Adventures of Sherlock Holmes

3.1 INTRODUCTION

That this chapter comes after the chapter devoted to gel structure and morphology is deliberate. If one is in full cognition of the molecular structure, one may understand the deformation mechanisms and eventually succeed in quantifying the mechanical properties of a gel. Conversely, in no case should the molecular structure be inferred from the mechanical behaviour.

Investigation of the mechanical properties of a thermoreversible gel certainly constitutes one of the most important tasks, especially as a knowledge of them may point to their potential applications.

Investigations into the mechanical properties consist essentially of evaluating the elastic response (elastic modulus) of a gel as well as its viscoelastic behaviour. These are usually measured by two methods: compression or submitting the gel to oscillatory perturbations.

In the first case, from the measurement, on a cylindrically shaped sample, of the true stress, σ, as a function of deformation, λ (λ = deformed length/initial length = l/l_0), one obtains the compression modulus, E, through the so-called Mooney–Rivlin relation derived from continuum mechanics arguments (Mooney, 1940, 1948; Rivlin, 1948):

$$\sigma = E[\lambda - (1/\lambda^2)] \qquad (3.1)$$

in which E amounts to one-third of the Young's modulus for small deformations. This relation is usually valid for small deformations (neohookean case).

Strictly speaking, Mooney and Rivlin have derived the following relation:

$$\sigma = 2C_1(\lambda - \lambda^{-2}) + 2C_2(1 - \lambda^{-3})$$

in which C_1 and C_2 are constants. It turns out that in gels, C_2 is usually equal to zero which yields $2C_1 = E$.

This relation is independent of time, provided that the system under study is purely elastic. In reality, this is seldom strictly true, or even not true at all, so that the stress is a function of time $\sigma(t)$. One then defines an isochronal modulus, $E(t_E)$, obtained by determining $\sigma(t = t_E)$ at constant deformation.

Sometimes the expression of $\sigma(t)$ as a function of time takes on a simple power form (Thirion and Chasset, 1967):

$$\sigma(t) \text{ or } E(t) \approx t^{-m} \tag{3.2}$$

The exponent m is to some extent a measure of the easiness with which a system relaxes at constant deformation. The value of m may also vary so that different regimes may be observed. Alternatively, $\sigma(t)$ can still obey a more complex behaviour.

The compression modulus determination is probably the easiest method to carry out once the gel has formed, can be handled safely and, above all, can retain its cylindrical shape (the gel is then said to be self-supporting).

Conversely, this method is evidently not suitable for study of gel formation and/or gel evolution kinetics, and recourse is made to a plate–plate or cone–plate rheometer or to Couette systems. Generally speaking, a layer of the potentially gelling solution is submitted to shear through a periodical strain. The modulus, G^*, is then a complex number, dependent upon the frequency of the periodical strain which reads:

$$G^* = G'(\omega) + iG''(\omega) = G(\omega)(\sin\delta + i\cos\delta) \tag{3.3}$$

in which $G'(\omega)$ is the elastic part (named storage modulus) and $G''(\omega)$ the viscous component (named loss modulus). The ratio of $G''(\omega)/G'(\omega) = tg\delta$ is designated as the loss tangent. For a gel-like material, one should observe $G'(\omega) \gg G''(\omega)$, and $G'(\omega)$ and $G''(\omega)$ should vary little with frequency. As has been emphasized in the introduction of this book, the rigorous definition for a gel with solid-like behaviour corresponds to $G'(0) \neq 0$.

$G(t)$ is related to $G'(\omega)$ and $G''(\omega)$ through the following relation:

$$G(t) = G'(\omega) - 0.4G''(\omega) + 0.014G''(\omega) \tag{3.4}$$

for $\omega = t^{-1}$. For a gel, where $G'(\omega) \gg G''(\omega)$, relation (3.4) reduces to:

$$G(t) \simeq G'(\omega) \tag{3.5}$$

One of the most common experiments carried out on thermoreversible gels consists of determining the relation between the gel modulus and the polymer concentration. Walter (1954) carried out this type of investigation in the early fifties on PVC gels.

Over the past twenty years theoretical models, essentially based on the scaling approach, have been developed to relate modulus to concentration in chemical gels (see for instance de Gennes, 1979). As a result, the relations are of the type:

$$E \approx C^n \tag{3.6}$$

where n is an exponent which depends upon the conformation of the chains linking junction points.

This type of relation is sometimes found with thermoreversible gels. This has led some scientists to infer from the mechanical properties that thermoreversible gels and chemical gels display similar molecular structure (flexible chains connecting small crystallites) and also to advocate that any theory derived for the latter can be applied without modification to the former (see for instance Clark and Ross-Murphy, 1987).

As many physical gels display a rigid fibre-like structure, one may seriously question such a notion as 'molecular weight between cross-links' or 'functionality' as would be calculated from the theory of ideal chemical networks.

Still, thanks to the topological resemblance between an array of cross-linked chains and an array of intertwined fibres, theoretical relations derived for chemical gels may also be relevant to physical gels as far as the overall behaviour is concerned. There exist, however, conspicuous differences between the two types of gel which cannot be ignored. In chemical gels, only one length, i.e. the mesh size, characterizes the system. As far as physical gels are concerned, this no longer holds, as another length, namely the fibre diameter, must also be taken into account. Also, polymer concentration and network concentration may differ. In fact, in most cases two phases constitute a thermoreversible network: the polymer-poor and the polymer-rich phases. The mechanical properties of the gel arise only from the polymer-rich phase. Ignoring this fact may lead one to determine only apparent exponents and eventually to draw erroneous conclusions. Here again a knowledge of the phase diagram proves to be of prime importance.

All these points are detailed and tentatively discussed in Appendix 4 in the light of theories originally developed for chemical gels which may yet be applicable to physical gels under given conditions (Jones and Marques, 1990).

Unfortunately, power law variations are not always found. Clearly, while theories developed for the elasticity of chemical gels may be regarded as a starting basis, the need in the near future for theories specifically suited to thermoreversible gels will probably grow.

As in the case with the first two chapters, Chapter 3 is organized into two major sections, **synthetic polymers** and **biopolymers**. Similarly, the section devoted to synthetic polymers will comprise two subsections, **solvent-induced gels** and **crystallization-induced gels**.

3.2 SYNTHETIC POLYMERS

Whereas some thermoreversible gels have been extensively studied, many systems have only received little attention so far. It therefore remains difficult to establish from the existing data general principles which may eventually be invalidated with the production of new data.

Still, it seems worth distinguishing again between solvent-induced gels and crystallization-induced gels as they apparently show different stress-relaxation phenomena. The exponent m of relation (3.2) turns out experimentally to be significantly larger for the former ($m > 0.1$) than for the latter ($m \simeq 0.01$–0.02).

Should this trend be confirmed, the stress–relaxation experiment would become a simple mechanical test to find out whether the gel junctions are crystalline or not.

3.2.1 Solvent-induced gels

(a) Isotactic polystyrene

The mechanical properties of isotactic polystyrene gels have been studied as a function of polymer concentration, solvent type and preparation temperature (Guenet and McKenna, 1986; McKenna and Guenet, 1988*a*,*b*).

First of all, these authors have shown by means of stress–relaxation experiments that iPS gels are not purely elastic but display a certain amount of relaxation (Figure 3.1). The relaxation behaviour obeys relation (3.2) at the early stage ($t < 10^3$ s) with an exponent m varying from 0.1 (*cis*-decalin) to 0.15 (*trans*-decalin and 1-chlorodecane). At longer times the relaxation exhibits a more or less steep downturn (Figure 3.1).

The value of m is rather large compared with that measured in swollen chemical gels ($m \simeq 0.02$) (see for instance Janacek and Ferry, 1969*a*,*b*). Also, it is found that the gel does not recover its initial height but retains permanent deformation ($\lambda_p/\lambda_a > 0.55$ where λ_p is the permanent strain and λ_a the applied one) as if the physical junctions had been destroyed and then re-formed.

This behaviour is quite consistent with the present picture of the molecular structure where the junctions are said to be non-crystalline. The

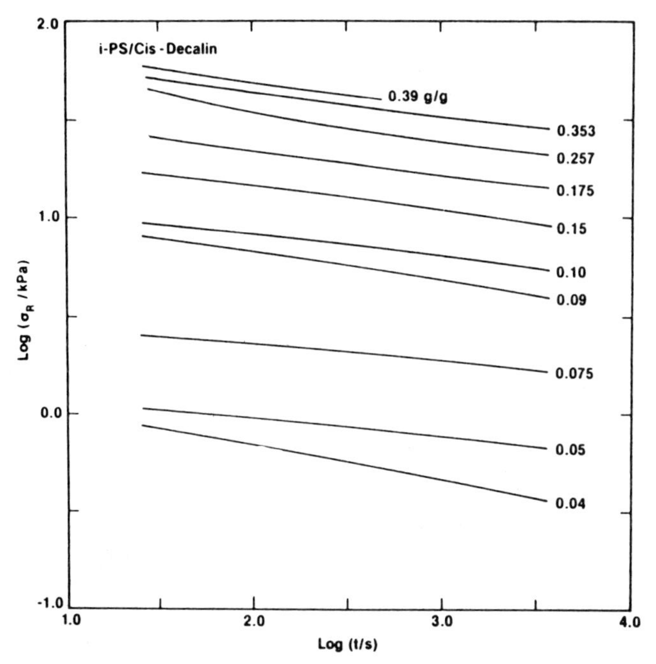

Figure 3.1 Double logarithmic representation of the reduced stress *vs* time in compression for iPS–*cis*-decalin gels formed at -20°C of different concentrations (w/w), as indicated (with permission from Guenet and McKenna, 1986).

ladder-like model as described by Klein *et al.* (1990*a*) does suggest a propensity of the physical knots to be labile by breaking and re-forming, particularly under stress.

As the relaxation turns out to be quite important, the isochronal compression modulus at $t = 120$ s has been considered for establishing a relation with polymer concentration. It has been observed that the gels display neo-hookean behaviour for $\lambda^{-1} < 1.5$, as the isochronal modulus does not depend, within experimental uncertainties, upon the applied strain.

The modulus–concentration relation does not follow a simple power law and depends markedly upon the solvent type.

Figure 3.2 shows the behaviour in *cis*-decalin. The isochronal modulus does not vary linearly with concentration. In addition, the modulus remains virtually constant in the range $0.2 \leqslant C_{pol} \leqslant 0.35$. The same phenomenon, although not so prominent, seems to occur for $0.12 \leqslant C_{pol} \leqslant 0.17$ so that there is a sharp jump between $C_{pol} = 0.17$ and $C_{pol} = 0.2$.

In *trans*-decalin and 1-chlorodecane the variation again does not obey a power law (Figure 3.3). Unlike gels obtained in *cis*-decalin, there are no large concentration domains in which the modulus remains practically invariant. However, as in *cis*-decalin, there are also 'features' due to a rapid

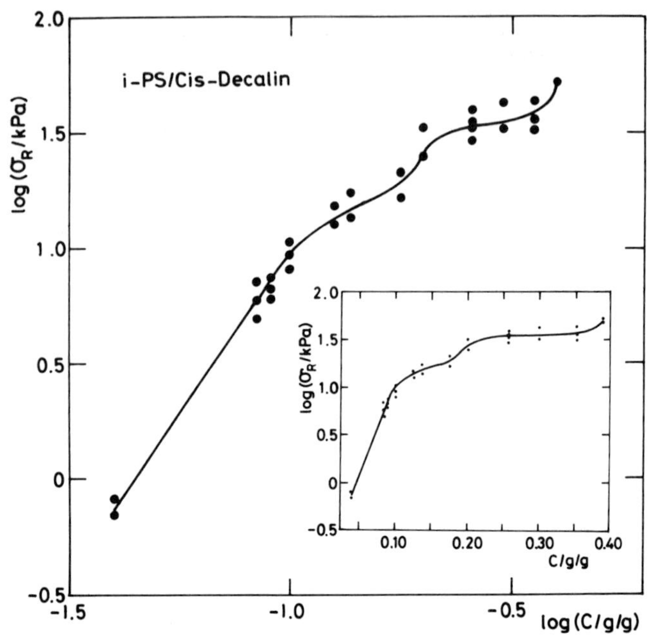

Figure 3.2 Double logarithmic plot of the isochronal modulus (at 120 s) as a function of the polymer concentration (w/w) for iPS–*cis*-decalin gels formed at -20°C. Insert shows a semi-logarithmic plot of the same data (with permission from Guenet and McKenna, 1986).

change in modulus over a narrow range of concentration followed by a domain of concentration for which the modulus does not vary as much. These features are emphasized in Figure 3.3 by dotted lines.

For concentrations up to about $C_{pol} = 0.14$, modulii in *cis*-decalin and *trans*-decalin are larger than those in 1-chlorodecane. Above $C_{pol} = 0.14$ the gels in *cis*-decalin possess a much lower modulus than those in *trans*-decalin and 1-chlorodecane, the latter two displaying values that can be regarded similar within experimental uncertainties.

The fact that significant departure from power laws are found with these gels, especially at high concentrations, may not be surprising if one bears in mind the occurrence of polymer–solvent compounds. In fact, the polymer concentration, C_{pol}, ought to be rescaled by the stoichiometric concentration, C_{γ}, since at this concentration the system is theoretically supposed to behave as a single entity (see Appendix 4). In other words it is as if one expected to observe a continuous power law from the semi-dilute state to the dried state with a chemical gel.

Alternatively, if $C_{pol} < C_{\gamma}$, then the variation may approach a power law which seems to be approximately the case in all three solvents (C_{γ} in *cis*-

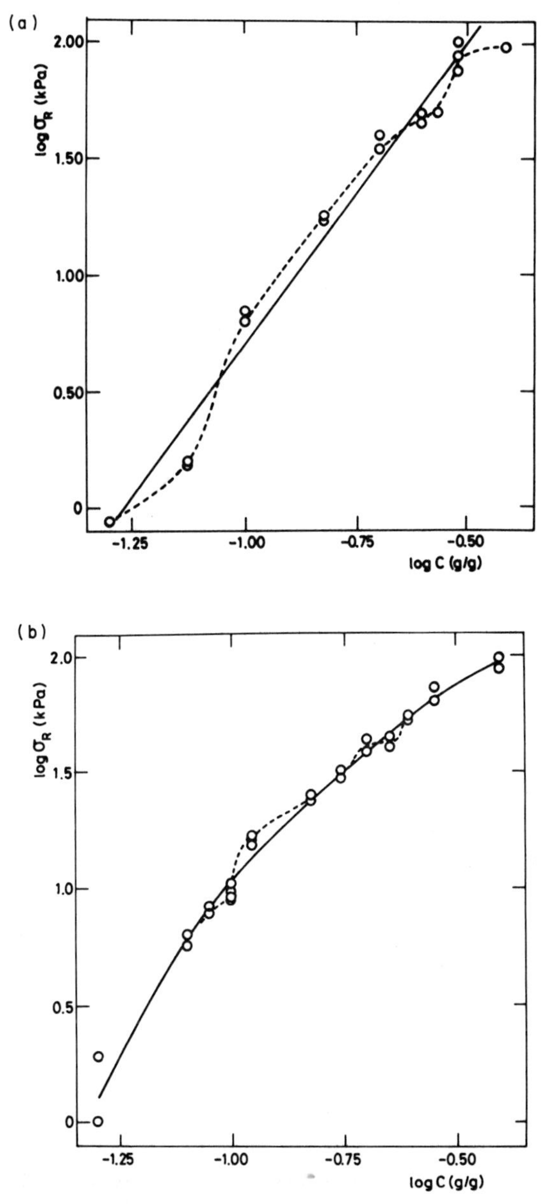

Figure 3.3 Double logarithmic representation of the isochronal modulus (at 120 s) as a function of the polymer concentration (w/w) for (a) iPS–1-chlorodecane gels and (b) for iPS–*trans*-decalin gels formed at -20°C. The dotted lines emphasize the possible 'features' of the variation, although the data can be fitted with monotonic curves (solid lines) (with permission from McKenna and Guenet, 1988*a*).

decalin $\simeq 0.28$ in *trans*-decalin $\simeq 0.38$ and in 1-chlorodecane $\simeq 0.46$). This argument may provide one sensible reason for the curvature at high concentration but fails to account for the invariance of the modulus over large concentration domains in *cis*-decalin or the features observed by Guenet and McKenna (1986).

The results discussed so far have been obtained on gel samples prepared by a quench at low temperature ($-20°C$) and on one polymer fraction. McKenna and Guenet (1988*b*) have noticed that ageing effects appear at higher temperature. If quenching temperatures, T_p, below $0°C$ are produced, no noticeable change of the modulus can be measured beyond 5 min (which is roughly the minimum preparation time). Conversely, above $0°C$ significant evolution is observed. For instance, at a quenching temperature of $T_p = 15°C$ the modulus increases from 71 kPa to 97 kPa after $t_a = 100$ min ageing.

Not only does the quenching temperature affect the gel formation kinetics, but it also influences the magnitude of the gel modulus. McKenna and Guenet (1988*b*) report a factor of about 4 for $C_{pol} = 0.15$ between gels prepared at $-20°C$ and $+15°C$ (25 kPa for the former against 97 kPa for the latter). They, however, impress upon the reader that the gelation temperature does not have any effect on the shape of the variation of the modulus with polymer concentration. They consider that the modulus may be expressed as:

$$E(C_{pol}, T_p, t_a) = H(T_p, t_a)G(C_{pol}) \tag{3.7}$$

in which G is a function of concentration alone and H a function of the preparation temperature and ageing time. Then, these authors were able to draw a master curve by simply shifting the curves vertically.

Curiously enough, the modulus variation with concentration and, to a lesser extent, its magnitude depend upon the polymer fraction. Yet, in no case is power law behaviour found. There seems to be no clear-cut relation between the fraction molecular weight and the modulus. The exact origin of this phenomenon remains unknown.

(b) Atactic polystyrene

Clark *et al.* (1983) have shown that atactic polystyrene solutions in carbon disulphide display the characteristics of a gel once cooled to the appropriate temperature. Using cone–plate or plate–plate geometry, these investigators have obtained the rheogram in Figure 3.4 which shows that G' is a constant over a wide range of frequencies while $G'' < G'$ in the same range.

The storage modulus $G' \approx 6 \times 10^3$ dynes/cm^2 which they report for a 9.5% solution at $-22°C$ ($M_w = 9 \times 10^5$) is unexpectedly high if one considers that this polymer was thought to possess no gelation capability.

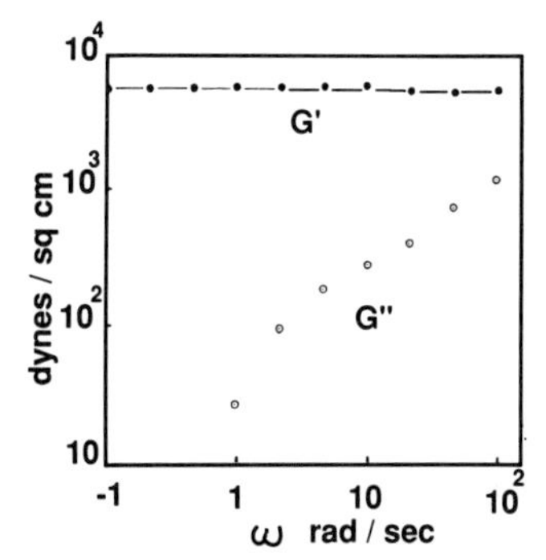

Figure 3.4 Storage (G') and loss (G'') modulii as a function of frequency run between parallel plates at 3% strain of a 9.5% (w/w) polystyrene gels in CS_2 ($M_w = 9 \times 10^5$) at $-22\,^\circ C$ (with permission from Clark, Wellinghoff and Miller, 1983).

Measurements of shear modulus as a function of polymer concentration have been carried out by Koltisko *et al.* (1986) by means of a coaxial cylinder. They have found a power law dependence with an exponent $n \approx 2$. It is worth stressing, however, that the modulii are taken not at constant temperature but at constant reduced temperature $T_R = T/T_{gel}$. T_{gel} is the extrapolated temperature at which the solution exhibits zero shear modulus. As discussed in Section 1.2.1.a.ii, this temperature increases with increasing concentration. While this approach is commonly used for amorphous rubbery systems, one may wonder about its relevance in the present case. There is now little doubt about the first-order character of aPS gelation. The gel is therefore a two-phase entity which consists of a polymer-rich phase, the network, and a dilute phase, the latter having little effect on the elastic properties. What should be of interest is establishing a relation between the modulus and the 'network concentration'. This can be achieved at constant temperature as the amount of polymer-rich phase (the 'amount' of network) is directly proportional to the polymer concentration. Conversely, at constant reduced temperature no simple equation relates the two quantities.

Re-examining the results of Koltisko *et al.* at constant temperature also yields a near C^2 behaviour at $-98\,^\circ C$ whereas the modulus varies as $C^{3.1}$ at $-50\,^\circ C$. These differing exponents may indicate a modification of gel morphology with temperature. Also, one has to realize that at $-98\,^\circ C$ the system aPS–CS_2 lies for the highest concentrations just a few degrees above its glass transition which probably affects the modulus behaviour.

(c) Poly(methyl methacrylate)

Early experiments by Pyrlik and Rehage (1975, 1976) have shown that gels formed from stereocomplex solutions in *o*-xylene (i/s = 4/5,

$C_{pol} = 5.6 \times 10^{-2}$ g/cm^3) possess a storage modulus virtually constant over the range $10^{-3} < \omega$ (s^{-1}) $< 10^2$.

Berghmans *et al.* (1987) have investigated the effect of ageing of sPMMA gels prepared in *o*-xylene in combination with temperature jumps. The results obtained on G' are drawn in Figure 3.5. Cooling the solution to 63°C leads to an increase in G' of about 4 decades within 20 min at almost constant rate. Cooling to 58°C accelerates the gelation process. Heating at 63°C leads to a decrease in G' but the ageing rate proceeds as if no temperature decrease to 58°C had been brought about. The same is observed when the system is further cooled to 23°C.

This experiment shows that ageing is relatively important and that there is a fair amount of reversibility in the gelation process if one assumes that the ageing rate observed at 63°C at the early stage remains constant. Although reasonable, this assumption is not borne out by an experiment carried out at 63°C only.

Fazel *et al.* (1992*b*) have also reported ageing effects on the isochronal modulus in sPMMA gels. At 0°C in *o*-xylene or bromobenzene, 1 h is required to reach a constant value.

These authors have also observed that the relaxation behaviour of sPMMA gels in these solvents resemble those of isotactic polystyrene. As a matter of course at constant compressive deformation, the relaxation behaviour can be described for relaxation times $t < 1000$ s by means of relation (3.2) for which: $m \simeq 0.13$ in bromobenzene and $m \simeq 0.15$ in *o*-xylene. At longer times the rate increases up to $m \simeq 0.46$ in *o*-xylene while it remains almost the same in bromobenzene.

These results altogether may suggest non-crystalline junction domains which is consistent with the absence of any report on gel crystallinity (see Section 2.3.1.c.ii).

Like isotactic polystyrene gels, sPMMA gels exhibit neohookean behaviour within the investigated range of deformations, i.e. for $\lambda^{-1} \leqslant 1.25$.

Fazel *et al.* (1992*b*) have determined the isochronal modulus at 120 s against the polymer concentration (in g/cm^3) in the two aforementioned solvents. They have found a power law with exponents differing quite markedly with the solvent type:

$$E \approx C^{1.84 \pm 0.1} \text{ in bromobenzene} \tag{3.8}$$

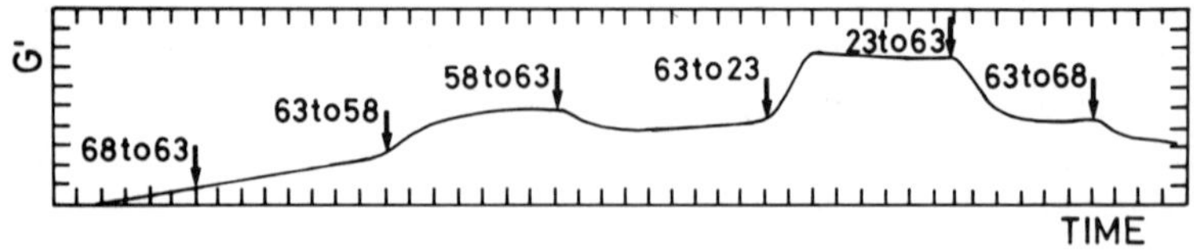

Figure 3.5 Variation as a function of time and of the thermal treatment of G' in sPMMA–*ortho*-xylene gels. Each scale tick represents 10 mm (with permission from Berghmans *et al.*, 1987).

$$E \approx C^{2.66 \pm 0.1} \text{ in } o\text{-xylene} \tag{3.9}$$

In bromobenzene the exponent close to 2 is quite consistent with the rod-like structure of the gels as demonstrated by small-angle neutron scattering (Section 2.3.1.c.ii). Jones and Marques (1990) have derived a general relation for infinitely thin rigid systems where the 'cross-links' are frozen (see Appendix 4):

$$E_r \approx C^{(3\nu+1)/(3\nu-1)} \tag{3.10}$$

in which ν^{-1} is the fractal dimension of the object between the junctions ($R \approx N^{\nu}$, R = end-to-end distance and N number of units). In this case the elastic behaviour is said to be of **enthalpic** (energetic) type. In the present case $\nu = 1$ which yields, for rod-like objects:

$$E \approx C^2 \tag{3.11}$$

Relation modulus–concentration is, however, different if one considers a freely hinged network where the rod-like chains are freely hinged (which means that the angle between two chains starting from the same junction point can fluctuate). Here the elasticity of the system is considered **entropic** (Boué *et al.*, 1988). Under these conditions Jones and Marques arrive at:

$$E_e \approx C^{3\nu/(3\nu-1)} \tag{3.12}$$

Then for $\nu = 1$ relation (3.12) becomes:

$$E \approx C^{3/2} \tag{3.13}$$

It is worth noting that relation 3.12 can be derived by considering, as was done by de Gennes (1979) for networks of chains with excluded volume, that the modulus is proportional to the inverse of the cube of the screening length ξ. As this length is related to polymer concentration through (Daoud *et al.*, 1975):

$$\xi = C^{-\nu/(3\nu-1)}$$

one finally retrieves relation (3.13) for rod-like objects.

The fact that an exponent slightly lower than 2 is determined experimentally may suggest an intermediate situation between the two types of elasticity. This assumption seems to be borne out by the fact that gels prepared in bromobenzene swell when immersed in an excess of solvent. They could not exhibit such a property if the 'cross-links' were frozen.

Jones and Marques suggest that the resulting modulus E is then given by:

$$G^{-1} = G_r^{-1} + G_e^{-1} \tag{3.14}$$

which eventually implies an apparent exponent somewhere between 1.5 and 2.

In Chapter 2 it is briefly mentioned how the molecular structure in bromobenzene depends slightly upon whether sPMMA has been used from the nascent state or once dissolved in chloroform prior to use. A significant effect is also observed on the mechanical properties. While the modulus–concentration relationship is characterized by nearly the same exponent, the front factor increases which gives (Fazel *et al.*, 1992*b*):

$$\text{nascent PMMA } E = 1.6 \times 10^2\, C^{1.86 \pm 0.1}\ (\text{kPa}) \tag{3.15}$$

$$\text{treated PMMA } E = 2.08 \times 10^2\, C^{1.86 \pm 0.1}\ (\text{kPa}) \tag{3.16}$$

When immersed in an excess of bromobenzene, these gels swell readily. For gels swollen at equilibrium the experimental modulus–concentration relation becomes (Fazel *et al.*, 1992*b*):

$$E = 1.94 \times 10^3\, C^{1.99 \pm 0.1}\ (\text{kPa}) \tag{3.17}$$

The nearly tenfold increase in the modulus magnitude would normally imply that the number of physical junctions per unit volume has been raised considerably. Such an assumption may seem rather paradoxical since gel swelling entails 'dilution' of the number of existing junctions and, besides, is thought to prevent the formation of new ones. However, the exponent has also increased and is now close to two, which means, in the light of Jones and Marques' approach, that the gel displays purely enthalpic elasticity. This suggests that the 'cross-links' are frozen which may probably also affect the magnitude of the modulus. That the 'cross-links' are frozen after swelling is not surprising, as higher solvation is liable to restrict the degrees of freedom. Further, neutron scattering results indicate that the cross-section of a part of the rod elements constituting the network has nearly doubled which is also consistent with modulus reinforcement.

Concerning gels in *o*-xylene, the exponent $n = 2.66$ may be taken as indicative of a fractal dimension of about $\nu^{-1} = 1.38$ in the framework of relation (3.10) or $\nu^{-1} = 1.87$ in the framework of relation (3.12). Apparently, both of these values disagree with the neutron scattering data gained for these systems (Fazel *et al.*, 1992*b*) which would imply $\nu^{-1} > 2$ if interpreted in terms of fractal dimension. However, if one considers, as Fazel *et al.* have done, fibre-like structures between 'cross-links' with a large cross-section but no longer strictly straight (that is possessing a longitudinal length with the required fractal dimension), the conclusions drawn from the two types of experiment can be reconciled. Indeed, mechanical properties are chiefly sensitive to the fractal dimension of the longitudinal length, provided that the cross-section remains unchanged when the concentration is changed, while neutron scattering only probes the cross-section size in the range of momentum transfer explored by Fazel *et al.*

Also, it might be worth asking whether relations (3.10) and (3.12) still hold when the distance between junction domains becomes of the same order of magnitude as the fibre cross-section, which may be the case here. Alternatively, the fibre cross-section may vary with polymer concentration which eventually makes the analysis of the experimental data more complex (see Appendix 4).

It should be noted that, as far as non-swollen gels are concerned, not only is the exponent larger in *o*-xylene but also the front factor ($E = 3.6 \times 10^3\ C^{2.66 \pm 0.1}$ kPa).

(d) Poly(γ-benzyl-L-glutamate)

Murthy and Muthukumar (1987) have carried out an extensive study of the PBLG gels in benzyl alcohol. They report that dynamic measurements can only be performed at very low strain ($\gamma < 3\%$). They also note that it takes several hours for the system to reach equilibrium.

Interestingly enough, the storage modulus, $G'(\omega)$, and the loss modulus, $G''(\omega)$, show the same frequency dependence whatever the range of temperatures investigated some of which correspond to the sol state (68°C for instance) (Figure 3.6). Murthy and Muthukumar accordingly suggest that

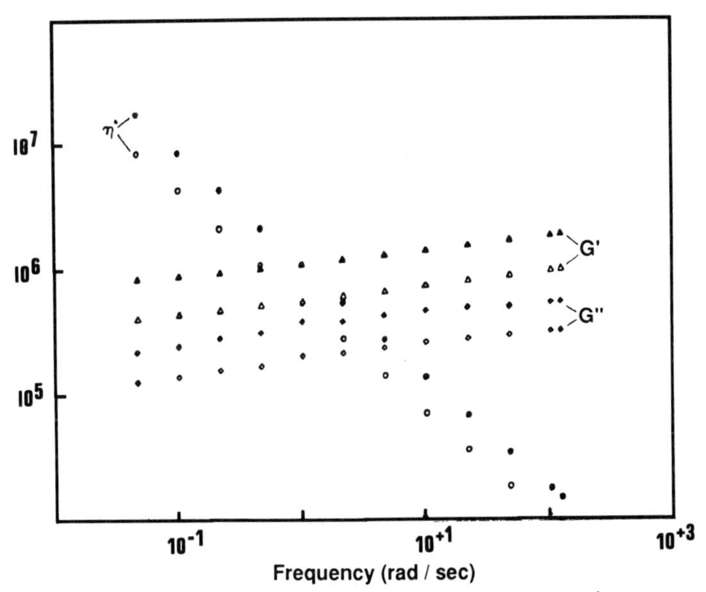

Figure 3.6 Double logarithmic plot of the complex viscosity, η^*, G' and G'' as a function of frequency for poly(γ-benzyl-L-glutamate)–benzyl alcohol gels. The solid points represent measurement carried out 3 h from the initial measurements (hollow points). Concentration 8.3% (w/w), $T = 68°C$, $\omega = 1$ rad/s and $\gamma = 1\%$. The modulii G' and G'' are in dynes/cm^2 and the complex viscosity expressed in poise (with permission from Murthy and Muthukumar, 1987).

the gel state retains high-temperature structures. The storage modulus $G'(\omega)$ does not, however, reach a plateau at low frequencies. Instead, $G'(\omega)$ varies approximately as ω^m with $m \simeq 0.05$ to 0.07 depending upon temperature. Using relations (3.2) and (3.5), this suggests that these gels exhibit a relaxation behaviour similar to the one of isotactic polystyrene gels.

These authors have also determined the variation of the storage modulus as a function of polymer concentration. The concentration shows the same dependence at all frequencies. Their results are plotted in Figure 3.7 for $\omega = 1$ rad/s and $\gamma = 1\%$. The modulus is seen to display a shoulder around 7% (w/w) and a maximum around 8.5%. The maximum is identified with the narrow biphasic region of the phase diagram (see Section 1.2.1.e) while the shoulder is believed to be an indication of the liquid crystalline phase dispersed in the isotropic matrix (Aharoni, 1980). Finally, Murthy and Muthukumar point out that at higher concentrations the modulus falls quite

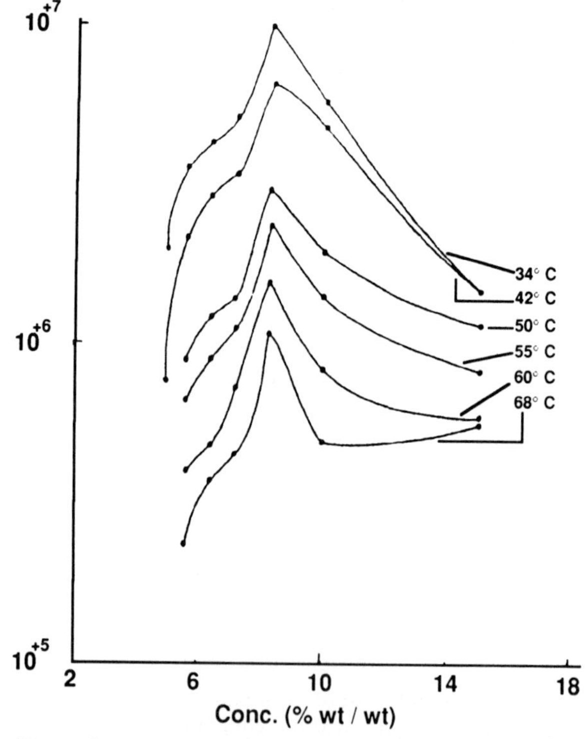

Figure 3.7 Dependence of the storage modulus, G' (dynes/cm^2) on concentration, covering all three phases of the poly(γ-benzyl-L-glutamate)–benzyl alcohol systems. $\omega = 1$ rad/s and $\gamma = 1\%$ (with permission from Murthy and Muthukumar, 1987).

steeply at low temperatures and more smoothly at high temperatures. This is a most remarkable feature of this polymer.

A few years earlier, Sasaki *et al.* (1982) also studied the mechanical properties of PBLG–benzyl alcohol gels by means of a Couette system. Since their study was restricted to concentrations between 0.1% and 8%, they did not observe any maximum. Otherwise, their results are qualitatively similar to those of Murthy and Muthukumar.

Here again, as with isotactic polystyrene, no simple power law can fit the experimental results.

Sasaki *et al.* (1982) also found that the storage modulus depends markedly upon molecular weight at low concentrations ($C \approx 0.1$–0.3%, w/w). This dependence becomes less and less conspicuous as the concentration is increased. The molecular weight dependence occurs in the vicinity of the critical gel concentration C_{gel} where overlap problems determine gel formation. It is probably early days to find out what actually takes place in this domain of concentration. A detailed knowledge of the molecular structure will be required to derive reliable theories.

Conversely, once the concentration lies well above C_{gel}, the chain molecular weight is supposed to play no major role whatsoever. This statement generally holds true.

3.2.2 Crystallization-induced gels

(a) Polyethylene

The mechanical properties of the gel state of polyethylene solutions have received no particular attention, unlike those of the fibres spun from it. While this book should, in principle, be restricted to the gel state, the high-modulus polyethylene fibres deserve a short paragraph to emphasize the technical achievement they represent and to highlight the potentiality of the gel state for preparing from 'old' polymers new materials with unusual properties.

Smith and Lemstra (1979, 1980*a*,*b*) have shown that the Young's modulus, Y, of the fibre varies, within experimental uncertainty, linearly with the draw ratio, Δ:

$$Y = 2.86\Delta - 0.46 \text{ (GPa)} \tag{3.18}$$

At the highest draw ratio (31.7), a Young's modulus of about 90 GPa is found which is almost as high as for Kevlar fibres.

These authors also report that the tensile strength increases as much as the Young's modulus from 0.06 GPa to 3.04 GPa at the maximum draw ratio.

Smith and Lemstra further mention that the presence or absence of solvent and/or the drawing temperature affect the maximum attainable draw ratio but do not significantly alter the modulus at comparable draw ratio.

(b) Poly(vinyl chloride)

Walter carried out a systematic study on PVC thermoreversible gels as early as 1954. Most of the solvents considered consist of common PVC plasticizers (molecules possessing high boiling points).

He found that in most cases the modulus is related to polymer concentration by a power law as long as the concentration was taken lower than about 30%. Above 30%, noticeable departures from the power law are seen. The exponents found at concentrations below 30% for plasticizers of similar properties towards PVC range from about 3.2 up to 3.9. Similar exponents have been measured by Mutin (1986) in diesters, bromobenzene and bromonaphthalene ($n = 3.1 \pm 0.2$). These exponents when examined in the light of Jones and Marques approach hint at fractal dimensions $\nu^{-1} = 1.53 \pm 0.07$ for 'frozen cross-links' and $\nu^{-1} = 2.03 \pm 0.07$ for freely hinged ones. However, this approach may not necessarily hold in this case as neutron scattering reveals structures differing noticeably from one solvent to another while the modulus–concentration exponent remains virtually identical (diethyl oxalate and bromobenzene for instance).

In fact, PVC ought to be regarded as a copolymer of highly syndiotactic sequences alternating with irregular ones. Accordingly, the approach through fractal dimension which implies self-similarity of the elastic element may not be appropriate. Possibly, the approach derived by He *et al.* (1989) for block copolymers is better suited to this system (see below relation 3.33).

As has been emphasized in the previous chapters, ageing is an important phenomenon in PVC gels. Ageing effects have already been observed by Walter (1954) and also by te Nijenhuis and Dijkstra (1975). The latter have shown, by studying the evolution of the storage modulus, G', as a function of ageing time, t_a, that the slope $dG'/d \log t_a$ is independent of frequency. They have considered this quantity as the 'rate of ageing'. After reduction of the data to a single reference temperature, a plot as a function of ageing temperature yields a maximum near 10°C (Figure 3.8). te Nijenhuis and Dijkstra regard this shape as indicative of ageing originating in crystallization.

According to Mutin and Guenet (1989), gel ageing originates chiefly in the formation of the so-called 'weak links' which differ from the 'strong links', the latter being thought to consist of crystallites of predominantly syndiotactic sequences.

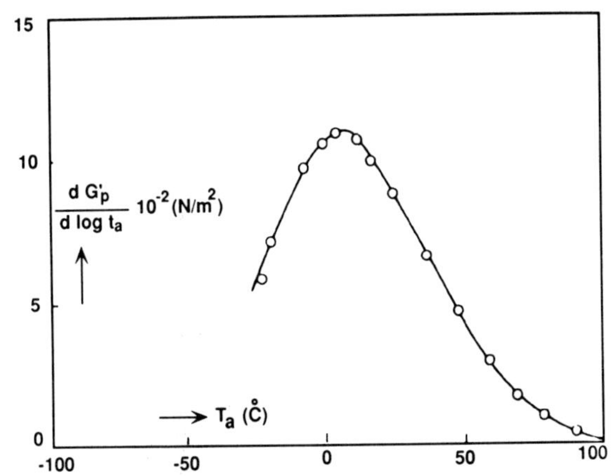

Figure 3.8 Variation of the derivative of the storage modulus G'_p to the logarithm of ageing time as a function of temperature (with permission from te Nijenhuis and Dijkstra, 1975).

DSC experiments have shown a melting endotherm near 50°C for aged gels (see Section 1.2.2.b.ii). Mutin and Guenet (1989) have shown that a gel prepared in diethyl malonate is rejuvenated by heating at 65°C since the modulus found after heat treatment is virtually equal to the one measured on the nascent gel.

The ageing process depends strongly upon both solvent type and polymer syndiotacticity. These effects are summarized in Figure 3.9 for gels prepared at the same concentration ($C_{pol} = 0.175$ g/cm^3) and aged at room temperature while immersed in an excess of preparation solvent. The evolution of the gel modulus is plotted as a function of the concentration. Swelling and deswelling entail a decrease and an increase in polymer concentration respectively.

As can be seen, gels of atactic PVC (HTPVC) in bromonaphthalene swell quite markedly while the modulus remains virtually constant. Conversely, a gel prepared from higher syndiotactic PVC (LTPVC) deswells in the same solvent while its modulus increases noticeably. Gels of atactic PVC in diethyl malonate display a behaviour quite similar to the one observed for the highly syndiotactic PVC in bromonaphthalene. From this, Mutin and Guenet conclude that the solvent plays a role similar to tacticity which is possibly an indication of the formation of a polymer–solvent compound with the less syndiotactic sequences of the atactic PVC.

It seems worth noting that the formation of weak links most probably also occurs for atactic PVC gels in bromonaphthalene, as the modulus remains constant while the gel swells.

The effect of PVC chain molecular weight has also been examined by Mutin and Guenet with gels prepared from HTPVC in either bromonaphthalene

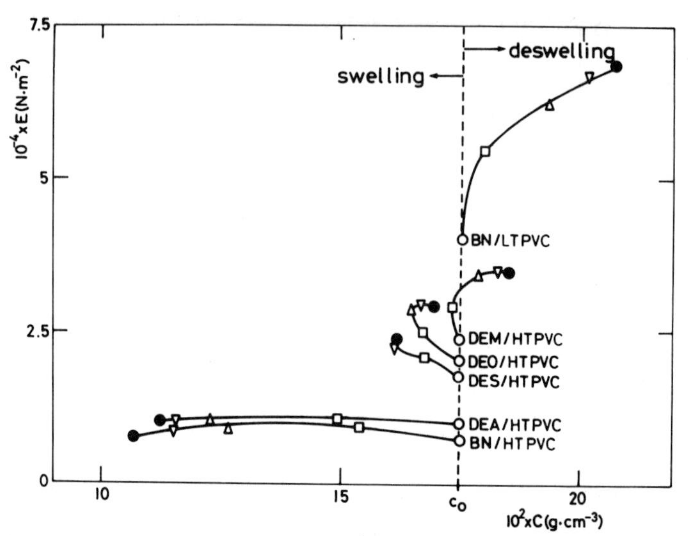

Figure 3.9 Variation of the compression modulus as a function of both the concentration and ageing time for PVC gels in various solvents. Original concentration of the gel is $C_0 = 17.5\%$ (w/w). All samples were aged in an excess of preparation solvent. HTPVC stands for PVC synthesized at 50°C (atactic), while LTPVC stands for PVC synthesized at −40°C (more syndiotactic). Solvents as indicated: BN = bromonaphthalene, DEA = diethyl adipate, DES = diethyl succinate, DEO = diethyl oxalate and DEM = diethyl malonate. Ageing times = ○, 1 h; □, 1 day; △, 7 days; ▽, 14 days and •, 21 days (with permission from Mutin and Guenet, 1989).

or diethyl malonate. In the former solvent, the modulus increases with molecular weight whereas it remains virtually constant in the latter. As the probability of finding sequences with the right size for forming crystals augments with increasing molecular weight (Mutin, 1986), the number of junctions is also liable to grow, hence the increase in modulus. Conversely, the virtual invariance in PVC–diethyl malonate gels may be viewed again, as above, as an indication of the involvement of the solvent in the formation of physical junctions.

However, as has been realized by Mutin and Guenet, the solvent may also play a negative role. For instance the effect of solvent quality on the modulus magnitude is not assessed: a bad solvent is liable to promote the formation of crystallites acting as junction domains, unlike a good solvent. As a result, the modulus in the former will be larger than the one in the latter. To decide whether solvent A rather than solvent B enhances the mechanical properties becomes meaningful only if solvent quality towards PVC is the same.

Determining solvent quality of a solvent towards PVC at room temperature, i.e. measuring χ_1 from dilute solutions, is not straightforward on account of aggregation. Another approach has been taken by Najeh *et al.* (1992) to overcome this difficulty. It consists of preparing gels in different solvents at the same concentration and then letting them swell up to equilibrium in an excess of preparation solvent. These authors reason that gels reaching the same swelling ratio, $G_v = V_\infty / V_0$ (V_∞ = volume at equilibrium and V_0 = initial volume), are directly comparable since their preparation solvent displays the same quality towards PVC. Therefore, if only crystallization of the syndiotactic sequences takes place, these gels must have, within experimental uncertainty, the same or nearly the same modulus.

Such an approach has allowed them to test the polymer–solvent compound structure proposed by Abied *et al.* (1990). As a reminder, these authors consider that a diester can interact through the C=O group with two neighbouring chains so as to establish a bridge (see Section 2.3.2.b). This effect is thought to be cooperative and accordingly to give birth to weak links. If this assumption is true, monoesters should give different results, as they can only interact with one chain.

Najeh *et al.* have studied the gel modulus in different solvents as a function of the swelling equilibrium ratio. And indeed, gels prepared in diesters, on the one hand, and monoesters on the other, do not behave the same (Figure 3.10).

In diesters the following relation is found:

$$E = 18.6 G_V^{-1.54 \pm 0.1} \text{ kPa} \tag{3.19}$$

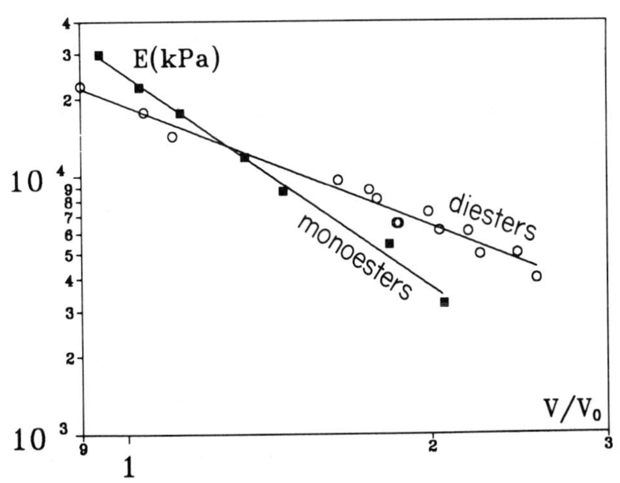

Figure 3.10 Variation of the compression modulus (E in kPa) as a function of the swelling ratio (vol/vol) for PVC gels prepared in diesters (○) or monoesters (■) and swollen to equilibrium in an excess of preparation solvent (initial concentration of all the gels: 0.175 g/cm^3 (from Najeh, Munch and Guenet, submitted for publication).

while in monoesters, one obtains:

$$E = 24.5 G_v^{-2.72 \pm 0.1} \text{ kPa} \tag{3.20}$$

Furthermore, for relatively high swelling ratios ($G_v > 1.5$), the modulus is unquestionably larger in diesters than in monoesters, which indicates unambiguously the involvement of the solvent in the formation of additional physical links. Conversely, for low swelling ratios ($G_v < 1.5$), the modulus becomes larger in monoesters. Najeh *et al.* consider the fiber-like model proposed by Abied *et al.* (1990) in order to account for the latter behaviour. For PVC/diester gels they still suggest that some kind of polymer-compound is formed with the less syndiotactic sequences which eventually gives rise to the 'weak' links. Conversely, for the PVC/monoester gels they note that those monoesters leading to low swelling ratio possess high molar volume. They accordingly suggest that these monoesters will meet with considerable difficulty in entering the amorphous part of the fiber. This part being less solvated is then liable to lie under its glass temperature. As a result, modulus reinforcement of enthalpic nature is expected which can possibly be far larger than the effect of additional links. Najeh *et al.* provide support to this interpretation by means of DSC experiments that do reveal a glass transition in gels prepared from these monoesters ($T_g \approx 30°C$) while a melting endotherm is observed with their diester counterparts ($T \approx 55$–$60°C$). This altogether implies that modulii of PVC/monoester gels should drop markedly above 30°C.

(c) Multiblock copolymers

He *et al.* have reported in a series of papers, the variation of the modulus of block copolymer gels as a function of temperature, concentration, crystalline sequence composition and solvent type (He *et al.*, 1987*a*, 1988, 1989).

First of all, relaxation experiments such as those performed on isotactic polystyrene gels have revealed a type of behaviour depicted by relation (3.2). Values of m range from 0.01 to 0.03 whatever the concentration or the crystalline sequence composition. It is worth keeping in mind that these gels possess crystalline junctions.

As a function of temperature, He *et al.* (1987*a*) have noticed that for low concentrations there is a sudden drop in modulus near 40°C while for high concentrations the modulus remains virtually constant (Figure 3.11). These investigators have interpreted these results with the notion of partial melting of the gel structure occurring at a given temperature corresponding to the monotectic temperature, as determined by differential calorimetry (see Section 1.2.2.d). They have derived a theoretical approach for which they chose to express the modulus as follows:

$$E \approx f(C)g(T)h(T) \tag{3.21}$$

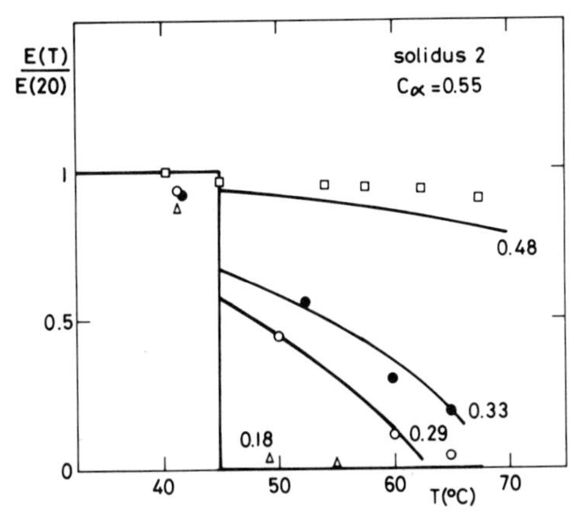

Figure 3.11 Reduced experimental compression modulus ($E(T)/E(20)$) as a function of temperature for various polymer concentrations (w/w) as indicated. The solid line stands for theoretical calculations achieved by means of equations (3.25) to (3.27) by considering the schematic phase diagram drawn in Figure 3.12 (with solidus 2) and using the values of C_α, C_M and T_M found experimentally (with permission from He *et al.*, 1987*a*).

in which $f(C)$ is the concentration-dependence, $g(T)$ the effect of temperature on the flexible sequences ($g(T)$ would be proportional to kT for a rubber-like material) and $h(T)$ the function linked to the number of effective links. The latter function is the one of interest. Taking into account the temperature–concentration phase diagrams of monotectic type established experimentally for these gels, they have expressed $h(T)$ as follows:

$$h(T) \propto C_{\text{prep}}/C_\alpha \qquad \text{for } T < T_M \tag{3.22}$$

$$h(T) = 0 \qquad \text{for } C < C_M \qquad \text{for } T > T_M \tag{3.23}$$

$$h(t) \propto \frac{C_{\text{prep}} - C_{M1}(T)}{C_{\alpha s}(T) - C_{M1}(T)} \qquad \text{for } T > T_M \tag{3.24}$$

These relations simply express that the modulus is proportional to the amount of polymer-rich phase. $C_{\alpha s}(T)$ is the concentration defined by the solidus line at T ($C_{\alpha s} = C_\alpha$ at $T = T_M$) and $C_{M1}(T)$ the one defined by the liquidus line ($C_{M1}(T) = C_M$ at $T = T_M$ and $C_{M1}(T) \simeq 0$ for $T < T_M$). By regarding $g(T)$ and $f(C)$ as constants in relation (3.21) within the range of temperature investigated and by normalizing $E(T)$ to the modulus at 20°C, one finally ends up with:

$$E(T)/E(20) = 1 \qquad \text{for } T < T_M \tag{3.25}$$

$$E(T)/E(20) = 0 \qquad \text{for } C < C_M \qquad \text{for } T > T_M \tag{3.26}$$

$$E(T)/E(20) = \frac{C_{\text{prep}} - C_{\text{M1}}(T)}{C_{\alpha s}(T) - C_{\text{M1}}(T)} \times \frac{C_{\alpha}}{C_{\text{prep}}} \quad \text{for } T > T_{\text{M}} \tag{3.27}$$

As can be seen in Figure 3.11, this set of simple equations combined with the schematic phase diagram of Figure 3.12 accounts for the experimental results quite satisfactorily.

Originally, He *et al.*'s assumption that $h(T)$ is proportional to the network concentration was purely empirical. This assumption finds justification if one considers that only the 'functionality' decreases with increasing temperature. Jones and Marques have shown that the modulus must then vary as the network concentration independent of whether the elasticity is of entropic or enthalpic origin (see Appendix 4). This may therefore suggest that partial melting observed in these gels consists chiefly of altering the number of fibrillar structures that merge at the same junction domain.

These outcomes provide further support to the statement of He *et al.* that the observation of two endotherms by differential calorimetry does not necessarily imply the presence of two distinct, unconnected structures, as was often inferred from the case of isotactic polystyrene gels. In the iPS

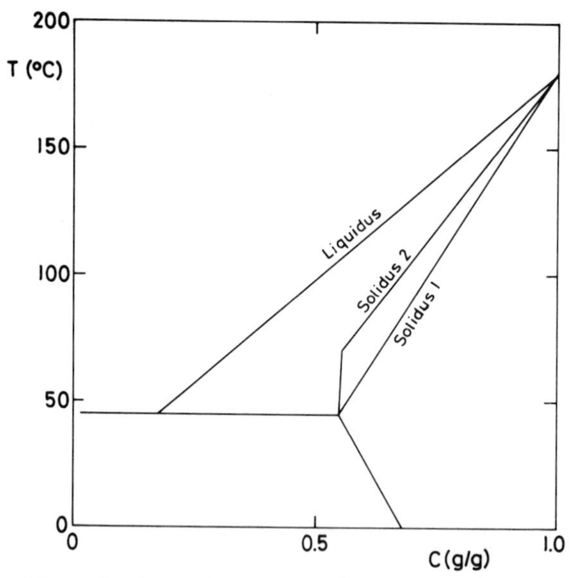

Figure 3.12 Schematic phase diagram used for calculating the variation of the compression modulus as a function of temperature. Here, the liquidus line has been approximated to a straight line and corresponds to $C_{\text{M1}}(T)$ in equation (3.27). Solidus lines are only tentative: 'solidus 1' increases monotonically with temperature, while 'solidus 2' is a constant up to 75°C. Both solidus lines correspond to $C_{\alpha s}(T)$ in equation (3.27) (with permission from He *et al.*, 1987*a*).

gels, only one structure forms the network while the other consists of single crystals scattered throughout the system. If this model pertained to the present case, the modulus should drop abruptly at T_M even for $C > C_M$, as the low-melting endotherm is attributed to the network melting. He *et al.*'s results eventually call into question the excessive tendency, which occurred at one time, of regarding isotactic polystyrene gels as a paradigm for synthetic polymers.

As a function of concentration and copolymer composition in crystallizable sequence, X_C, He *et al.* (He *et al.* 1987*a*, 1988, 1989) have found, within experimental uncertainty, a power law variation (Figure 3.13):

$$E = E_0 X_C^2 C^{4.5 \pm 0.3} \tag{3.28}$$

This relation has been found to be independent of the solvent type (*trans*-decalin, bromobenzene and 1-phenyldodecane). However, the front factor, E_0, is about three times larger in *trans*-decalin ($E_0 = 14 \times 10^4$ kPa) than in bromobenzene and 1-phenyldodecane ($E_0 = 4.9 \times 10^4$ kPa). This discrepancy may originate in the solvent quality towards the non-crystalline

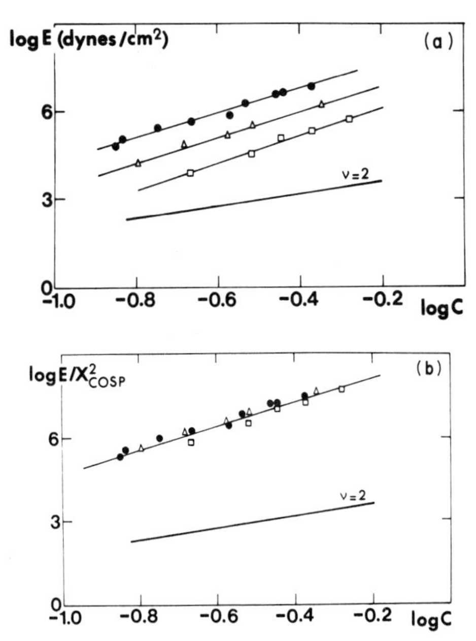

Figure 3.13 (a) upper figure: double logarithmic representation of the variation of the compression modulus as a function of polymer concentration for Copo gels prepared in *trans*-decalin. • = Copo50, ▵ = Copo20 and □ = Copo10 (for details on the copolymers see Table 1.3). (b) lower figure: reduced modulus (modulus divided by the square of the copolymer composition in crystallizable sequence, X_{COSP}) *vs* concentration. Symbols as above (with permission from He *et al.*, 1988).

sequence which is better with *trans*-decalin than the other two solvents. This sequence, being more swollen in *trans*-decalin, may therefore be more rigid.

In bromobenzene and 1-phenyldodecane, no swelling occurs when the gel is immersed in an excess of preparation solvent. Conversely, in *trans*-decalin, which is a good solvent of the non-crystallizable sequence (PDMS), subsequent swelling takes place together with an alteration of the power law. The latter becomes:

$$E \sim C^{6 \pm 0.3} \tag{3.29}$$

To account for the variation given by relation (3.28), He *et al.* (1989) have derived a short theory based on the proportionality between the modulus of the gel and the osmotic pressure developed by an equivalent solution, as has been used to account for the mechanical properties of chemical gels (de Gennes, 1979). This approach assumes that the elasticity is of entropic origin. The osmotic pressure, π, and the scattering function of the solution, $S(q)$, are related through:

$$\partial\pi/\partial C = C \lim_{q \to 0} S^{-1}(q) \tag{3.30}$$

Furthermore, $\partial\pi/\partial C$ may be expressed as:

$$\partial\pi/\partial C = \partial\pi/\partial C_{cry} \times \partial C_{cry}/\partial C \tag{3.31}$$

in which $C_{cry} = CX_C$ and in which the first term in relation (3.31) is assumed to be:

$$\partial\pi/\partial C_{cry} = C_{cry} \lim_{q \to 0} S_{cry}^{-1}(q) \tag{3.32}$$

To calculate $S_{cry}(q)$, He *et al.* (1989) argued that the crystalline character of the gel entails sharp variations in polymer density. As a result, they have used the Debye–Bueche scattering function (Debye and Bueche, 1949) derived for two-density systems (see relation (2.23)). For $q = 0$, this function becomes proportional to the cube of the correlation length a. Then the modulus finally reads:

$$E \sim X_C^2 C^2 a^{-3} \tag{3.33}$$

They estimate the variation in a from the results gathered in atactic polystyrene ($a \sim C^{-0.95}$, relation (2.51)) which eventually yields:

$$E \sim X_C^2 C^5 \tag{3.34}$$

This relation agrees quite well with the experimental results.

The approach of Jones and Marques (1990) may also be contemplated. For the experimental exponent $n = 4.5$, relations (3.10) and (3.12) give fractal dimensions of $\nu^{-1} = 1.88 \pm 0.08$ or $\nu^{-1} = 2.33 \pm 0.03$ respectively. However, the concept of fractal dimension implies objects characterized by self-similarity which may be somehow inappropriate for the system under study.

3.3 BIOPOLYMERS

3.3.1 Gelatin gels

In his review on protein gels, Ferry (1948*a*) mentions that the mechanical properties of gelatin gels were investigated as early as the turn of this century (Leick, 1904; Rankine, 1906; Hatschek, 1921).

Rankine (1906) and later Hatschek (1921) have observed that these gels relax noticeably at constant strain and that the stress can drop to zero after some time. More precise figures of the relaxation rate can be obtained from the paper by Pines and Prins (1973). For a 15% (w/w) gel they found a relaxation of the type given in relation (3.2) which yields from their raw data $m \simeq 0.08$ for $t = 10^2$ to $t \simeq 5 \times 10^2$ s and for strains between 2.87% and 22.8%. As noted by Pines and Prins, this value is larger than in chemical gels. Conversely, it is quite reminiscent of values obtained with iPS or PBLG.

The variation of the modulus with gelatin concentration has been the subject of many studies (Leick, 1904; Ferry, 1948*b*). For concentrations ranging from about 1% to 30%, it is found that E varies as the square of the concentration in a wide range of temperature (wide for gelatin). Ferry (1948*b*) has obtained the following relations for C varying between 10^{-2} g/cm^3 and 5.8×10^{-2} g/cm^3:

$$\left.\begin{array}{ll} E(\mathrm{kPa}) = 4.5 \times 10^3 C^2 & \text{at } 0^\circ\mathrm{C} \\ E(\mathrm{kPa}) = 2.3 \times 10^3 C^2 & \text{at } 15^\circ\mathrm{C} \\ E(\mathrm{kPa}) = 1.2 \times 10^3 C^2 & \text{at } 21.8^\circ\mathrm{C} \end{array}\right\} \quad (2.35)$$

This square dependence of the modulus as a function of concentration was interpreted earlier in the framework of rubber elasticity, as gelatin was thought to resemble chemical gels (point-like junctions instead of three-dimensional ones) (Ferry, 1948*a*). Then the exponent of two was believed to arise from some kind of 'physical dimerization'.

However, as has been discussed above, an exponent of two can be found for many different textures, e.g. assemblies of fibres (see Section 3.3.2) or rigid cellular structures (see for instance Ashby and Gibson, 1988). In fact this result is quite in accord with the morphology of gelatin gels, which display a cellular arrangement.

To account for this exponent, another approach has been taken by Clark and Ross-Murphy (1985) on the basis of the treatment developed by Hermans (1965). It essentially rests upon the analogy between a physical gel and a chemical gel, particularly the growth process. By defining a concentration C_0 below which no infinite gel can be formed, Clark and Ross-Murphy arrive at the general, dimensionless relation between the modulus E and the gelatin concentration C:

$$\frac{EK}{aRT} = \frac{C}{C_0}\,[h(f, C/C_0)] \tag{3.36}$$

in which f is some functionality, and K and a are constants.

For $C > C_0$, the function in brackets tends to C, which eventually gives the desired power law but predicts departure at concentrations of the order of C_0. Yet, this theory contains a sufficient number of adjustable parameters, particularly the functionality, to permit good fit of the experimental data. Also, it seems conceptually wrong to treat a physical gel, even gelatin, on the basis of chemical gels. As has been suggested in the introduction of this chapter and discussed in more detail in Appendix 4, there may exist different reasons, related to physical gel structure in general, of why the power law is not found all along.

One reason may lie with the considerable mechanical changes displayed by these gels on ageing, a phenomenon particularly enhanced at low concentrations. Gel ageing has been thoroughly studied by a careful investigation of the evolution with time of the gel modulus by te Nijenhuis (1981*a*,*b*).

A typical example of the evolution of the storage modulus with ageing

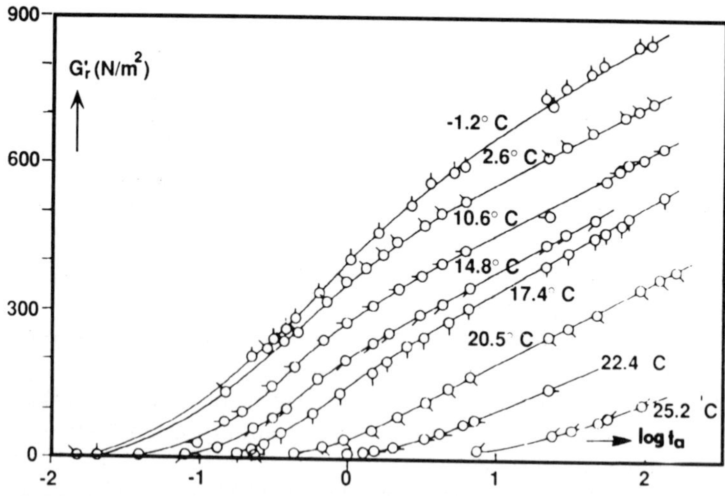

Figure 3.14 Reduced storage modulus of 1.95% (w/w) gelatin–water plotted against ageing time for different ageing temperatures (as indicated). $\omega = 0.393$ rad/s. Reference temperature $-1.2°C$ (with permission from te Nijenhuis, 1981*a*).

time at different temperatures is given in Figure 3.14. As can be seen, equilibrium is not reached even after some 100 h (according to te Nijenhuis, even 150 h ageing is still not sufficient to reach an equilibrium modulus). The derivative of the curve $G = f(\log t_a)$, $dG/d \log t_a$, reaches a constant value beyond $t_a \simeq 1$ h, a result much reminiscent of the variation of the amount of helix, χ_H, reported by Djabourov *et al.* (1988*a*).

The rate of ageing at constant temperature increases with increasing gelatin concentration. As already observed, as a function of the ageing temperature, the derivative $dG/d \log t_a$ becomes a constant beyond $t_a \simeq 1$ h. Its slope depends, however, upon the concentration: the higher the concentration, the greater the slope.

Another interesting experiment reported by te Nijenhuis deals with the effect on the modulus of a pre-quench at higher temperature than the final gelation temperature. In the present case, the sample was first quenched to 17.4°C for time t_a and then cooled down to −1.2°C for an ageing time t_{ea}. The results (Figure 3.15) indicate that the longer t_a, the higher the modulus. However, beyond $t_a \simeq 3$ h, the pre-ageing process at 17.4°C does not bring much additional effect. For the largest discrepancy a factor of about 2 is obtained between the modulus of a sample pre-aged at 17.4°C and a sample directly cooled down to −1.2°C.

This type of experiment involving different thermal histories shows that gelatin gels cannot be treated with the concept of chemical equilibrium as advocated by Eldridge and Ferry (1954).

Curiously enough, reversibility is observed when the reverse thermal history is brought about, i.e. ageing first at low temperature and then at higher temperature. Under these conditions the evolution of the modulus with ageing time does not display, beyond a certain time, any memory effect of the annealing at low temperature.

Djabourov *et al.* (1988*b*) have put the emphasis on the gelation behaviour near the gelation threshold. Combining polarimetry with modulus determination, they have observed that the onset of gelation is located near $\chi_{H_c} = 7\%$. This value virtually does not depend upon the gelation temperature (at least between 24°C and 28°C). They have tentatively examined their data in the framework of de Gennes' percolation theory (de Gennes, 1976). This theory particularly predicts the behaviour of the gel modulus as a function of the amount of reacted bonds p. The gel point is characterized by a critical value p_c. Then the following relation should pertain near the gel point ($p - p_c \to 0$):

$$E \sim (p/p_c - 1)^t \qquad \text{for } p > p_c \tag{3.37}$$

where the exponent t is estimated to amount to 1.8 to 2 (de Gennes, 1976; Derrida *et al.*, 1983).

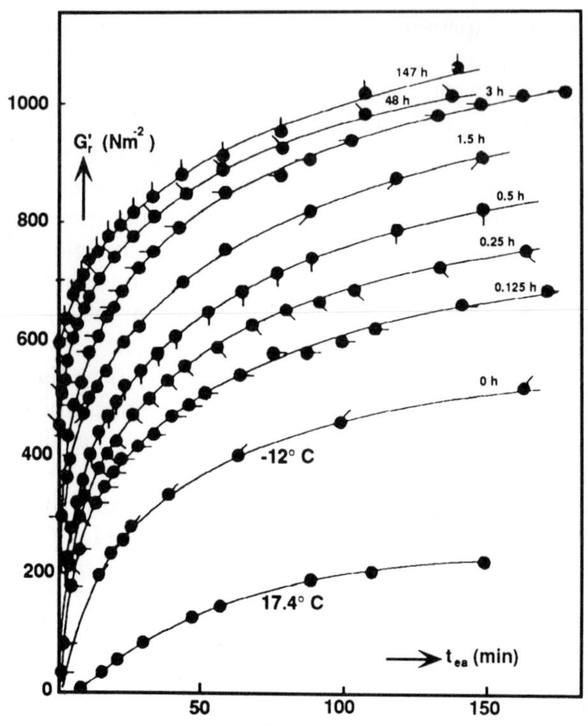

Figure 3.15 Reduced storage modulus of 1.95% (w/w) gelatin–water plotted against excess ageing time elapsed at -1.2°C after a previous ageing at 17.4°C during several pre-ageing times (as indicated). The ageing curves of -1.2°C and 17.4°C are also given; $\omega = 0.393$ rad/s and reference temperature -1.2°C (with permission from te Nijenhuis, 1981*a*).

To analyse their data, Djabourov *et al.* (1988*b*) have postulated:

$$p_c = \chi_{H_c} \tag{3.38}$$

where $6.95\% \leqslant \chi_{H_c} \leqslant 7.1\%$. With this assumption they show that E obeys relation (3.38) in the range $0.1 \leqslant (\chi_H - \chi_{H_c})/\chi_H \leqslant 0.4$ ($7\% < \chi_H < 9.8\%$) with $t = 1.82 \pm 0.15$ (Figure 3.16). In this very narrow range the modulus is augmented about tenfold.

This type of analysis relies heavily upon the precise determination of p_c. Djabourov *et al.*'s plot (Figure 3.16) emphasizes that a small variation in χ_{H_c} has a considerable effect on the slope.

Higgs and Ross-Murphy (1990) consider that the actual number of 'cross-links' is a very complex function of concentration, temperature and time. They further argue that the power laws observed experimentally may well arise from mutual cancellation of different effects. Do gelatin gels really form through a percolation mechanism or are the exponents simply

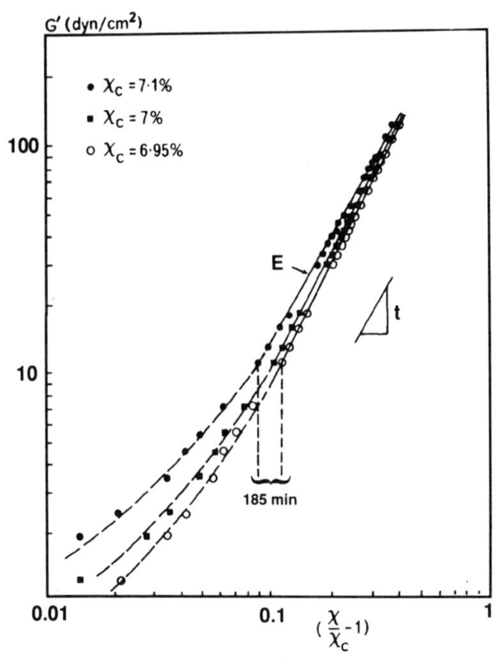

Figure 3.16 Double logarithmic representation of the shear modulus G' as a function of the reduced degree of helicity ($\chi - \chi_c/\chi_c$ in which χ_c is the critical helix content at which gelatin sets in) for gelation–water gels. Different values of χ_c are tested (as indicated). The slope $t = 1.82$ (with permission from Djabourov *et al.*, 1988*b*).

fortuitous? These may remain unanswered questions for some time. Alternatively, this might not be the essential question at the moment as long as the molecular structure is not better elucidated.

Creep measurements by Higgs and Ross-Murphy (1990) have allowed determination of the variation of the instantaneous compliance $J(0)$ with concentration. Their experiments have been achieved on gels prepared at 15°C and aged for 19 h at that temperature for concentrations ranging from 2.4% to 15%.

For $J(0)$ they have obtained by fitting their data with a power law:

$$J(0) \approx C^{-1.7 \pm 0.1} \tag{3.39}$$

In principle, $J(0) \approx G^{-1}(0)$ and thus both quantities should show the same variation with concentration. As G varies as C^2, one would expect the same for $J(0)$. Yet, according to Higgs and Ross-Murphy, the discrepancy between 1.7 and 2 is not significant in the light of the scatter of their experimental data.

Also, they have measured ΔJ, which is the difference between $J(0)$ and the intercept of the linear region of $J(t)$ observed at large t values extrapolated to $t = 0$. This quantity may also be expressed as a function of concentration with a power law:

$$\Delta J \approx C^{-1.1} \tag{3.40}$$

ΔJ decreases thus less rapidly with concentration than $J(0)$. This Higgs and Ross-Murphy interpret by considering the effect of disentanglements of dangling ends as in chemical gels. It remains to be defined precisely what are dangling ends (dangling chains?) and what is their relevance to the mechanical behaviour of a network which, based upon morphological evidence, is rather fibre-like with a mesh size much larger than the coil dimension.

As already stressed above, the lack of sufficient knowledge of the molecular structure inevitably leads to speculations based on concepts that, although relevant to other systems, do not necessarily apply to these gels.

3.3.2 Polysaccharides

(a) Agarose

Agarose gels display considerable relaxation, as was noticed by Pines and Prins (1973), the rate of which is strain dependent. At the early stage of relaxation, the stress is roughly proportional to $t^{-0.4}$. The system seems to reach an equilibrium state after a few hours.

Pines and Prins consider that the fast relaxation arises from the breaking of the weakest gel junctions while the strongest that remain are responsible for the steady state.

The variation in the modulus G' as a function of agarose concentration has been determined by Watase *et al.* (1989) in the range 3–30% (w/w) at a constant frequency of 2.5 Hz. Their results, once replotted in a double logarithmic scale, show that $G' \approx C^2$ as with gelatin gels, confirming earlier results obtained by Smidsrød (1972). Watase *et al.* use Clark and Ross-Murphy's equation (3.37) to fit these data. Yet, knowing that agarose gels consist essentially of intertwined, rigid fibres, it seems simpler to use Jones and Marques' (1990) relation (2.11). This allows prediction of the correct value of the exponent with a minimum of assumptions and by taking into account the actual structure of the network.

Watase *et al.* have also investigated the modulus behaviour with temperature. They found it to increase slightly (about 10% or less) up to $T \simeq 35\,^\circ\mathrm{C}$ and then to decrease rapidly (their measurements were carried out up to 65°C). They discuss this result in the light of a hypothetical model where flexible chains connect crystalline junction zones, a model somewhat in contradistinction with the fibrillar morphology. One may wonder whether the modulus decrease might not be accounted for by partial melting, as has been discussed for block copolymers (see Section 3.2.2.c).

(b) Carrageenans

As can be expected, the mechanical properties of carrageenans depend upon the counter-ion concentration. Rochas and Landry (1988) have shown that the modulus varies with the square of the polymer concentration with a front factor increasing with salt (KCl) concentration (Figure 3.17):

$$\left.\begin{aligned} E &= 0.1994\, C^{2.02} && 0.05\ \text{M} \\ E &= 0.389\, C^{1.98} && 0.1\ \text{M} \\ E &= 0.745\, C^{1.95} && 0.25\ \text{M} \end{aligned}\right\} \tag{3.41}$$

It should be noted that, in the case of high salt concentration, there appears to be a significant departure from the square dependence at high polymer concentration. Again, the square dependence can be simply accounted for by the fibrillar morphology of these gels.

Rochas *et al.* (1990) have extensively studied the effect of molecular weight of carrageenan samples on both the compression modulus and the yield stress, also determined by compression. They have found that the modulus is molecular weight dependent up to a certain critical value of about $M_c \simeq 10^5$. This value does not depend upon salt concentration (KCl) (Figure 3.18).

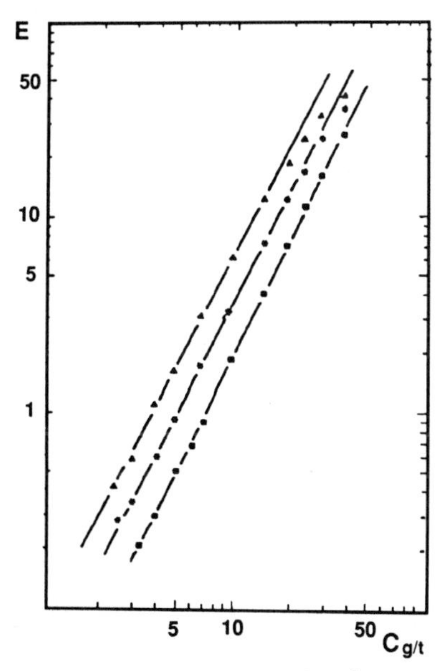

Figure 3.17 Variation of the elastic modulus (10^5 dyne/cm^2) as a function of polymer concentration for $\varkappa$-carrageenan gels in water with different salt concentration (KCl): $\triangle$ = 0.25 M, $*$ = 0.1 M and $\blacksquare$ = 0.05 M (with permission from Rochas and Landry, 1988).

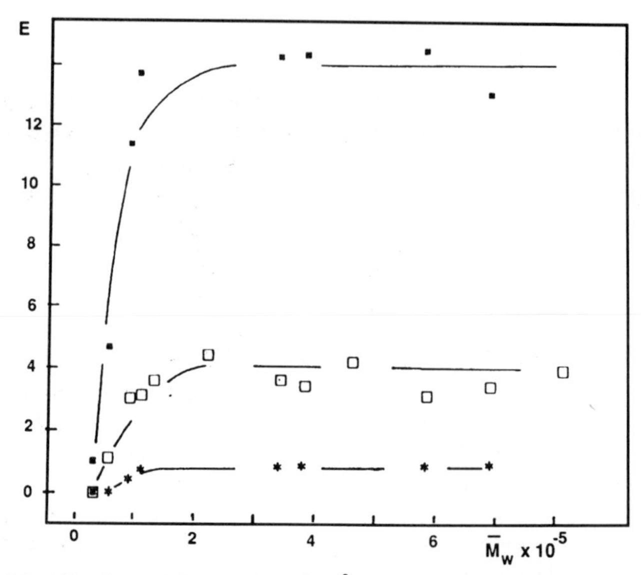

Figure 3.18 Elastic modulus E (dynes/cm^2) as a function of molecular weight for gels of different fractions of κ-carrageenans in 0.1 M KCl. Concentrations: $* = 0.5 \times 10^{-2}$ g/cm^3, ■ = 10^{-2} g/cm^3, □ = 2×10^{-2} g/cm^3 (with permission from Rochas *et al.*, 1990).

Alternatively, the yield stress varies as the molecular weight, which has been postulated by Mitchell (1980) for polysaccharides. Whereas the modulus is markedly dependent upon salt concentration, the yield stress does not significantly depend on it in the range 0.1–0.05 M. At higher salt concentration, the yield stress starts to decrease.

(c) Amylose and amylopectin

Ellis and Ring (1985) have carried out a systematic study of the mechanical properties of amylose–water gels as a function of amylose concentration and temperature.

Amylose gels do not obey the usual square dependence found with common biopolymers. Instead, Ellis and Ring report a seventh power relation for gels matured for 24 h at 20°C (in fact as is apparent from Figure 3.19, the relation is $E \sim C^{5.45}$). For gels aged for 15 h, Doublier and Choplin (1989) have further shown that gel rigidity depends upon the ionic force: gels are stiffer in 0.2 M-KCl than in 0.5 M-KCl.

Ellis and Ring (1985) have found that the modulus decreases slightly with increasing temperature. On these bases, Ellis and Ring have seriously questioned the use of theories based on rubber elasticity to account for the mechanical properties of amylose gel.

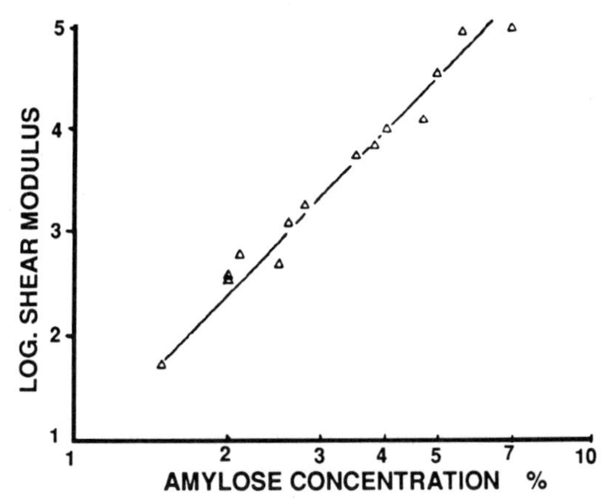

Figure 3.19 Double logarithmic plot of the shear modulus of pea amylose aqueous gels as a function of amylose concentration (with permission from Ellis and Ring, 1985).

Clark *et al.* (1989) have found similar results which slightly depend upon the degree of polymerization:

$$\left.\begin{array}{ll} E \sim C^{5.9} & \text{for } DP = 1100 \\ E \sim C^{4.7} & \text{for } DP = 660 \\ E \sim C^{4.4} & \text{for } DP = 300 \end{array}\right\} \quad (3.42)$$

These authors claim, however, that these experimental results can be accounted for with the model developed by Clark and Ross-Murphy (1985) (see relation (3.36)) based on rubber elasticity. To do so they have to consider very low functionalities ($f = 3$ to 10). They have obtained an approximate agreement in a rather narrow range of concentrations (from about 1% to 3%).

In view of the amylose gel morphology (see Section 2.2.2.c.ii), which proves to be rather fibre-like with a mesh of micrometric size (≈ 1 μm, i.e. far larger than the chain dimension), one may wonder what is the meaning of such low functionalities and accordingly question the use of this theory. (In Clark and Ross-Murphy's theory, the functionality is the number of chains merging at the same 'cross-link'.)

Amylose gels may not behave as common biopolymer gels, but similar types of variation are found with synthetic polymer gels (block copolymers see Section 3.2.2.c). If the amylose molecule is partially crystallized in such a way as to resemble a multiblock copolymer made up of crystallized sequences alternating with non-crystallized ones, then one may apply relation (3.33). Now, if one considers further that the value of a in relation (3.33) varies as the mean pore radius (relation (2.6)), one would obtain for amylose gels:

$$E \sim C^{5.66} \tag{3.43}$$

a variation in good agreement with the experimental results.

However, the exponents, although similar to those measured with block copolymers, may not originate from the same phenomenon. It seems that a knowledge of the temperature–concentration phase diagram for amylose–water gels could be invaluable in throwing light on the origin of this high exponent (see Appendix 4).

Amylopectin gels also display an unusual modulus–concentration behaviour. In contrast with the high exponent found for amylose gels, Ring *et al.* (1987) report a linear relation between modulus and concentration in the range 10–25% for gels stored for 6 weeks at 1°C (Figure 3.20). So far, amylopectin gels represent the only case for which there exist no theories that predict such an exponent (the lowest exponent predicted is 3/2 (see Appendix 4)). In fact, this concentration-dependence is not even a strict power law variation, as extrapolation to $C = 0$ yields a value differing from 0 (the actual law is of the type: $E = kC - E_0$).

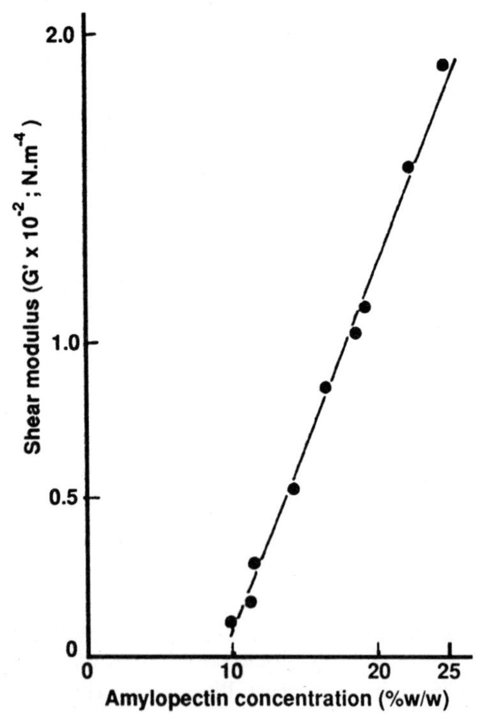

Figure 3.20 Shear modulus *vs* amylopectin concentration (linear scale) in waxy-maize amylopectin gels (with permission from Ring *et al.*, 1987).

One should bear in mind that the chief difference between amylose and amylopectin is the presence of branching in the latter. Offhand, one would have thus expected amylopectin gels to exhibit close rubber-like behaviour which is apparently not the case. The huge effect of branching on the mechanical properties emphasizes the important role played by the chemical architecture.

(d) Gellan

Mechanical properties of gellan gels are strongly influenced by the ionic strength as well as by the nature of the ions present in solution. The effect of ionic strength is illustrated in Figure 3.21 where results obtained by Grasdalen and Smidsrød (1987) are plotted as a function of the molarity of KCl salt. As can be seen, modulus reinforcement can reach one order of magnitude.

Like most biopolymer gels, the modulus of rigidity of gellan gels varies as the square of the concentration. Yet, the front factor depends strongly on the counter-ion (Figure 3.22). With monovalent cations the strength increases with atomic number ($Li^+ < Na^+ < K^+ < Cs^+$) while there is little difference with alkaline-earth metal ions. The transition elements give stronger gels with modulus increasing in the order $Zn^{2+} < Cu^{2+} < Pb^{2+}$. The largest modulii are obtained, however, with HCl. A comparison between different ions is given in Table 3.1 for gels prepared at 2%.

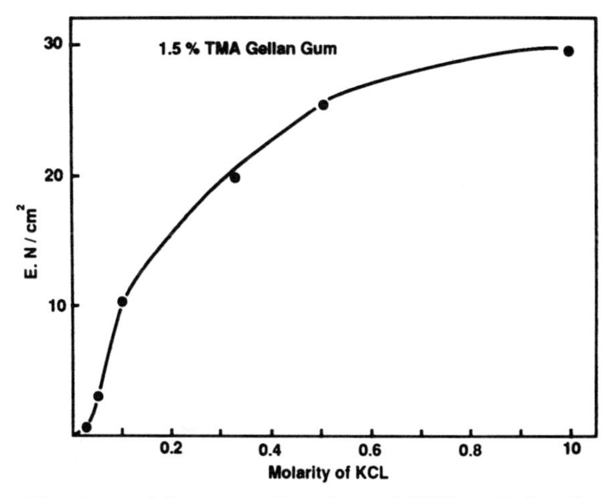

Figure 3.21 Elastic modulus as a function of KCl molarity for 1.5% (w/v) aqueous gellan gels. Temperature = 22 ± 2°C (with permission from Grasdalen and Smidsrød, 1987).

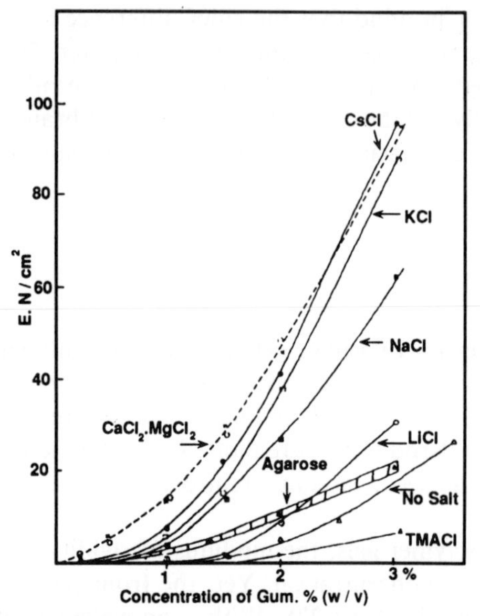

Figure 3.22 Elastic modulus as a function of gellan concentration in aqueous gellan–water gels (hatched area represent the variation of agarose gels). △ = native gellan gum (i.e. no salt addition); ▲ = 0.1 M TMACl; ◯ = 0.1 M LiCl; ■ = 0.1 M NaCl; □ = 0.1 M KCl; • = 0.1 M CsCl; ○ = 0.033 M $MgCl_2$ and x = 0.033 M $CaCl_2$ (with permission from Grasdalen and Smidsrød, 1987).

Table 3.1 Comparison of the modulus of rigidity for 2% gellan–water gels in different salt environments for an ionic strength $I = 0.1$ mol/l. After Grasdalen and Smidsrød (1987). (TMA = tetramethylammonium).

Salt	Modulus (kPa)	Molarity (M)
TMACl	18	0.1
No	49	—
LiCl	87	0.1
NaCl	267	0.1
KCl	377	0.1
CsCl	412	0.1
$CaCl_2$	480	0.033
$MgCl_2$	455	0.033
$SrCl_2$	385	0.033
$BaCl_2$	429	0.033
$ZnCl_2$	500	0.033
$CuSO_4$	570	0.025
$Pb(NO_3)_2$	650	0.033
HCl	1300	0.1

(e) Cellulose derivatives

Mechanical properties of cellulose derivative gels have usually been limited to testing phenomenological theory for semi-cystalline systems (Ninamiya and Ferry, 1967) or to determining ultimate properties such as yield stress (Sarkar, 1979).

Sarkar (1979) has observed in the case of methylcellulose that the gel strength of 2% aqueous gels (at 65°C) is molecular weight dependent up to $M_w \simeq 10^5$ and then remains virtually independent of this parameter. The degree of hydroxypropyl substitution influences considerably the gel strength. A substitution of 0.15 mol produces a fivefold drop in gel strength. Also, the presence of additives such as salt alter the mechanical properties of the gel. The gel strength of methylcellulose gels can increase by about 40% by adding 6% NaCl to the solution.

Hermann (1965) has studied the variation of the gel modulus of carboxymethylcellulose as a function of concentration. Gels prepared from a high-molecular-weight sample ($M_w = 3.7 \times 10^5$) rapidly reach the square dependence ($G \approx C^2$ from about 3 g/100 cm^3 to about 20 g/100 cm^3 whereas deviation is seen between about 1.2 g/100 cm^3 and 3 g/100 cm^3). For samples prepared from lower-molecular-weight carboxymethylcellulose, the square dependence is either reached at higher concentrations ($M_w = 10^5$) or not reached at all ($M_w = 4.6 \times 10^4$) (Figure 3.23). It is worth mentioning that the three carboxymethylcellulose samples used by Hermans do not possess the same degree of substitution so that they are not strictly comparable.

3.3.3 Gels from blends of biopolymers

A large variety of mixtures of biopolymers give homogeneous solutions in water above a well-defined temperature, unlike synthetic polymers which usually exhibit liquid–liquid phase separation in organic solvents under similar conditions. As a consequence, investigations have been carried out on the mechanical properties of aqueous gels prepared from such blends (see for instance Dea, 1979; Clark *et al.*, 1983; Zhang and Rochas, 1990). Such studies have revealed in many instances the existence of synergistic effects.

Clark *et al.* (1983) have investigated blends of gelatin and agar and observed microscopic phase separation. Depending upon the concentration of gelatin, the agar-rich phase is the supporting phase ($C_{gelatin} \leqslant 1\%$) or forms inclusions in the gelatin-rich phase. In all cases, the resulting modulus is higher than for pure gelatin but never exceeds that of pure agar. Clark *et al.* account for these results by using a theory developed by Takayanagi *et al.* (1963) for composite materials which fixes the upper boundary and

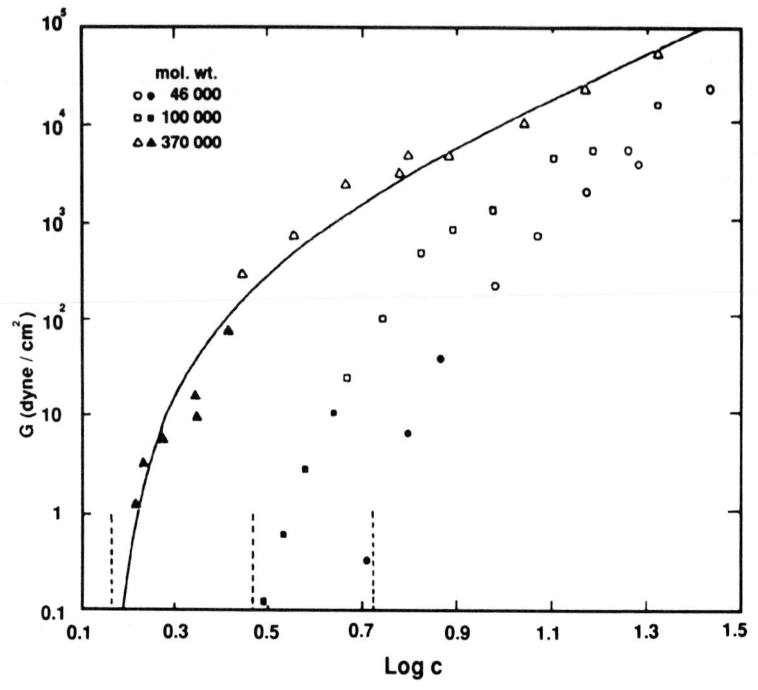

Figure 3.23 Shear modulus of aqueous gels of three samples of carboxymethyl cellulose as a function of polymer concentration (expressed here in g/100 cm^3). Molecular weight as indicated. The curve drawn is derived from a theory of the type given in relation 3.36. Solid and hollow symbols represent data obtained with two different types of experimental set-up (with permission from Hermans, 1965).

the lower boundary of the expected variation of the modulus through:

$$G = \Phi_1 G_1 + \Phi_2 G_2 \qquad \text{for the upper boundary} \tag{3.44}$$

$$G^{-1} = \Phi_1 G_1^{-1} + \Phi_2 G_2^{-1} \qquad \text{for the lower boundary} \tag{3.45}$$

in which Φ_1 and Φ_2 are the volume fractions of phase 1 and 2 respectively ($\Phi_2 = 1 - \Phi_1$).

Zhang and Rochas (1990) have reported an investigation on agarose + $\varkappa$-carrageenan blends. They have observed that the resulting modulus, E_B, is, to a very good approximation, a linear combination of the modulus of each biopolymer gel measured at the same concentration:

$$E_B(C_{Ag} + C_\varkappa) = E_{Ag}(C_{Ag}) + E_\varkappa(C_\varkappa) \tag{3.46}$$

In which C_{Ag}, $C_\varkappa$ and $C_{Ag} + C_\varkappa$ are the concentrations of agarose, $\varkappa$-carrageenan and the blend respectively. From this, Zhang and Rochas suggest that this type of gel probably consists of interpenetrated networks of agarose and $\varkappa$-carrageenan. This assumption is backed up by their optical rotation measurements which demonstrate the independent coil–helix transition of $\varkappa$-carrageenan first and then, at lower temperature, of agarose.

In some cases the mixing of a gelling biopolymer with a biopolymer that does not gel on its own can significantly improve the mechanical properties but also allows the gelation phenomenon to proceed in more diluted

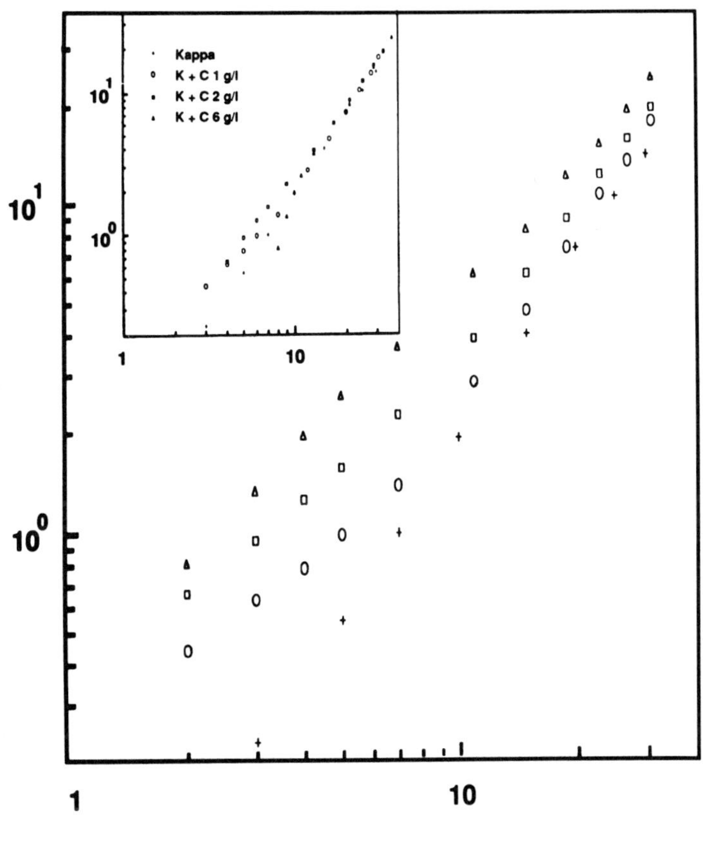

Figure 3.24 Elastic modulus as a function of polymer concentration in $\varkappa$-carrageenan + galactomannan gels in 0.05 M KCl: (+) $\varkappa$-carrageenan alone; (○) $\varkappa$-carrageenan + galactomannan (1 g/l); (□) $\varkappa$-carrageenan + galactomannan (2 g/l); (△) $\varkappa$-carrageenan + galactomannan (6 g/l). Inset represents the variation of the same modulus but this time as a function of the total polysaccharides concentration ($\varkappa$-carrageenan + galactomannan). Here, a master curve can be seen from $C_{total} = 10$ g/l up to the highest concentration, while there is significant departure from the master curve below $C_{total} = 10$ g/l (with permission from Turquois, Rochas and Taravel, submitted for publication).

solutions. Agarose + galactomannan or carrageenan + galactomannan mixtures are typical examples for which this situation is encountered (Ainsworth and Blanshard, 1978; Cairns *et al.*, 1986; Dea *et al.*, 1986; Tako and Nakamura, 1986; Damasio *et al.*, 1990; Turquois *et al.*, 1991).

Turquois *et al.* (1991) have extensively studied the system $\varkappa$-carrageenan + galactomannan by considering three different situations:

(1) The polysaccharide concentration ($\varkappa$-carrageenan + galactomannan) is kept constant.

They have found that the gel modulus first increases when the galactomannan concentration is increased, then decreases with a maximum around 35% galactomannan + 65% carrageenan. At the maximum, the modulus is about 30% higher than the modulus of $\varkappa$-carrageenan.

(2) The $\varkappa$-carrageenan concentration is kept constant.

Gel rigidity is continuously enhanced up to a galactomannan content of about 40%, but it decreases beyond. However, up to a galactomannan content of 100% the modulus is still larger than with $\varkappa$-carrageenan alone.

(3) The galactomannan concentration is kept constant.

The results are summarized in Figure 3.24. Interestingly, the variation of the modulus as a function of $\varkappa$-carrageenan concentration depends upon the galactomannan content. Below a given concentration, $C_\varkappa$, of $\varkappa$-carrageenan, which increases with increasing galactomannan content, the modulus does not follow the usual square dependence but rather a C-dependence. Conversely, above this 'critical' concentration, $C_\varkappa$, the C^2 dependence is retrieved. A master curve can also be drawn by plotting the modulus as a function of total polysaccharide concentration (carrageenan + galactomannan) (see inset, Figure 3.24). Above $C_{pol} > 10$ g/l, a master curve is obtained whereas at lower concentrations there is significant deviation. It is interesting to note that carrageenan can be replaced by a non-gelling polysaccharide without affecting the gel mechanical properties that much. Obviously, such behaviour remains to be understood at the molecular level.

4 Concluding remarks

'Mehr Licht'

von Goethe's last words

As with every domain of science, but especially such a relatively new one, there exist a certain number of well-established facts that have received sound and firm explanations, former explanations that are now seriously called into question and, finally, a large body of investigations which, together with their interpretations have still to be confirmed or invalidated. These concluding remarks will chiefly attempt to highlight what might possibly become the future trends in this field.

Investigations into the gelation mechanism will probably still constitute an important research activity. The concept of **liquid–liquid** phase separation (LLPS for short) became popular at one time to account for the phenomenon of physical gelation and it was widely taken for granted that it was the 'mechanism'. This belief probably originated in the theoretically predicted occurrence of a network structure at the early stage of phase separation, provided that the latter proceeds via spinodal decomposition. In other words, the final gel structure was thought to arise from a macroscopic phenomenon involving distances far larger than the chain molecular dimension. Clearly in this view, thermoreversible gelation can only occur in bad solvents (bad in the sense of Flory, that is with a gelation temperature below the θ-temperature) wherein some miscibility gap exists so as to yield two liquid phases. However, a number of contradictory examples have been now discovered. Polymer solutions in good solvents can also produce gels such as atactic polystyrene in carbon disulphide or PMMA in bromobenzene. Obviously, for gels to be produced some **liquid–solid** phase separation (LSPS for short) must take place, but unlike LLPS, LSPS is not necessarily governed by solvent quality, and one can create a high degree of organization, such as crystallization, in good solvents (good in the sense of Flory, that is gelation temperature located well-above the θ-temperature).

It is worth emphasizing that this questioning of LLPS as the 'gelation mechanism' occurred as a result of investigations on synthetic polymers with which a large variety of solvents were employed. This has not so far been done with biopolymers. As with synthetic polymers, it is believed that investigations with solvents other than water would be more than profitable in casting additional light on the gelation mechanism.

Investigation into the molecular structure has also revealed that, as opposed to what was inferred when considering only the LLPS mechanism, the gel structure, and ultimately the gel morphology, stem from the particular chain structure and not the reverse situation. As a rule, one succeeds in preparing fibre-like networks not because some macroscopic mechanism has stretched the chains, but because extended, rigid chains, unable to fold, are obtained in the first place. So far, only the case of polyethylene gels seems to require a macroscopic effect, i.e. the shearing of the solution.

Thermodynamic studies have thus revealed that no universal mechanism produces thermoreversible gels. Several mechanisms have been discovered and others remain to be found. The quest for universality at the level of the gelation mechanism is evidently doomed to failure. Universality does, in fact, appear at the level of the gel morphology and to some extent at the level of the chain conformation. There is no sharp distinction between synthetic polymers and biopolymers as far as morphology is concerned: in all cases, fibre-like morphology is observed. Furthermore, in all the cases that have been thoroughly investigated, it has been shown that this morphology probably arises from enhanced chain rigidity. To achieve this goal, Nature uses a number of ploys, hence the variety of gelation mechanisms: the mechanism and the conditions to form a polymer–solvent compound are different from those required to form a double helix, and so on. However, in all cases the resulting effect is the enhancement of chain rigidity. It seems that this is essential to prevent chains from folding and, consequently, from giving birth to chain-folded crystals or to spherulites.

These recent discoveries show that a comprehensive understanding of the molecular structure of the gel and, whenever achievable, of the chain trajectory within the gel matrix are of prime importance in understanding and predicting gel properties. For example, a knowledge of the gel molecular structure together with the gel morphology are necessary to account for the mechanical properties such as the stress relaxation or the modulus-concentration relation.

Paradoxically, a great deal remains to be discovered about the nature and, correspondingly, the structure of the physical junctions. Indeed, available data show the existence of organized domains in the gel, and yet there is no way that we can claim with absolute certainty that these domains are effectively the gel junctions. To illustrate this, it is worth recalling an early model proposed by Arnott *et al.* (1974*b*). In this model the physical junctions, that is to say the domains wherein chains originating from two different bundles cross one another, are regarded as amorphous (see figure). Disordered domains, provided that they are located between two organized domains, are apparently just as effective and efficient as organized junctions are assumed to be. Such a model could more than adequately explain the high relaxation rates observed for some gels.

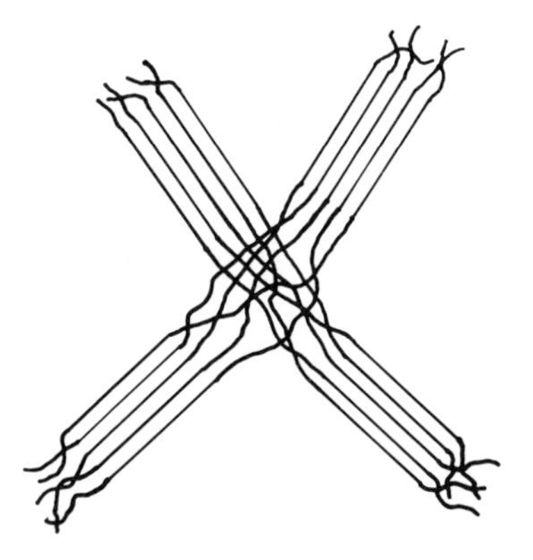

Finally, another future development will undoubtedly concern the elaboration of theories specifically suited to thermoreversible gels of the kind discussed in this book. While some existing theories primarily developed for chemical gels can be used under some conditions, they generally rely on concepts that are not always relevant to the problem.

With regard to theories, the phenomena which occur near the gelation threshold might also be of interest, although admittedly, in many systems further investigation is still needed to confirm the experimental knowledge of what is happening. In some systems such as isotactic polystyrene solutions, experiments have, already, unveiled aspects such as the difference of chain trajectory in *cis*-decalin on either side of the temperature gelation threshold (T_{gel}): worm-like below while hair-pin-like above. A theoretical approach is expected to account for why this occurs in *cis*-decalin but not in *trans*-decalin. Clearly, in the solvent-induced gels the way in which the solvent interacts with the chain to promote gel formation is expected to be modelized so as to enable one to predict the behaviour of a polymer–solvent systems.

The gelation threshold also depends on the polymer concentration at a given temperature. While in some cases this concentration gelation threshold, C_{gel}, turns out to be close to the value either determined or calculated for the chain overlap concentration, C^*, there seems little doubt that this is purely coincidental. The availability of some theoretical predictions for C_{gel} as a function of parameters such as the chain length may provide us with useful hints as to the gelation mechanism, the growth process and gel structure.

These are but a few suggestions for further studies in the field of physical gelation. There will, needless to say, be many other approaches not yet envisaged, in the years to come. In any event, thermoreversible gelation obviously constitutes a new branch of its own in polymer science with rules often differing from those of the polymer crystallization and chemical network. Conversely, the fields of synthetic polymers and biopolymers that were once believed to be far apart in the end seem to obey the same principles.

The ever-increasing number of fundamental studies and new concepts should boost the potential applications of the phenomenon of thermoreversible gelation. At present these applications are relatively numerous for biopolymers, although mainly restricted to the food industry, but still comparatively few for synthetic polymers. Hopefully, this book will give more scientists involved in fundamental or applied research the incentive to investigate this promising field.

Appendix 1

Notions on phase diagrams

In this appendix are presented some types of temperature-concentration (T, C) phase diagrams that may be encountered in physical gel systems.

In most cases, phase separation occurs during the gelation process. Two types of phase separation will be considered: **direct liquid–solid phase separation** and **indirect liquid–solid phase separation**. The latter is preceded by a **liquid–liquid phase separation**. Whereas the former may occur in good solvents, the latter only occurs with poor solvents since liquid–liquid phase separation takes place within a so-called **miscibility gap**.

Six main schematic types of phase diagram relevant to the understanding of thermoreversible gelation can be recognized for binary systems:

(1) Eutectic system with formation of a solid solution.
(2) Monotectic system with formation of a solid solution.
(3) Polymer–compound system, either congruently or incongruently melting.
(4) Peritectic system.
(5) Metatectic system.
(6) Liquid–crystalline systems with miscibility gap.

In practice, several types can be observed with the same system.

GENERAL RULES

Variance

These different types of phase diagram all obey the Gibbs phase rule. This rule gives the degree of freedom of the system, its variance v_G, and particularly the maximum number of coexisting phases for a given number of components. It is worth emphasizing that this rule holds even for systems formed under non-equilibrium conditions (Koningsveld *et al.*, 1990).

The variance, v_G, of a system may be regarded as the dimension in the $T, X_1, X_2, \ldots, X_i$ diagram of the locus where a given transformation takes place. If Φ is the number of possible phases and c the number of components of the system, then the variance reads at constant pressure:

$$v_G = c - \Phi + 1 \tag{A1.1}$$

For a two-component system ($c = 2$), if $v_G = 2$, then $\Phi = 1$ (one phase in equilibrium with itself) and this equilibrium occurs within an area of the T, X diagram. Now, if $v_G = 1$, one phase is in equilibrium with another one and the locus of the transition is a line ($T = f(C)$). Finally, for $v_G = 0$ three phases coexist, an event that can only appear on a zero-dimension locus, i.e. a point. Correlatively, three phases cannot coexist at different temperatures or concentrations even for non-equilibrium conditions without violating Gibbs phase rule.

Temperature-invariant transitions

For a two-component system, when three phases coexist there is a so-called **temperature invariant transition** which is first-order and, accordingly, produces a thermal event. It is possible to determine experimentally the temperature at which this invariant occurs, and also the composition of each of the three phases: X_T, the composition at which the three phases coexist (eutectic, peritectic, etc.), X_α, the polymer-poor phase composition and X_β, the polymer-rich phase composition. At X_T, only one endotherm will be observed corresponding to the transformation of two phases into one. If ΔH_T is the heat associated with this transition, the following relations are derived from the lever rule:

$$\text{for } X \leqslant X_T \qquad \Delta H_T(X) = \frac{|X - X_\alpha|}{|X_T - X_\alpha|} \times \Delta H_T \tag{A1.2}$$

$$\text{for } X \geqslant X_T \qquad \Delta H_T(X) = \frac{|X - X_\beta|}{|X_T - X_\beta|} \times \Delta H_T \tag{A1.3}$$

As a result, for $X \leqslant X_T$, $\Delta H_T(X)$ increases linearly with X up to X_T, then decreases for $X \geqslant X_T$. $\Delta H_T(X) = 0$ for $X = X_\alpha$ and $X = X_\beta$. As a result, plotting $\Delta H_T(X)$ *vs* X allows one to determine X_T, X_α and X_β. Such a diagram is known as Tamman's diagram (see for instance Carbonnel *et al.*, 1970; Rosso *et al.*, 1973).

So far in this section, it has been taken for granted that Gibbs phase rule cannot be violated. There are, however, at least two cases for which this rule may break down: for finely dispersed matter for which the surface free energy is no longer negligible and for kinetic reasons which appear when the completion of transformation is significantly delayed on account of diffusion impediments.

Effect of finite size

There are known examples where the finite size of the phases leads to alteration of the variance. For instance, in a pressure–temperature phase diagram the variance can be $v_G = 2$ at the triple point instead of $v_G = 0$ (Defay *et al.*, 1966; Brun *et al.*, 1977).

If one phase (minor phase) is finely divided into the other one and if its size depends upon the composition, X, then the temperature-invariant transition will no longer occur at a well-defined temperature.

The effects due to finite size can be calculated. Gibbs has shown that, if the particle or the minor phase are of finite size, then the free energy, ΔF_0, should read:

$$\Delta F_0 = S\sigma - V\Delta f \tag{A1.4}$$

in which S and V are the surface and the volume of the particle (or of the minor phase) respectively, σ the surface free energy and Δf the free energy per unit volume. For crystalline systems, relation (A1.4) may be rewritten by considering the undercooling ΔT, the temperature of fusion of a particle of infinite size, T^0_m, and the enthalpy of fusion, Δh_f. By assuming the latter quantity to be invariant with temperature, one can replace the free energy by the free energy of fusion near the melting point. One finally ends up with:

$$\Delta F_0 = S\sigma - (V\Delta h_f \Delta T / T^0_m) \tag{A1.5}$$

From this relation one can deduce the actual temperature of fusion due to the finite size of the particle:

$$T_m = T^0_m[1 - (S\sigma / V\Delta h_f)] \tag{A1.6}$$

Now, provided that both the size of the particles or the minor phase and the surface free-energy term remain constant in the whole range of existence, it is then clear from relation (A1.6) that the so-called temperature-invariant transitions will be effectively temperature-invariant. This explains why temperature-invariant transitions can still be observed with some physical gels in spite of microphase separation (see for instance He *et al.*, 1987*a*, 1988, 1989).

Kinetic effects

Kinetic effects are often observed with polymers for which high viscosities and, correlatively, low diffusion rates are encountered. This gives significant, measurable transition rates such as crystallization rates, etc. For example, this results in the coexistence of two phases in a one-component slowly crystallizing material, the amorphous phase and the crystalline phase (the crystalline phase usually consisting of crystalline lamellae alternating with amorphous domains), in spite of Gibbs phase rule which allows the presence of only one. Obviously, this two-phase state is metastable and transforms spontaneously after a finite period of time. It ought to be mentioned that, if the system lies under a glass transition, then the transformation is in practice impossible to within a reasonable period of time.

One may wonder whether a semi-crystalline polymer, in which the extra-spherulitic amorphous phase no longer exists, ought to be regarded as a two-phase system or as a single phase. In practice, it will be considered as one phase containing defects (the amorphous layers), since the amorphous material is not a free entity but is intimately linked to the crystalline part. Often a polymer chain participates in both domains which renders physical separation between either domain impossible without severe degradation.

Critical size and nucleation step

Relations (A1.4) and (A1.5) also contain the notion of critical size. In fact, it can be shown that, at the early stage of particle growth, ΔF_0 first increases and is therefore positive, which implies that the particle is unstable. By further increasing the particle's size, ΔF_0 starts to decrease to become zero for a critical size, ρ_c, given by:

$$\rho_c = \frac{V}{S} = \frac{\sigma T_m^0}{\Delta h_f \Delta T} \tag{A1.7}$$

This specifically means that, as long as the particle has not reached this critical size, it remains unstable and may accordingly vanish. As can be seen, the higher the surface free energy, the higher the critical size. Conversely, the higher the fusion enthalpy or the higher the undercooling, the lower this critical size.

If the growth proceeds via the formation of a critical nucleus, then it is designated as **homogeneous nucleation**. Conversely, if impurities play the rôle of the critical nucleus, then it is designated as **heterogeneous nucleation**. In the absence of impurities capable of initiating the crystal growth, large undercoolings can be observed.

As a rule, first-order transitions require a nucleation process, unlike second-order transitions. This accounts for why the former very often show hysteresis between formation and fusion, unlike the latter.

PHASE DIAGRAMS

As recalled by Koningsveld *et al.* (1990), a few simple rules must be followed to establish phase diagrams. Two one-phase domains must be separated by a two-phase domain. Metastability extensions of domain lines cannot enter one-phase domains.

Eutectic system

A eutectic system (Figure A1.1) corresponds to the following reaction on heating:

$$\text{solid 1} + \text{solid 2} \Rightarrow \text{liquid}$$

The eutectic transformation is defined at a given temperature and concentration, T_E and C_E. T_E is an invariant. Solid 1 is the polymer-poor phase (or solvent-rich phase) while solid 2 is the polymer-rich phase and is here a solid solution. For a semi-crystalline polymeric system, the solid solution may be viewed as crystalline regions alternating with amorphous ones, where solvent molecules are randomly interspersed in the latter.

For polymer–solvent systems, Flory (1953) has derived the equations of the liquidus lines. Making use of the general relation:

$$\frac{\Delta\mu}{T_m} = \frac{\mu(T_m) - \mu^0}{T_m} = \Delta H_m\left(\frac{1}{T_m^0} - \frac{1}{T_m}\right) \qquad \text{(A1.8)}$$

in which $\Delta\mu$ is the variation of chemical potential, T_m the new melting point due to this variation and T_m^0 the melting point for $\Delta\mu = 0$. ΔH_m is the melting enthalpy. This expression is valid provided that specific heat capacity does not vary from the solid to the liquid.

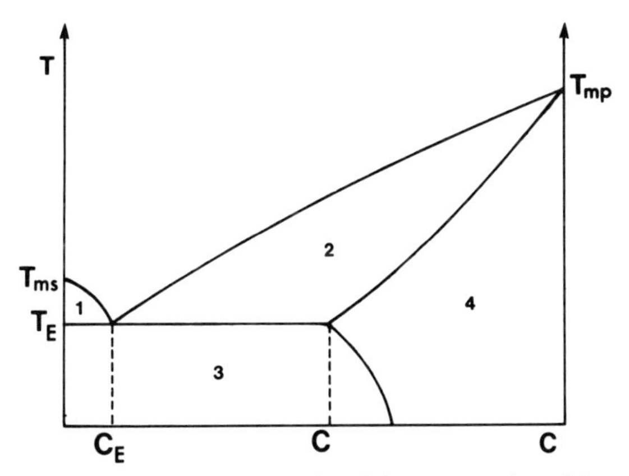

Figure A1.1 Example of phase diagram involving a eutectic melting: C_E = eutectic concentration, T_E = eutectic temperature. In 1 = crystallized solvent + liquid, 2 = solid solution + liquid, 3 = crystallised solvent + solid solution and 4 = solid solution.

The chemical potential derived for polymer and solvent in a polymer–solvent mixture from the Flory–Huggins theory are expressed as follows for long polymer chains:

$$\Delta\mu_s = RT[\ln v_s + (1 - v_s) + \chi_1(1 - v_s)^2] \quad \text{(A1.9)}$$

$$\Delta\mu_p = RT(V_p/V_s)(-v_s + \chi_1 v_s^2) \quad \text{(A1.10)}$$

in which R is the gas constant, ΔH_{0_p} and ΔH_{0_s} the polymer melting enthalpy per mol in the pure crystalline state and the solvent melting enthalpy respectively, χ_1 the Flory's interaction parameter and V_p and V_s the monomer and the solvent molar volumes respectively.

One then derives the melting point depressions of the solvent and the polymer crystals as a function of the solvent volume fraction, v_s:

for the polymer liquidus line:

$$\frac{1}{T_{m_p}} - \frac{1}{T^0_{m_p}} = \frac{RV_p}{\Delta H_{0_p} V_s}(v_s - \chi_1 v_s^2) \quad \text{(A1.11)}$$

for the solvent liquidus line:

$$\frac{1}{T_{m_s}} - \frac{1}{T^0_{m_s}} = -\frac{R}{\Delta H_{0_s}}[\ln v_s + (1 - v_s) + \chi_1(1 - v_s)^2] \quad \text{(A1.12)}$$

These relations have been applied successfully to some polymer–solvent systems by Smith and Pennings (1974), Wittmann and StJohn Manley (1977) and Klein and Guenet (1989).

In polymer–solvent systems for which the solvent melting point lies well below that of the polymer, the eutectic point shifts towards lower polymer composition.

Monotectic system

A monotectic transition is identified by the following reaction (Figure A1.2):

$$\text{liquid 1} + \text{solid} \Rightarrow \text{liquid 2}$$

As with eutectic melting, the monotectic transformation occurs at a given temperature and concentration, T_M and C_M respectively. Liquid 1 is the polymer-poor phase and solid the polymer-rich phase. A monotectic transition results from a liquid–liquid phase separation followed by liquid–solid transformation.

This case is most often considered for gelation due to the particular morphological features that are said to arise from the liquid–liquid phase separation (Cahn and Hilliard, 1959, 1965). Of particular interest is the

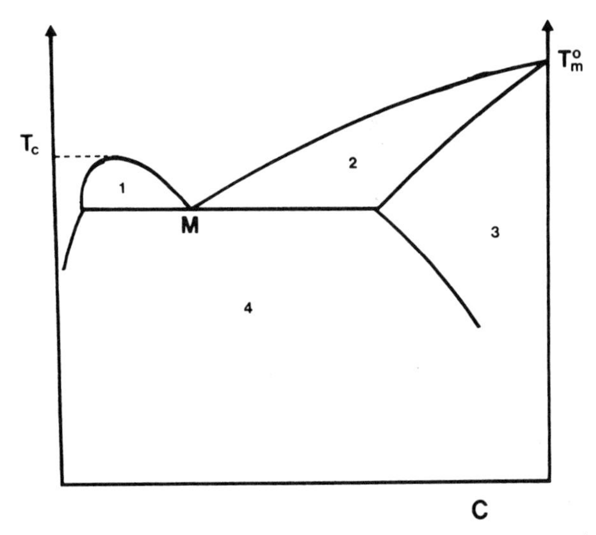

Figure A1.2 Example of a phase diagram involving a monotectic transition: T_c = critical temperature, T_m^0 = melting temperature of the pure polymer, M is the monotectic point. 1 = liquid 1 + liquid 2, 2 = liquid 2 + solid solution, 3 = solid solution and 4 = solid solution + liquid 1.

mechanism whereby phase separation occurs at its early stage. Below the miscibility gap, also termed **binodal** for homodisperse polymer, two regions are considered as defined by the second derivative of the free energy of mixing with respect to concentration:

For $\partial^2 \Delta G_m / \partial X^2 > 0$ the metastable region in which liquid–liquid phase separation proceeds via nucleation and growth.

For $\partial^2 \Delta G_m / \partial X^2 < 0$ the unstable region in which liquid–liquid phase separation takes place via diffusion and is designated as **spinodal decomposition.**

These regions are delimited by the so-called **spinodal** curve which is the locus where $\partial^2 \Delta G_m / \partial X^2 = 0$. With homodisperse polymers, the binodal and spinodal merge at the critical temperature, T_c. Below the spinodal curve, the early stage of the phase separation process is said to produce a network structure of two interconnected phases (Cahn and Hilliard, 1959, 1965). It is thought that the polymer-rich phase keeps its structure after 'freezing' by liquid–solid transformation, hence producing a permanent physical network. Such assumptions rely mainly upon morphological observations (see for instance Tohyama and Miller, 1981). However, one does not know to what extent the solidification process (liquid 1 + liquid 2 ⇒ liquid 1 + solid) alters the network structure built up by the spinodal decomposition.

On heating, part of the solid solution will melt at the monotectic temperature, T_M. There a discontinuous change of liquidus curve occurs which can be quantified by means of Flory's relation (A1.8) with two different χ_1 parameters. This therefore means that there is an abrupt change of solvent quality. Here it should be emphasized that the melting at T_M does not necessarily imply the existence of polymer crystals of lower perfection. Conversely, it is most likely that such crystals will melt, if present, at T_M.

Such types of phase diagram have been obtained by He *et al.* (1987*a*, 1988, 1989) for physical gels of a multiblock copolymer.

Polymer–solvent compound

The polymer–solvent compound is characterized by the following reactions at the stoichiometric composition, X_γ (Figure A1.3):

Compound ⇒ liquid — for a congruently melting compound where the compound and the liquid have the same composition (Figure A1.3a)

Compound ⇒ liquid + solid — for an incongruently melting compound where the liquid and the solid have both compositions, which differs from the compound (Figure A1.3b)

The intermediate case can also be found and is designated as a compound possessing a singular point (Figure A1.3c).

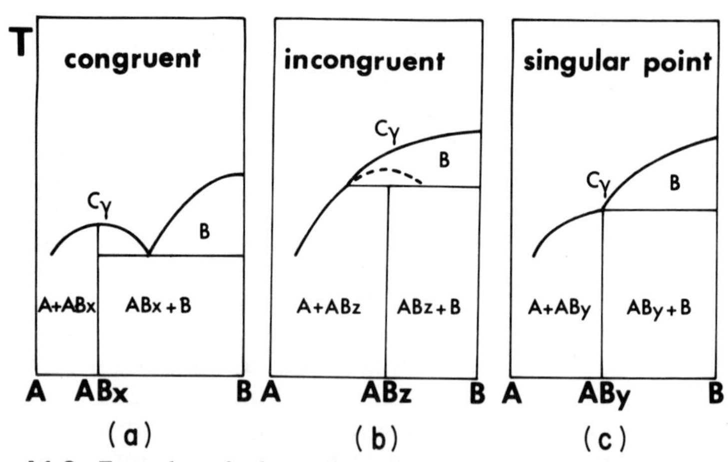

Figure A1.3 Examples of phase diagrams involving polymer + solvent compounds. C_γ = stoichiometric composition. The dotted line in 'incongruent' represents the metastable extension (with permission from Guenet and McKenna, 1988).

Also, an incongruently melting compound can show congruent melting at high heating rates, as one then observes the metastable extensions of the phase diagram (dashed line, Figure A1.3*b*).

A congruently melting compound displays the melting property of a pure component which implies intimate co-ordering of the polymer and the solvent at the molecular level such as cocrystallization.

Typical examples of temperature–concentration phase diagrams involving congruently melting polymer–solvent compounds have been established for poly(oxyethylene glycol)–*para*-halogenodisubstituted benzene (Point *et al.*, 1985) and for iPS–decalin physical gels (Guenet and McKenna, 1988).

The liquidus line located on the left-hand side of the diagram may be theoretically expressed from Flory's relation by making some assumptions. In Flory's theory, it is said that the crystalline phase is in equilibrium with a solution composed of polymer chains + solvent. Here it is assumed that the organized polymer–solvent compound gives solvent and disordered polymer–solvent compound. This can be considered as a new entity, polymer chains with bound solvent molecules, which may be termed amorphous polymer–solvent compound of the same composition as the ordered form. The molten part accordingly consists of amorphous compound + solvent. Then all one needs is to rescale Flory's relation with the solvent volume fraction, v_{cs}, corresponding to the compound stoichiometry and to consider an amorphous compound–solvent-interaction parameter, Λ_1:

$$\frac{1}{T_{m_p}} - \frac{1}{T^0_{m_p}} = \frac{RV_c}{\Delta H_{0_c} V_s} [(v_s/v_{cs}) - \Lambda_1 (v_s/v_{cs})^2] \tag{A1.13}$$

in which V_c and ΔH_{0_c} are the molar volume and the heat of fusion, respectively, associated with the polymer–solvent compound.

Obviously, if on melting, the compound gives polymer chains on the one hand and solvent on the other hand, the above relation is not appropriate since one has to derive the chemical potential of the solvent located in the ordered compound which does not appear as a trivial calculation.

As regards incongruently melting compounds, the final liquidus line can simply be expressed through Flory's relation (A1.8).

Peritectic system

A peritectic melting (Figure A1.4) is seen when:

$$\text{solid 1} \Rightarrow \text{liquid} + \text{solid 2}$$

with $X_{\text{solid 1}} < X_{\text{solid 2}}$.

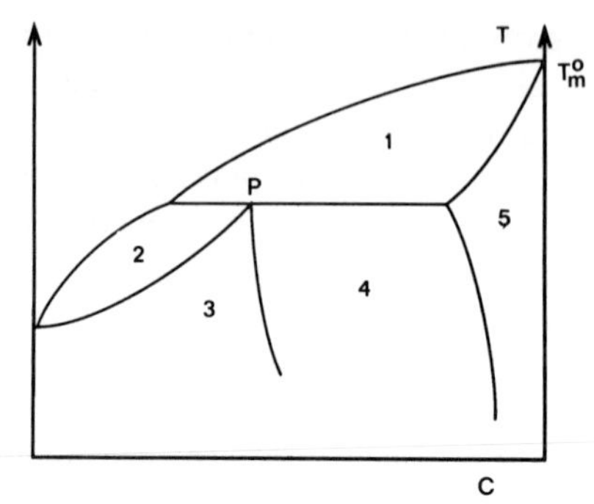

Figure A1.4 Example of phase diagram involving a peritectic transition: P is the peritectic point; T_m^0 is the melting of the pure polymer. 1 = solid solution + liquid, 2 = liquid + peritectic, 3 = peritectic domain, 4 = solid solution + peritectic, 5 = solid solution.

The incongruent melting of a polymer–solvent compound is a limiting case of peritectic reaction.

Metatectic system

This type of phase diagram can be observed when a crystalline system possesses two crystalline forms, one form being more stable than the other, at low temperature (Figure A1.5). The reaction at the metatectic point, T_{MT}, is:

$$\text{liquid} + \text{solid } 1 \Rightarrow \text{solid } 2$$

with $X_{\text{solid }1} > X_{\text{solid }2}$.

PVC gels may fit into either peritectic or metatectic systems.

Liquid–crystalline systems with miscibility gap

As pointed out by Flory (1956), rigid polymers possessing a rod-like structure ought to show special behaviour in solution. For one thing the rod-like structure is liable to induce ordering in concentrated solutions. This led Flory to propose the diagram drawn in Figure A1.6. This diagram exhibits a so-called chimney which delimits the dilute isotropic phase from the semi-dilute ordered phase. Below there is a miscibility gap with a kind of monotectic transition at T_L, where:

$$\text{liquid 1 (isotrope)} + \text{liquid 2 (anisotrope)} \Rightarrow \text{liquid 3 (anisotrope)}$$

It must be stressed that liquid 2 (anisotrope) may well be a solid phase in the case of thermoreversible gels. As with the diagram displaying a miscibility gap, spinodal decomposition is expected to be possible and is

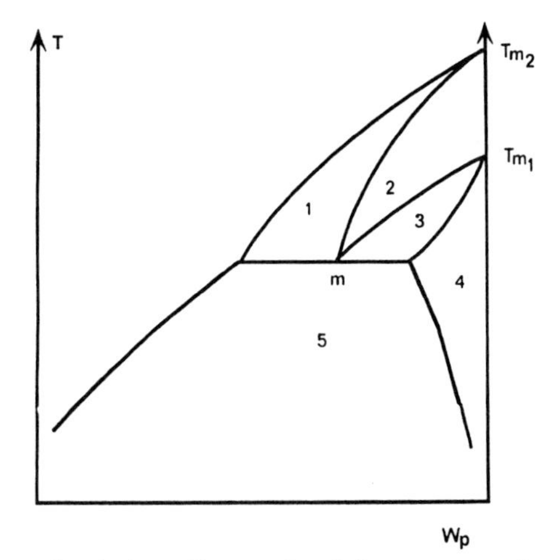

Figure A1.5 Example of phase diagram involving a metatectic transition: m is the metatectic point, T_{m_1} represents the transformation of one crystalline form, stabler at lower temperature (form 1), into another, stabler at higher temperature (form 2), while T_{m_2} represents the final melting temperature of form 2. 1 = liquid + solid solution of form 2, 2 = solid solution of form 2, 3 = solid solution of form 2 + solid solution of form 1, 4 = solid solution of form 1 and 5 = liquid + solid solution of form 1.

much invoked in these systems (see for instance Tohyama and Miller, 1981).

Flory (1956) has given an expression for the critical volume fraction v_2^* at which the chimney should begin:

$$v_2^* \approx (8/x)[1 - (2/x)] \tag{A1.14}$$

in which x is the axial ratio.

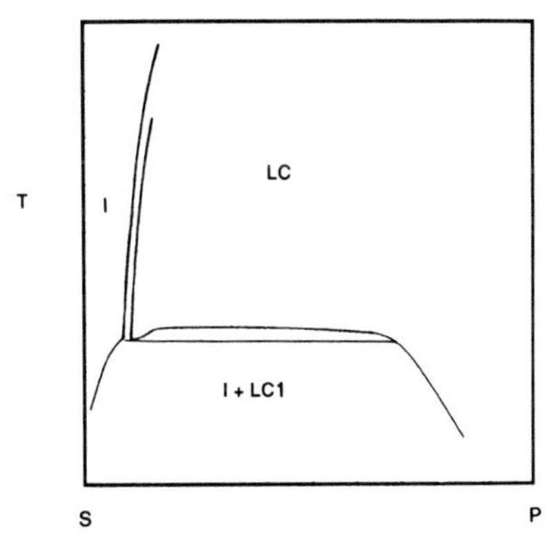

Figure A1.6 Example of a phase diagram involving a liquid crystalline polymer with miscibility gap: I = isotrope domain, LC = liquid crystalline domain, in the miscibility gap = I + LC1, in the chimney = I + LC2.

THERMAL ANALYSIS

Thermal analysis, such as differential scanning calorimetry, is the most convenient tool for establishing phase diagrams. Very often some conclusions are drawn from the shape of the endotherm that are not necessarily correct.

Here, typical endotherms that are expected with a simple eutectic system are portrayed in Figure A1.7. Their shapes are actually valid whatever the system. As can be seen, the transition at the eutectic temperature is quite narrow whereas the rest of the melting is broader and ends abruptly at the liquidus line. The broadness of the second endotherm depends upon the difference between the eutectic temperature and the melting temperature of the pure component. The latter also displays a narrow melting.

It is often concluded with polymers that broad melting endotherms indicate a broad range of crystal perfection. While this statement is true with a single component, it is not necessarily accurate with two components or more as exemplified in Figure A1.7. In particular, the lower the preparation temperature, the broader the melting endotherm. Still, for a population of identical or nearly identical crystals the endotherm displays high asymmetry, in that a sharp decrease follows its maximum, a pattern that is less likely to be encountered with a population of crystals differing in perfection.

Finally, exotherms recorded on cooling are always narrower than the corresponding endotherms obtained on heating. This can be illustrated, for example, by comparison in a polymer solution of the solvent crystallization exotherm and the solvent melting endotherm, the latter being several times broader than the former (Klein and Guenet, 1989).

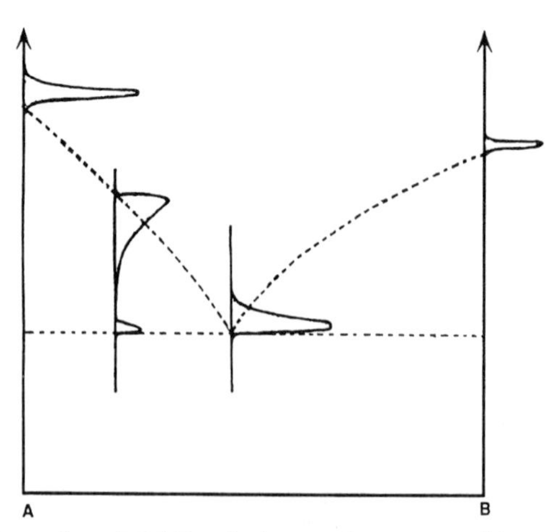

Figure A1.7 Examples of DSC endotherms that occur with an eutectic system.

ORDER OF THE TRANSITION

In most cases gel melting or gel formation proceeds via a first-order transition. Very often the Eldridge–Ferry relation (Eldridge and Ferry, 1954) is used to describe the variation of the gel melting point as a function of polymer concentration. Such a use is irrelevant, as this theory, based on a chemical equilibrium, concerns only transitions of second or higher orders (actually, it also works for first order transitions in ideal systems which cannot be the case for polymer–solvent mixtures). Originally, the Eldridge–Ferry relation was developed at a time when physical gel junctions, or even crystalline polymers and crystalline domains in general, were thought to be tiny enough to be regarded as point-like (fringed-micelle model). This misconception did not escape Flory's attention who wrote: 'it (the fringed-micelle model) unfortunately has engendered the collateral assumption that boundaries between crystalline and amorphous regions ... are inherently so diffuse as to negate consideration of two distinct phases. Melting in polymers was thus placed outside the purview of thermodynamic theory of phase changes' (Flory, 1961).

For a first-order transition, a discontinuity of the order parameter, m_T, occurs at the transition. In other words, the value of m_T drops from m_0 to 0 at the transition temperature. This also means that the derivative dm_T/dT is infinite at the transition temperature. Conversely, for a second-order transition, the order parameter decreases continuously to reach 0 at the transition temperature. Similarly, the derivative dm_T/dT always has a finite value.

Another property differentiates first-order from second-order transitions: undercooling or superheating phenomena often take place in the former whereas they do not in the latter. This means that first-order transitions proceed via nucleation and growth unlike second-order transitions (for further reading see for instance Landau and Lifshitz, 1967*a*).

The Eldridge–Ferry approach consists of defining an equilibrium constant K_{EF} which is the ratio between the number of actual junctions and the number of potential junctions. There is no mention of a two-phase system but the system is believed to transform continuously from the sol to the gel state. K_{EF} is assumed to be proportional to the polymer concentration. Then, applying van t' Hoff's equation, one ends up with:

$$\partial \log K_{EF} = \Delta H/\mathbf{R}(\partial \log 1/T) \tag{A1.15}$$

Integrating (A1.9) leads to:

$$K_{EF} = K_{EF_0} \exp \Delta H/\mathbf{R}\left(\frac{1}{T} - \frac{1}{T_0}\right) \tag{A1.16}$$

in which T_0 is the melting temperature.

K_{EF}, which is proportional to the concentration, is a kind of order parameter and it should thus be realized that it will not show discontinuity at T_0 nor will the derivative dK_{EF}/dT be infinite.

In fact, the Eldridge–Ferry relation does not represent the melting point variation as a function of concentration but the number of junctions as a function of temperature. This relation is simply not symmetrical.

Similarly, Flory's melting point equation (A1.7) gives the variation of the melting point with concentration, but inverting it would not give the amount of crystallites as a function of temperature. The latter quantity can only be calculated from the lever rule (Gibbs phase rule) provided that the solidus line is known experimentally or theoretically (which is seldom the case).

Finally, any time gel melting is shown or is thought to be first order, Flory's equation (A1.11) or more generally relation (A1.8) must be employed.

Incidentally, combining equations derived from a theory based on a first-order transition, on the one hand, with equations based on a second-order transition, on the other, is simply irrelevant. (Only in the case of a tricritical point is this possible, but, as is apparent from the name, it is restricted to one point so that it cannot describe gel melting over a domain of concentration.)

Appendix 2

Diffraction by helices

A helix is a three-dimensional object defined by the following equations in cylindrical polar coordinates (r, φ, z):

$$\left.\begin{aligned} r &= r_{\mathrm{H}} = \text{constant} \\ z &= P\varphi/2\pi \end{aligned}\right\} \tag{A2.1}$$

in which P represents the pitch of the helix and r_{H} its radius. If P is positive, the helix is said to be right-handed and, correspondingly, left-handed if P is negative.

The following mathematical derivation concerns helices that are all oriented in one direction, perpendicular to the incident beam.

Continuous helix

In cylindrical polar coordinates, the structure factor reads:

$$F(R, \Psi, \zeta) = \int_V \rho(r, \varphi, z) \exp[2\pi i(Rr \cos(\varphi - \Psi) + \zeta z)] \, \mathrm{d}v \tag{A2.2}$$

in which $\rho(r, \varphi, z)\mathrm{d}v = \text{const} \times \mathrm{d}\varphi$ for a helix. Since a helix possesses periodicity along the z-axis, the Fourier transform will differ from zero only when $\zeta = n/P$ where n is an integer. After some algebraic manipulation, it can be shown that the diffracted intensity is:

$$I(R, n/P) = \langle F(R, \Psi, \zeta) F^*(R, \Psi, \zeta) \rangle = J_{\mathrm{n}}^2(2\pi R r_{\mathrm{H}}) \tag{A2.3}$$

in which J_n is a Bessel function of first kind and order n.

Discontinuous helix

In practice, however, helices are made up of a collection of atoms and are, accordingly, discontinuous. A discontinuous helix can be regarded as the product of a continuous helix of pitch P and radius r_H and a set of planes perpendicular to the z-axis and spaced h apart, where h is the axial rise (Cochran *et al.*, 1952). The axial rise is the distance between the projections on to the z-axis of identical scatterers lying on two consecutive units. In other words, the axial rise is directly proportional to the length of the helix unit. One can then define a screw-axis h/P which can also be expressed in terms of integers u/v (u is always taken as positive while v is positive for a right-handed helix and negative for a left-handed one). A helix is then often defined as u units occurring in v turns.

Under these conditions it can be shown that the Fourier transform is zero everywhere except for a set of planes defined by:

$$\zeta = (m/h) + (n/P) \tag{A2.4}$$

in which m and n are integers. These planes are expressed in terms of layer lines with index l ($l = vP\zeta$):

$$l = um + vn \tag{A2.5}$$

It should be noted that $vP = c$ represents the distance over which the structure repeats itself. In contrast with the continuous helix, relation (A2.5) implies that for a given l, there exist several different values of n. As a result, for a discontinuous helix, after averaging F over Ψ, the intensity is finally expressed as:

$$I(R, l/c) = \sum_n J_n^2(2\pi R r_H) \tag{A2.6}$$

where the summation is carried out on all the integral values of n, satisfying relation (A2.5). (Here n must not be confused with the layer line in A2.3.)

Klug *et al.* (1958) have shown that equation (A2.6) can be rewritten in a more generalized form which takes into account that all atoms do not possess the same diffraction power. Then the diffracted intensity reads:

$$\langle I(R, l/c) \rangle_\Psi = \sum_n G_{n,1} G_{n,1}^* \tag{A2.7}$$

in which, once the atomic scattering factor f_j of atom j is introduced, $G_{n,1}$ reads:

$$G_{n,1} = \sum_j f_j J_n(2\pi R r_{H_j}) \exp[i(-n\varphi_j + 2\pi l z_j/c)] \tag{A2.8}$$

Provided that atomic structure factors are not too different (which can be the case with most deuterated polymers if neutron diffraction is used) or that the atomic structure factor of one type of atom predominates (which is the case with X-ray diffraction in most polymers for instance for which $f_C \gg f_H$), relation (A2.6) is usually a good approximation.

If helices are oriented along the z-axis, then spots of diffraction will be observed according to the rules given by relations (A2.5) and (A2.6). Interestingly, only the Bessel function of zero order displays a maximum at $R = 0$. The consequence of this property is that the first meridional reflection (located directly on the z-axis) will occur for a value of l for which $n = 0$ is possible. This is true for $l = um$, that is for $l = u$ for the first one (except $l = 0$ which is in the direct beam). Other layer lines for which $l \neq u$ will show spots off the z-axis.

To give a practical example, consider two helices: 3_1 ($u = 3$ and $v = 1$) and 7_2 ($u = 7$ and $v = 2$). For the 3_1 helix, relation (A2.5) is:

$$l = 3m + n \tag{A2.9}$$

To have $n = 0$ entails $l = 3m$, which means that the first meridional reflection will occur at $l = 3$.

For the 7_2 helix, relation (A2.5) is:

$$l = 7m + 2n \tag{A2.10}$$

Here, $n = 0$ entails $l = 7m$ which means that the seventh layer line will have a meridional spot. However, it can be shown that, except for $l = 7$, all other odd layer lines satisfy relation (A2.5) with large values of n. As a result all the odd layer lines with spots off the z-axis will show only weak diffraction. Relation A2.6 indicates that the first intense layer line will occur for $l = 2$, that is more generally for $l = v$ as $n = 1$ is allowed (with $m = 0$). As a result, the type of helix (the ratio u/v) can be determined from the ratio of the number of the layer line containing the first meridional reflection to the number of the first intense layer line with spots off the z-axis (in present case 7/2).

Helices in a crystalline lattice

These are simple cases. When the helices are included in a crystalline lattice, some layer lines may turn out to be absent because of the symmetry of the crystal.

In this case the structure factor reads:

$$F(R, \Psi, l/c) = \sum_q \sum_n G_{n,1} \exp[in(\Psi + \pi/2 - \varphi_q] \times \exp[2i\pi(hx_q + ky_q + lz_q)] \tag{A2.11}$$

in which x_q, y_q and z_q are the fractionate coordinates of the qth macromolecule of the crystalline unit cell.

For example, isotactic polystyrene crystals, although made up of 3_1 helices, do not show any reflection at $l = 3$ since the space group ($R3c$) implies $l = 2n$.

Complex helices

The case of a coiled-coil, where a simple helix of radius r_{H_1} repeats itself after N_1 turns and the major helix of radius r_{H_0} ($r_{H_0} > r_{H_1}$) repeats itself after N_0 turns on a repeat distance c, can be very complex. If M is the number of units in the repeat distance and if M and N_1 have a common factor C, then the coiled-coil may be regarded as a simple helix with M/C units. Then the selection rule is:

$$\zeta = l/c = (m/h) + (n/P_c) \tag{A2.12}$$

in which h is the axial rise of M/C units and P_c is the pitch of the major helix.

Concerning double helices, they can be thought of as single helices with a pitch $P/2$. This is the reason why a diffraction pattern can be ambiguous and one then needs to supplement the investigation with conformational analysis to find out whether the helical conformation is energetically feasible or not. This was the case for polysaccharides which explains why there are still some controversies. This is typically the case for agarose for which Arnott *et al.* (1974*a*,*b*) dismissed single helices on the basis of their sterical infeasibility.

Appendix 3

Scattering by semi-rigid macromolecules

Once extrapolation to zero concentration has been achieved in the appropriate manner, the reduced intensity, $I(q)/C$, for diluted species reads:

$$(I(q)/C)_{C=0} = KMP(q) \tag{A3.1}$$

in which C is the concentration, K a constant including the apparatus calibration constant and the contrast factor, M the molecular weight of the species and $P(q)$ their form factor. For species consisting of N scattering centres and randomly oriented, the general expression of the form factor reads:

$$P(q) = N^{-2} \sum_{i=1}^{N} \sum_{j=1}^{N} \overline{\frac{\sin qr_{ij}}{qr_{ij}}} \tag{A3.2}$$

$P(q)$ can also be rewritten in a continuous form by using the correlation function $\gamma(r)$:

$$P(q) = \frac{\int_0^\infty \gamma(r) \frac{\sin qr}{qr} 4\pi r^2 \, dr}{\int_0^\infty \gamma(r) \, 4\pi r^2 \, dr} \tag{A3.3}$$

FORM FACTOR $P(q)$ FOR A WORM-LIKE CHAIN

The first studies on the form factor $P(q)$ of a worm-like chain were carried out by Heine *et al.* (1961). More recent theoretical approaches have been achieved by Des Cloizeaux (1973) and Yoshisaki and Yamakawa (1980). Des Cloizeaux calculated the asymptotic regimes of the form factor while Yoshisaki and Yamakawa worked out an approximated analytical expression which is valid up to $qb = 10$, where b is the statistical segment. For the Kratky–Porod worm-like chain, they express $P(q)$ as the product of two functions:

$$P(q) = P_0(q)\Gamma(q) \tag{A3.4}$$

$P_0(q)$ is expressed as follows ($\varkappa$ is defined in relation (A3.8)):

$$P_0(q) = [1 - \varkappa(q)]\, P_{\mathrm{p}}(q) + \varkappa(q) P_{\mathrm{B}}(q) \tag{A3.5}$$

in which $P_{\mathrm{p}}(q)$ is the Debye form factor for a statistical coil:

$$P_{\mathrm{p}}(q) = 2[\exp(-u) + u - 1]/u^2 \tag{A3.6}$$

with $u = \langle R_{\mathrm{G}}^2\rangle q^2 b^2$, $\langle R_{\mathrm{G}}^2\rangle$ is the mean square radius of gyration and $P_{\mathrm{B}}(q)$ is the form factor for a rod:

$$P_{\mathrm{B}}(q) = [2Si(\beta)/qL] - [4\sin^2(\beta/2)/(qL)^2] \tag{A3.7}$$

in which $Si(\beta) = \int t^{-1} \sin t \,\mathrm{d}t$ with $\beta = qL$ and L the contour length.
$\varkappa(q)$ reads:

$$\varkappa(q) = \exp(-\zeta^{-5}) \tag{A3.8}$$

with $\zeta = \pi\langle R_{\mathrm{G}}^2\rangle q b^2/2L$
$\Gamma(q)$ is expressed as follows:

$$\Gamma(q) = 1 + (1 - \varkappa) \sum_{i=2}^{5} A_i \zeta^i + \sum_{i=0}^{2} B_i \zeta^{-i} \tag{A3.9}$$

where:

$$A_i = \sum_{j=0}^{2} a_{1,ij} L^{-j} \mathrm{e}^{-40/4L} + \sum_{j=1}^{2} a_{2,ij} L^j \mathrm{e}^{-2L} \tag{A3.10}$$

$$B_i = \sum_{j=0}^{2} b_{1,ij} L^{-j} + \sum_{j=1}^{2} b_{2,ij} L^j \mathrm{e}^{-2L} \tag{A3.11}$$

$a_{1,ij}$, $a_{2,ij}$, $b_{1,ij}$ and $b_{2,ij}$ being numerical coefficients (see Table A3.1).
At the limit of very long chains, Des Cloizeaux (1973) has given the asymptotic expressions in the two different domains:

$$\text{regime 1} \quad qa < 1: \qquad (L/a)P_1(q) = 6/q^2a^2 \tag{A3.12}$$

$$\text{regime 2} \quad qa > 1: \qquad (L/a)P_2(q) = \pi/qa + 2/3q^2a^2 \tag{A3.13}$$

in which a is the so-called persistence length related to the statistical segment by $a = b/2$ and L is the chain contour length. The intersect between the asymptotes of these regimes in a Kratky plot ($q^2 I(q)$ *vs* q) defines a momentum transfer q^* which amounts to:

$$q^* = 16/(3\pi a) \approx 1.7/a \tag{A3.14}$$

Table A3.1 Values for the different coefficients of equations (A3.10) and (A3.11) (from Yoshisaki and Yamakawa, 1980).

$a_{1,20} = 0.17207$	$a_{1,30} = 0.077459$	$a_{1,40} = 0.9633$	$a_{1,50} = -1.1307$
$a_{1,21} = -7.0881$	$a_{1,31} = 4.8101$	$a_{1,41} = 26.45$	$a_{1,51} = -23.971$
$a_{1,22} = 19.577$	$a_{1,32} = -200.99$	$a_{1,42} = 406.47$	$a_{1,52} = -224.71$
$a_{2,21} = 0.33157$	$a_{2,31} = -3.9383$	$a_{2,41} = 12.608$	$a_{2,51} = -9.7252$
$a_{2,22} = -1.0692$	$a_{2,32} = 11.279$	$a_{2,42} = -38.021$	$a_{2,52} = 33.515$
$b_{1,00} = 1.3489$	$b_{1,10} = -2.035$	$b_{1,20} = 1.3744$	
$b_{1,01} = 16.527$	$b_{1,11} = -30.016$	$b_{1,21} = 12.268$	
$b_{1,02} = -65.909$	$b_{1,12} = 112.9$	$b_{1,22} = -46.316$	
$b_{2,01} = 13.544$	$b_{2,11} = 32.504$	$b_{2,21} = -51.258$	
$b_{2,02} = 60.772$	$b_{2,12} = -138.36$	$b_{2,22} = 72.212$	

The mean square radius of gyration for a worm-like chain of persistence length a reads (Benoit and Doty, 1953):

$$\langle R^2 \rangle = La\left\{\frac{1}{3} - \frac{a}{L} + \frac{2a^2}{L^2} - \frac{2a^3}{L^3}\,[1 - \exp^{-1}(L/a)]\right\} \tag{A3.15}$$

If $L \gg a$, then relation (A3.15) reduces to:

$$\langle R^2 \rangle = La/3 = Lb/6 \tag{A3.16}$$

FORM FACTOR FOR A COPOLYMER-LIKE MODEL

The copolymer-like model consists of long rods that are connected by portions made up of smaller rods of the same chemical nature. Calculation of the form factor has been worked out by Muroga (1988).

Consider a chain of contour length L composed of N rods each of length Λ alternating with disordered sequences of n rods each of length λ. The n rods of length λ form a sequence that is assumed to be characterized by Gaussian statistics with a mean square end-to-end distance $n\lambda^2$. The following expression has thus been derived by Muroga:

$$P(q) = \frac{f^2}{N}\left[2\Xi(\beta) - \frac{4}{\beta^2}\sin^2\frac{\beta^2}{2}\right] + \frac{2f^2}{N}\,\Xi^2(\beta)\left[\frac{\exp(-\omega)}{1 - \nu\exp(-\omega)}\right]$$

$$- \frac{2f^2}{N^2}\,\Xi^2(\beta)\left[\frac{\exp(-\omega)}{1 - \nu\exp(-\omega)}\right]\left[\frac{1 - [\nu\exp(-\omega)]^N}{1 - \nu\exp(-\omega)}\right]$$

$$+ (1 - f)^2\left(\frac{2}{N\omega} - 2\left[\frac{1 - \exp(-\omega)}{N\omega}\right]^2\left[\frac{\nu}{1 - \nu\exp(-\omega)}\right]\right.$$

$$\times \left\{ \frac{1 - [\nu \exp(-\omega)]^N}{1 - \exp(-\omega)} \right\} - 2 \frac{[1 - \exp(-\omega)](1 - \nu)}{N\omega^2 [1 - \nu \exp(-\omega)]} \Bigg)$$

$$+ \frac{2f(1-f)\Xi(\beta)[1 - \exp(-\omega)]}{N^2 \omega [1 - \exp(-\omega)]}$$

$$\times \left\{ 2N - \frac{[1 + \exp(-\omega)]\{1 - [\nu \exp(-\omega)]^N\}}{1 - \exp(-\omega)} \right\} \qquad \text{(A3.17)}$$

in which ω, β, ν and $\Xi(\beta)$ have the following meanings:

$$\omega = qn\lambda^2/6, \qquad \beta = q\Lambda, \qquad \nu = (\sin \beta)/\beta$$

$$\Xi(\beta) = (1/\beta) \int_0^{q\Lambda} (\sin t)/t \, \mathrm{d}t = (1/\beta) Si(\beta)$$

The fraction of rods of length Λ, f, is given by:

$$f = N\Lambda/(nN\lambda + N\Lambda) = N\Lambda/L$$

If the proportion of small rods is taken equal to zero, then one obtains the expression of the form factor of freely jointed rods as derived by Hermanns and Hermanns (1958):

$$P(q) = \frac{f^2}{N} \left[2\Xi(\beta) - \frac{4}{\beta^2} \sin^2 \frac{\beta^2}{2} \right] + \frac{2f^2}{N} \Xi^2(\beta) \left[\frac{\exp(-\omega)}{1 - \nu \exp(-\omega)} \right]$$

$$- \frac{2f^2}{N^2} \Lambda^2(\beta) \left[\frac{1 - \nu^N}{(1 - \nu)^2} \right] \qquad \text{(A3.18)}$$

The asymptotic approximation of $P(q)$ in this case is for $qR > 1$ and $q\Lambda > 1$:

$$P(q) = \frac{\Pi}{qL} + \left(\frac{\Pi^2}{2} - 4 \right) \frac{1}{q^2 bL} \qquad \text{(A3.19)}$$

The radius of gyration of Muroga's model can be derived from the Garland model (Guenet, 1980*a*):

$$\langle R_G^2 \rangle = (1 - f)\langle R_A^2 \rangle + (Mf/6\mu_R)L^2 \qquad \text{(A3.20)}$$

in which $\langle R_A^2 \rangle$ is the mean square radius of gyration of a Gaussian chain of the same contour length, L, but possessing a statistical segment λ, μ_R the molecular weight of a rod of length Λ, f the fraction of these rods and M the chain molecular weight. Introducing $\zeta_r = \Lambda/\lambda$ and since $(M/L\mu_R) \simeq 1/\Lambda$, relation (A3.20) may be rewritten:

$$\langle R_G^2 \rangle = [(1 - f) + f\zeta_r] \lambda L/6 \qquad \text{(A3.21)}$$

EFFECT OF CHAIN CROSS-SECTION

The form factors derived above are strictly only valid for infinitely thin chains. In practice, the chain has a non-negligible cross-section. If the radius of gyration of the cross-sectional area, R_σ, is such that $R_\sigma \ll a$, then for $R_\sigma^{-1} < q < a^{-1}$, $P(q)$ may be written:

$$P(q) = P_{\mathrm{R}}(q) f(qr) + o(q^{-2}) \tag{A3.22}$$

in which $P_{\mathrm{R}}(q)$ is the form factor of a rod:

$$P_{\mathrm{R}}(q) = \frac{\pi}{qL} - \frac{2}{q^2L^2} \approx \frac{\pi}{qL} \tag{A3.23}$$

and $f(qr)$ is the cylindrical Fourier transform of the correlation function of the cross-sectional area:

$$f(qr) = \int_0^R \gamma^{\mathrm{R}}(r) J_0(qr) 2\pi r \, \mathrm{d}r \tag{A3.24}$$

and $o(q^{-2})$ the residual terms related to kinks, as in the Hermans–Hermans model, or to contacts (see Luzzati and Benoit, 1961).

If the cross-section possesses a cylindrical symmetry, then $P(q)$ reads:

$$P(q) = \frac{\pi}{qL} \times \frac{4J_1^2(qr)}{(qr)^2} \tag{A3.25}$$

in which J_1 is the Bessel function of first kind and order 1. If the experiment is carried out in a momentum transfer range where $qr < 1$, then the Bessel function can be developed which ultimately gives:

$$P(q) = (\pi/qL)\exp(-q^2R_\sigma^2/2) \tag{A3.26}$$

in which R_σ^2 is the mean square radius of gyration of the cross-sectional area. This approximation is actually valid whatever the cross-section symmetry.

For filament helices of radius r_{H} and pitch P, Schmidt (1970) has derived the following expression:

$$I(q) \simeq MP(q) \simeq (\pi\mu_{\mathrm{L}}/q) \sum_{n=0}^{\infty} \varepsilon_n [J_n(qr_{\mathrm{H}}\sqrt{1 - a_n^2})]^2 \tag{A3.27}$$

in which $\varepsilon_0 = 1$ and $\varepsilon_n = 2$ for $n \geqslant 1$ with:

$$\left.\begin{array}{lll} a_n = 2\pi n/qP & \text{for} & q > 2\pi n/P \\ a_n = 1 & \text{for} & q \leqslant 2\pi n/P \end{array}\right\} \tag{A3.28}$$

In the usual range of transfer momentum ($q \leqslant 2\pi/P$), only a_0 differs from unity so that relation (A3.27) reduces to:

$$I(q) \simeq MP(q) \simeq (\pi\mu_L/q) J_0^2(qr_H) \tag{A3.29}$$

If double or triple helices are dealt with, only the linear mass, μ_L, will be affected, otherwise relation (A3.29) holds under the same conditions of transfer momentum. Relation (A3.29) is identical with the scattering by a hollow cylinder with an infinitely thin wall (Mittelbach and Porod, 1961).

If $qr_H < 1$, then the Bessel function can be expanded in a power series which gives by considering the two first terms:

$$I(q) \simeq MP(q) \simeq (\pi\mu_L/q)[1 - (q^2 r_H^2/2)] \approx (\pi\mu_L/q)\exp(-q^2 r_H^2/2] \tag{A3.30}$$

Now if the chain cross-section is taken into account in the fashion portrayed in Figure A3.1, Pringle and Schmidt (1971) arrive at:

$$I(q) \simeq (\pi\mu_L/q) \sum_{n=0}^{\infty} \varepsilon_n \cos^2(n\varphi/2) \frac{\sin^2(n\omega/2)}{(n\omega/2)^2} [g_n(qr_H, \gamma)]^2 \tag{A3.31}$$

in which $g_n(qr_H, \gamma)$ reads:

$$g_n(qr_H, \gamma) = \frac{2}{r_H^2(1-\gamma^2)} \int_{\gamma r_H}^{r_H} rJ_n(qr\sqrt{1-a_n^2})\, \mathrm{d}r \tag{A3.32}$$

in which the same set of conditions as in relation (A3.28) holds. As above, in the usual transfer momentum range ($q \leqslant 2\pi/P$), $P(q)$ reduces to:

$$I(q) \simeq (\pi\mu_L/q)\left\{\frac{2}{(1-\gamma^2)qr_H} [J_1(qr_H) - \gamma J_1(q\gamma r_H)]\right\}^2 \tag{A3.33}$$

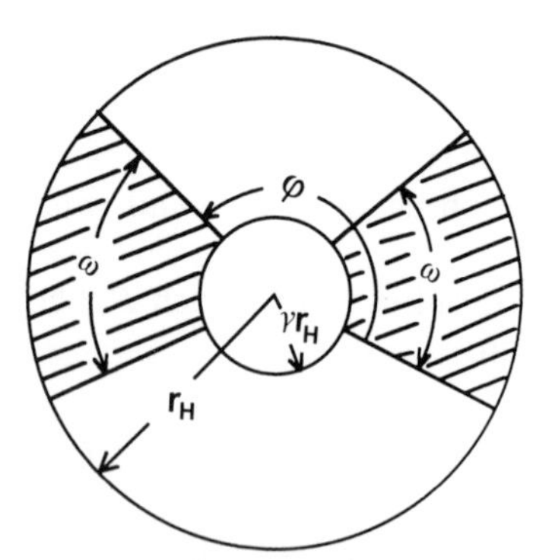

Figure A3.1 Schematic representation of the cross-section of a double helix (cross-hatching). r_H is the outer radius and γr_H the inner radius. The cross-section of a simple helix can be obtained by equating $\varphi = 0$ (after Pringle and Schmidt, 1971).

This expression is identical with the one derived by Mittelbach and Porod (1961) for a hollow cylinder with a wall of finite thickness $r_H(1-\gamma)$. A given system can be regarded as a hollow cylinder when the scattering power of the matter (liquid or solid) located inside and outside the cylinder is identical. Now, if the scattering powers differ, one should use the following relation:

$$I(q) \approx (\pi\mu_L/q)\left\{\frac{2A_c}{q\gamma r_H} J_1(q\gamma r_H) + \frac{2A_e}{(1-\gamma^2)qr_H}[J_1(qr_H) - \gamma J_1(q\gamma r_H)]\right\}^2 \tag{A3.34}$$

in which A_c and A_e stand for the scattering powers with respect to the external environment of the cylinder for $r \leqslant \gamma r_H$ and $\gamma r_H \leqslant r \leqslant r_H$ respectively.

Appendix 4

Elasticity of rigid networks

Many physical gels are networks of fibre-like structures, the latter being rigid objects. Tentative theories have already been developed for rigid objects (Litt, 1961). However, this appendix relies mainly upon the theoretical relations derived by Jones and Marques (1990) which are more recent and incorporate such notions as fractal dimension, ν^{-1}. Despite the fact that they were primarily intended for describing the mechanical behaviour of chemical gels, they may be used under certain conditions for physical gels. The limits of applicability of these relations to physical gels are therefore discussed.

Elasticity may arise from two different effects: **entropic effect** when a system can experience different conformations (as in chemical gels consisting of Gaussian chains) or **enthalpic effect** when the conformation is frozen. Rigid networks display essentially **enthalpic elasticity**.

Consider n elements of cross-sectional dimension r crossing at the junction point and the distance between junction points to be R. It is worth emphasizing that n represents some kind of functionality of the junction point.

If the elements are not freely hinged at the junction point, i.e. if any arbitrary angle formed by the vectors joining the junction point to two points located on two different elements is fixed, then elasticity will be enthalpic.

If F represents the force acting on each element and since there are n elements in a volume R^3, the stress σ reads:

$$\sigma \approx nF/R^2 \tag{A4.1}$$

If X is the deflection of the element by force F, then the strain $\varepsilon = X/R$ by assuming affine deformation. Young's modulus then reads:

$$G_r = \sigma/\varepsilon = nF/XR \tag{A4.2}$$

The force is simply given by $F = kX$ where k is a spring constant. For this constant, Jones and Marques arrive at:

$$k \approx Ba/NaR^2 \tag{A4.3}$$

in which Ba is the bend constant of the element and a the elementary length of the element. Using the fractal dimension $1/\nu$ one also has:

$$R \approx N^{\nu}a \tag{A4.4}$$

For a rod of cross-sectional dimension r and length L, the bending spring constant k is (see, for instance, Landau and Lifshitz, 1967*b*):

$$k \approx Er^4/L^3 \tag{A4.5}$$

where E is the rod material modulus. This leads to $Ba = Er^4$.

For an arbitrary fractal dimension, G_r reads:

$$G_r \approx \frac{Er^4 n}{N^{3\nu+1}a^4} \tag{A4.6}$$

Of particular interest is the relation between the modulus and element volume fraction, φ. In the simple case where all the polymer is contained in the elements, φ is simply the polymer volume fraction and is given by:

$$\varphi \approx nNar^2/R^3 \approx nN^{1-3\nu}r^2/a^2 \tag{A4.7}$$

Finally one obtains the following relation for n constant:

$$G_r \approx \frac{Er^4 n}{a^4}\left(\frac{\varphi a^2}{nr^2}\right)^{(3\nu+1)/(3\nu-1)} \tag{A4.8}$$

This gives the following relation for rods ($\nu^{-1} = 1$):

$$G_r \approx E\varphi^2/n \tag{A4.9}$$

If, on the contrary, the distance between junction points is kept constant (i.e. N = constant) but the 'functionality' n is varied, then G_r varies as:

$$G_r \approx Er^2\varphi/a^2N^2 \tag{A4.10}$$

Relation (A4.10) does not depend upon the fractal dimension of the element, unlike relation (A4.8).

Obviously these relations may be used for physical gels provided that $L \gg r$ and $\varphi_{pol} \simeq \varphi_{net}$.

Now, one may consider the case when $\varphi_{pol} \neq \varphi_{net}$ because the gel is a two-phase system. Then, by applying the lever rule, φ_{net} reads:

$$\varphi_{net} = \frac{\varphi_{pol} - \varphi_{poor}}{\varphi_{rich} - \varphi_{poor}} \tag{A4.11}$$

in which φ_{poor} and φ_{rich} are the volume fraction of the polymer-poor phase and the polymer-rich phase respectively. No assumptions are made as to the nature of the polymer-rich phase. It may be a solid solution or a polymer–solvent compound – this does not really matter. Relation (A4.11) emphasizes that, without the knowledge of the phase diagram, the interpretation of the modulus–polymer concentration relation can be tricky.

In many instances, the phase diagram reveals that the conditions $\varphi_{\text{rich}} \gg \varphi_{\text{poor}}$ and $\varphi_{\text{pol}} \gg \varphi_{\text{poor}}$ are fulfilled so that relation (A4.11) reduces to $\varphi_{\text{net}} \simeq \varphi_{\text{pol}}/\varphi_{\text{rich}}$. Evidently, the above relations hold for $\varphi \ll 1$ which implies that $\varphi_{\text{pol}} \ll \varphi_{\text{rich}}$ must be obeyed in the case of physical gels.

For chemical gels, two cases have been examined by Jones and Marques (1990): variation of the distance between cross-links (variation of N) and variation of 'functionality' (variation of n). Concerning physical gels, a third case may be contemplated. Suppose that the cross-section of the fibre is augmented through fostering by the chains of the polymer-poor phase (this implies a decrease in the polymer-poor phase concentration). If the φ_{rich} remains constant (which is the case for a polymer–solvent compound) then, as $\varphi_{\text{net}} \approx r^2$, one finally ends up with:

$$G_{\text{r}} \approx \varphi_{\text{net}}^2 E \tag{A4.12}$$

Such a case may occur on lowering the temperature which therefore gives a variation quite different from the one observed in rubber elasticity (decrease in the modulus as kT).

References

Abied, H., Brûlet, A. and Guenet, J.M. (1990) *Colloïd Polym. Sci.* **268**, 403
Adkins, G.K. and Greenwood, C.T. (1966) *Starch/Stärke* **18**, 213
Aharoni, S.M. (1980) *Polymer* **21**, 1413
Aharoni, S.M., Charlet, G. and Delmas, G. (1981) *Macromolecules* **14**, 1390
Aiken, W., Alfrey Jr., T., Janssen, A. and Mark, H. (1947) *J. Polym. Sci.* **2**, 178
Ainsworth, P.A. and Blanshard, J.M.V. (1978) *Lebensm.-Wiss.u.-Technol.* **11**, 279
Alfrey Jr., T., Wiederhorn, N., Stein, R.S. and Tobolsky, A. (1949) *Ind. Eng. Chem.* **41**, 701
Anderson, N.S., Campbell, J.W., Harding, M.M., Rees, D.A. and Samuel, J.W.B. (1969) *J. Mol. Biol.* **45**, 85
Arnauts, J. and Berghmans, H. (1987) *Polym. Commun.* **28**, 66
Arnott, S., Scott, W.E., Rees, D.A. and McNab C.G.A. (1974*a*) *J. Mol. Biol.* **90**, 253
Arnott, S., Fulmer, A., Scott, W.E., Dea, I.C.M., Moorhouse, R. and Rees, D.A. (1974*b*) *J. Mol. Biol.* **90**, 269
Arpin, M., Strazielle, C. and Skoulios, A.J. (1977) *J. Phys. (Paris)* **38**, 307
Ashby, M.F. and Gibson, L.J. (1988) in *Cellular Solids, Structure and Properties*, Pergamon Press, London
Atkins, E.D.T. (1986) *Int. J. Biol. Macromol.* **8**, 323
Atkins, E.D.T., Isaac, D.H. and Keller, A. (1980) *J. Polym. Sci. Polym. Phys. Ed.* **18**, 71
Atkins, E.D.T., Keller, A., Shapiro, J.S. and Lemstra, P.J.(1981) *Polymer* **22**, 1161
Atkins, E.D.T., Hill, M.J., Jarvis, D.A., Keller, A. Sarhene, E. and Shapiro, J.S. (1984) *Colloid Polym. Sci.* **262**, 22
Avrami, M. (1939) *J. Chem. Phys.* **7**, 1103
Avrami, M. (1940) *J. Chem. Phys.* **8**, 212
Avrami, M. (1941) *J. Chem. Phys.* **9**, 177
Baker, G.K., Carrow, J.N. and Woodmansee, C.W. (1949) *Food Ind.* **21**, 617
Banks, W. and Greenwood, C.T. (1975) in *Starch and its Components*, Edinburgh University Press, Edinburgh
Barham, P.J., Hill, M.J. and Keller, A. (1980) *Colloid Polym. Sci.* **258**, 899
Barteling, S.J. (1969) *Clin. Chem.* **15**, 1002
Bear, R.S. (1952) *Advances in Protein Chemistry*, Vol. VII, Academic Press, New York
Belnikevitch, N.G., Mrkvičkova, L. and Quadrat, O. (1983) *Polymer* **24**, 713
Benoit, H. and Doty, P. (1953) *J. Phys. Chem.* **57**, 958
Benoit, H. and Picot, C. (1966) *Pure Appl. Chem.* **12**, 545
Bensusan, H.B. and Nielsen, S.O. (1964) *Biochemistry* **3**, 1367
Berghmans, H., Govaerts, F. and Overbergh, N. (1979) *J. Polym. Sci. Polym. Phys. Ed.* **17**, 1251

Berghmans, H., Donkers, A., Frenay, L., De Schryver, F.E., Moldenaers, P. and Mewis, J. (1987) *Polymer* **28**, 97
Berry, G. C. (1966) *J. Chem. Phys.* **44**, 4450
Biros, J., Masa, Z. and Pouchly, J. (1974) *Eur. Polym. J.* **10**, 629
Boedtker, H. and Doty, P. (1954) *J. Phys. Chem.* **58**, 968
Boedtker, H. and Doty, P. (1955) *J. Am. Chem. Soc.* **77**, 248
Boedtker, H. and Doty, P. (1956) *J. Am. Chem. Soc.* **78**, 4267
Borchard, W., Pyrlik, M. and Rehage, G. (1971) *Makromol. Chem.* **145**, 169
Borisova, T.I., Lifshits, M.I., Chichagova, Ye., Shevelev, R.V.A. and Shibayev, V.P. (1980) *Polym. Sci. USSR* **22**, 967
Borisova, T.I., Burshtein, L.L., Stepanova, T.P. and Shibayev, V.P. (1984) *Polym. Sci. USSR* **26**, 2897
Bosscher, F., Ten Brinke, G. and Challa, G. (1982) *Macromolecules* **15**, 1442
Boué, F., Edwards, S.F. and Vilgis, T.A. (1988) *J. Phys.* **49**, 1635
Boyer, R.F., Baer, E. and Hiltner, A. (1985) *Macromolecules* **18**, 427
Brandt, D.A. and Dimpfl, W.L. (1970) *Macromolecules* **3**, 655
Broesma, S. (1960) *J. Chem. Phys.* **32**, 1626
Brun, M., Lallemand, A., Quinson, J.F. and Eyraud, Ch. (1977) *Thermochim. Acta* **21**, 59
Bunn, C.W. (1948) *Nature* **161**, 159
Burchard, W. (1963) *Makromol. Chem.* **59**, 16
Buter, R., Tan, Y.Y. and Challa, G. (1973) *J. Polym. Sci. Polym. Phys. Ed.* **13**, 1699
Cahn, J.W. and Hilliard, J.E. (1959) *J. Chem. Phys.* **31**, 688
Cahn, J.W. and Hilliard, J.E. (1965) *J Chem. Phys.* **42**, 93
Cairns, P., Morris, V.J., Miles, M.J. and Brownsey, G.J. (1986*a*) *Food Hydrocolloids* **1**, 89
Cairns, P., Miles, M. J. and Morris, V. J. (1986*b*) *Nature* **322**, 89
Cairns, P., Miles, M. J., Morris, V. J. and Brownsey, G. J. (1987) *Carb. Hydr. Res.* **160**, 411
Candau, S.J., Dormoy, Y., Mutin, P.H., Debeauvais, F. and Guenet, J.M. (1987*a*) *Polymer* **28**, 1334
Candau, S.J., Dormoy, Y., Hirsch, E., Mutin, P.H. and Guenet, J.M. (1987*b*) *ACS Symp. Ser.* **350**, 33
Carbonnel, L., Guieu, R. and Rosso, J.C. (1970) *Bull. Soc. Chim.* **8–9**, 2855
Carroll, V., Miles, M.J. and Morris, V.J. (1982) *Int. J. Biol. Macromol.* **4**, 432
Carroll, V., Chilvers, G.R., Franklin, D., Miles, M.J., Morris, V.J. and Ring, S.G. (1983) *Carbohydr. Res.* **114**, 181
Challa, G., de Boer, A. and Tan, Y.Y. (1976) *Int. J. Polym. Mater.* **4**, 239
Chandrasekaran, R., Millane, R.P., Arnott, S. and Atkins, E.D.T. (1988) *Carbohydr. Res.* **175**, 1
Charlet, G. and Delmas, G. (1982) *Polym. Bull. (Berlin)* **6**, 367
Charlet G., Phuong-Nguyen H. and Delmas G. (1984) *Macromolecules* **17**, 1200
Chatellier, J.Y., Durand, D. and Emery, J.R. (1985) *Int. J. Biol. Macromol.* **7**, 311
Clark A.H. and Ross-Murphy S.B. (1985) *Br. Polym. J.* **17**, 165
Clark A.H. and Ross-Murphy S.B. (1987) *Adv. Polym. Sci.* **83**, 57
Clark, A.H., Richardson, R.K., Ross-Murphy, S.B. and Stubbs, J.M. (1983) *Macromolecules* **16**, 1367
Clark, A.H., Gidley, M.J., Richardson, R.K. and Ross-Murphy, S.B. (1989) *Macromolecules* **22**, 346

Clark, J., Wellinghoff, S.T. and Miller, W.G. (1983) *Polym. Prep. (Am. Chem. Soc., Div. Polym. Chem.)* **24**, 86
Cochran, W., Crick, F.H.C. and Vand, V. (1952) *Acta Crystallogr.* **5**, 581
Conio, G., Bianchi, E., Ciferri, A., Tealdi, A. and Aden, M.A. (1983) *Macromolecules* **16**, 1264
Corradini, P., Guerra, G., Petracone, V. and Pirozzi, B. (1980) *Eur. Polym. J.* **5**, 1089
Cowie, J.M.G. (1961) *Makromol. Chem.* **42**, 230
Dagan, A., Avichai, M., Gartstein, E. and Cohen, Y. (1991) *Polym. Prep.* **32**, 459
Damasio, N.H., Fiszman, S.M., Costell, E. and Duran, L. (1990) *Food Hydrocolloids* **3**, 457
Daoud, M. and Jannink, G. (1976) *J. Phys. (Paris)*, 973
Daoud, M., Cotton, J.P., Farnoux, B., Jannink, G., Sarma, G., Benoit, H., Duplessix, R., Picot, C. and de Gennes, P.G. (1975) *Macromolecules* **8**, 804
Dautzenberg, H. (1970) *Faserforsch. Textiltech.* **21**, 117
Dea, I.C.M. (1979) in *Polysaccharides in Food* (Blanshard and Mitchell, eds), Butterworths, London, p. 229
Dea, I.C.M., McKinnon, A.A. and Rees, D.A. (1972) *J. Mol. Biol.* **68**, 153
Dea, I.C.M., Clark, A.H. and McCleary, B.V. (1986) *Carbohydr. Res.* **147**, 275
de Boer, A. and Challa, G. (1976) *Polymer* **17**, 633
Debye, P. and Bueche, A.M. (1949) *J. Appl. Phys.* **20**, 518
Defay, R., Prigogine, A., Bellemans, A. and Everett, D.H. (1966) in *Surface Tension and Adsorption*, Longmans Green, London
de Gennes, P.G. (1976) *J. Phys. Lett. (Paris)* **37**, L1
de Gennes, P.G. (1979) in *Scaling Concepts in Polymer Physics*, Cornell, University Press, New York
Derrida, B., Stauffer, D., Herrmann, H.J. and Vannimenus, J. (1983) *J. Phys. Lett. (Paris)* **44**, L701
Des Cloizeaux, J. (1973) *Macromolecules* **6**, 403
Djabourov (1986) Thesis, Paris
Djabourov, M., Leblond, A. and Papon, P. (1988*a*) *J. Phys. (Paris)*, **49**, 319
Djabourov, M., Leblond, A. and Papon, P. (1988*b*) *J. Phys. (Paris)* **49**, 333
Djabourov, M., Clark, A.H., Rowlands, D.W. and Ross-Murphy S.B. (1989) *Macromolecules* **22**, 180
Domszy, S.C., Alamo, R., Edwards, C.O. and Mandelkern, L. (1986) *Macromolecules* **19**, 310
Doolittle, A.K. (1946) *Ind. Eng. Chem.* **38**, 535
Dormoy, Y. and Candau, D.S.J. (1991) *Biopolymers* **31**, 109
Dorrestijn, A., Keijzers, A.E.M. and te Nijenhuis, K. (1981) *Polymer* **22**, 305
Djorrestijn, A., Lemstra, P.J. and Berghmans, H. (1983) *Polym. Commun.* **24**, 226
Doty, P., Bradbury, J.H. and Holtzer, A.M. (1956) *J. Am. Chem. Soc.* **78**, 947
Doublier, J.L. and Choplin, L. (1989) *Carbohydr. Res.* **193**, 215
Drake, M.P. and Veis, A. (1964) *Biochemistry* **3**, 135
Drifford, M., Tivant, P., Bencheikh-Larbi, F., Tabti, K., Rochas, C. and Rinaudo, M. (1984) *J. Phys. Chem.* **88**, 1414
Dybal, J. and Spěvaček, J. (1988) *Makromol. Chem.* **189**, 2099
Dybal, J., Stokr, J. and Schneider, B. (1983) *Polymer* **24**, 971
Dybal, J., Spěvaček, J. and Schneider, B. (1986) *J. Polym. Sci. Polym. Phys. Ed.* **24**, 657
Edwards, S.F. (1966) *Proc. Phys. Soc. London* **88**, 265
Eagland, D., Pilling, D. and Wheeler, R.G. (1974) *Disc. Farad. Soc.* **67**, 181

Ehrenfest, P. (1915) *Proc. Amst. Acad.* **17**, 1132
Eldridge, J.E. and Ferry, J.D. (1954) *J. Phys. Chem.* **58**, 992
Ellis, H.S. and Ring, S.G. (1985) *Carbohydr. Polym.* **5**, 201
Errede, L.A. (1989) *J. Phys. Chem.* **93**, 2668
Errede, L.A. (1990) *J. Phys. Chem.* **94**, 466
Errede, L.A. (1991) *Adv. Polymer Sci.* **99**, 1
Escaig, J. (1981) *J. Microsc.* **126**, 221
Fazel. N., Fazel, Z., Brûlet, A. and Guenet. J.M. (1992*a*) (submitted to *Macromolecules*)
Fazel, Z., Fazel, N. and Guenet, J.M. (1992*b*) (submitted to *Macromolecules*)
Feke, G.T. and Prins, W. (1974) *Macromolecules* **7**, 527
Ferry, J.D. (1948*a*) *Adv. Protein Chem.* **4**, 1
Ferry, J.D. (1948*b*) *J. Am. Chem. Soc.* **70**, 2244
Finer, E.G., Franks, F., Phillips, M.C. and Suggett, A. (1975) *Biopolymers* **14**, 1995
Flory P.J. (1953) *Principles of Polymer Chemistry*, Cornell University Press, Ithaca, NY
Flory, P.J. (1956) *Proc. R. Soc. London Ser. A* **234**, 73
Flory, P.J. (1961) *J. Polym. Sci. Polym. Symp.* **49**, 105
Flory, P.J. and Garrett, R.R. (1958) *J. Am. Chem. Soc.* **80**, 4836
Flory, P.J. and Weaver, E.S. (1960) *J. Am. Chem. Soc.* **82**, 4518
Foord, S.A. and Atkins, E.D.T. (1989) *Biopolymers* **28**, 1345
Fouradier, J. and Venet, A.M. (1950) *J. Chim. Phys.* **47**, 391
François, J., Gan, Y.S. and Guenet, J.M. (1986) *Macromolecules* **19**, 2755
François, J., Gan, Y.S., Sarasin, D. and Guenet, J.M. (1988) *Polymer* **29**, 898
Frank, F.C., Keller, A. and O'Connor (1959) *Phil. Mag.* **4**, 200
French, D. (1984) in *Starch; Chemistry and Technology* (R.L. Whistler, J.N. BeMiller and E.D. Paschall, eds), p. 183, Academic Press, New York
Fujii, M., Honda, K. and Fujita, H. (1973) *Biopolymers* **12**, 1177
Gan, Y.S., François, J., Guenet, J.M., Gauthier-Manuel, B. and Allain, C. (1985) *Makromol. Chem. Rapid Commun.* **6**, 225
Gan, Y.S., François, J. and Guenet, J.M. (1986) *Macromolecules* **19**, 173
Gidley, M.J. (1989) *Macromolecules* **22**, 351
Gidley, M.J. and Bulpin, P.V. (1989) *Macromolecules* **22**, 341
Ginzburg, B., Siromyatnikova, T. and Frenkel, S. (1985) *Polym. Bull. Berlin* **13**, 139
Girolamo, M., Keller, A., Miyasaka, K. and Overbergh, N. (1976) *J. Polym. Sci. Polym. Phys. Ed.* **14**, 39
Godard, P., Biebuyck, J.J., Daumerie, M., Naveau, H. and Mercier, J.P. (1978) *J. Polym. Sci. Polym. Phys. Ed.* **16**, 1817
Godard, P., Biebuyck, J.J., Barriat, P.A., Naveau, H. and Mercier, J.P. (1980) *Makromol. Chem.* **181**, 2009
Godovskii, Yu. K., Mal'tseva, I.I. and Slonimskii, G.L. (1971) *Vysokomol. Soed.* **13A**, 2768
Graessley, W. (1974) *Adv. Polym. Sci.* **16**, 1
Grant, G.T., Morris, E.R., Ress, D.A., Smith, P.J.C. and Thom, D. (1973) *FEBS Lett.* **32**, 195
Grasdalen, H. and Smidsrød, O. (1987) *Carbohydr. Polym.* **7**, 371
Guenet, J.M. (1980*a*) *Polymer* **21**, 1385
Guenet, J.M. (1980*b*) *Macromolecules* **13**, 387
Guenet, J.M. (1981) *Polymer* **22**, 313
Guenet, J.M. (1986) *Macromolecules* **19**, 1960
Guenet, J.M. (1987*a*) *Phys. Rev. Lett.* **58**, 1532

Guenet, J.M. (1987*b*) *Macromolecules* **20**, 2874
Guenet, J.M. and McKenna, G.B. (1986) *J. Polym. Sci. Polym. Phys. Ed.* **24**, 2499
Guenet, J.M. and McKenna, G.B. (1988) *Macromolecules* **21**, 1752
Guenet, J.M. and Klein, M. (1990) *Makromol. Chem. Symp.* **39**, 85
Guenet, J.M., Picot, C. and Benoit, H. (1979) *Macromolecules* **12**, 86
Guenet, J.M., Willmott, N.F.F. and Ellsmore, P.A. (1983) *Polym. Commun.* **24**, 230
Guenet, J.M., Lotz, B. and Wittmann, J.C. (1985) *Macromolecules* **18**, 420
Guenet, J.M., Klein, M. and Menelle, A. (1989) *Macromolecules* **22**, 494
Guenet, J.M., Klein, M., Schaffhauser, V. and Terech, P. (1992) (submitted to *Macromolecules*)
Guerrero, S.J. and Keller, A. (1981) *J. Macromol. Sci B* **20(2)**, 167
Guerrero, S.J., Keller, A., Soni, P.L. and Geil, P.H. (1980) *J. Polym. Sci. Polym. Phys. Ed.* **18**, 1533
Guinier, A. (1956) *Théorie et Technique de la Radiocrystallographie*, Dunod, Paris
Guiseley, K.B. (1970) *Carbohydr. Res.* **13**, 247
Guttmann, C.M., Hoffman, J.D. and Di Marzio, E.A. (1979) *Farad Disc.* **68**
Halperin, B.I. and Nelson, D.R. (1978) *Phys. Rev. Lett.* **41**, 121
Hardy, W.B. (1899) *J. Physiol.* **24**, 288
Hardy, W.B. (1900) *Proc. R. Soc.* **66**, 95
Harrington, W.F. and von Hippel, P.H. (1961) *Arch. Biochem. Biophys.* **92**, 100
Harrington, W.F. and Karr, G.M. (1970) *Biochemistry*, **9**, 3725
Harrington, W.F. and Rao, N.V. (1970) *Biochemistry* **9**, 3714
Harrison, M.A., Morgan, P.H. and Park, G.S. (1972) *Eur. Polym. J.* **8**, 1361
Hatschek, E. (1921) *Kolloïd-Z.* **28**, 210
Hayashi, A., Kinoshita, K. and Kuwano M. (1977) *Polym. J.* **9**, 219
Hayashi, A., Kinoshita, K. and Yasueda S. (1980) *Polym. J.* **12**, 447
He, X., Herz, J. and Guenet, J.M. (1987*a*) *Macromolecules* **20**, 2003
He, X., Fillon, B., Herz, J. and Guenet, J.M. (1987*b*) *Polym. Bull. (Berlin)* **17**, 45
He, X., Herz, J. and Guenet, J.M. (1988) *Macromolecules* **21**, 1757
He, X., Herz, J. and Guenet, J.M. (1989) *Macromolecules* **22**, 1390
Heine, S., Kratky, O., Porod, G. and Schmitz, J.P. (1961) *Makromol. Chem.* **44**, 682
Hengstenberg, J. and Schuch, E. (1964) *Makromol. Chem.* **74**, 55
Hermann, K., Gerngross, G. and Abitz, W.Z. (1930) *Z. Physik. Chem.* **B10**, 371
Hermans, P.H. (1949) in *Colloïd Science* (H.R. Kruyt, ed.), p. 483, Elsevier, Amsterdam
Hermans, J. Jr. (1965) *J. Polym. Sci. A* **3**, 1859
Hermans, J. and Hermans, J.J. (1958) *J. Phys. Chem.* **62**, 1543
Hermansson, A.M. (1989) *Carbohydr. Polym.* **10**, 163
Heymann, E. (1935) *Trans. Farad. Soc.* **31**, 846
Hickson, T.G.L. and Polson, N. (1968) *Biochim. Biophys. Acta* **165**, 43
Higgs, P.G. and Ross-Murphy, S.B. (1993) *Int. J. Biol. Macromol.* **12**, 233
Hikata, M., Sasaki, S. and Uematsu I. (1977) *Rep. Prog. Polym. Phys. Japan* **20**, 621
Hill, A. and Donald, A.M. (1988) *Polymer* **29**, 1426
Hjerten, S. (1962) *Biochim. Biophys. Acta* **62**, 445
Hoffmann, J.D. and Lauritzen, J.I. (1961) *J. Res. Natl. Bur. Stand.* **65A**, 4
Holtzer, A.M., Benoit, H. and Doty, P. (1954) *J. Phys. Chem.* **58**, 624
Hyde, A.J. and Taylor, R.B. (1963) *Makromol. Chem.* **62**, 204
Imberty, A. and Pérez, S. (1988) *Biopolymers* **27**, 1205

Imberty, A., Chanzy, H., Pérez, S., Buléon, A. and Tran, V. (1988) *J Mol. Biol.* **201**, 365
Immirzi, A., de Candia, F., Iannelli, P., Zambelli, A. and Vittoria, V. (1988) *Makromol. Chem. Rapid Commun.* **9**, 761
Izmailova, V.N., Pchelin, V.A. and Sabir Abu Ali (1965) *Dokl. Akad. Nauk. SSSR* **164**, 131
Janacek, J. and Ferry, J.D. (1969*a*) *Macromolecules* **2**, 397
Janacek, J. and Ferry, J.D. (1969*b*) *J. Polym. Sci. A-2* **7**, 1681
Jelich, L.M., Nunes, S.P., Paul, E. and Wolf, B.A. (1987) *Macromolecules* **20**, 1943
Jimenez-Barbero, J., Bouffar-Roupe, C., Rochas , C. and Pérez, S. (1989) *Int. J. Biol. Macromol.* **11**, 265
Jones, J.L. and Marques, C.M. (1990) *J. Phys. (Les Ulis)* **51**, 1113
Juijn, J.A., Gisolf, A. and de Jong, W.A. (1973) *Kolloïd Z.Z. Polym.* **251**, 456
Kang, K.S. and Veeder, G.T. (1982) *US Patent* 4,326,053
Katime, I.A. and Quintana, J.R. (1988*a*) *Polym. J.* **20** 459
Katime, I.A. and Quintana, J.R. (1988*b*) *Makromol. Chem.* **189**, 1373
Katime, I.A., Quintana, J.R., Valenciano, R. and Strazielle, C. (1986) *Polymer* **27**, 742
Katz, J.R. (1930) *Z. Phys. Chem. Abt. A* **150**, 37
Katz, J.R., Derksen, J.C. and Bon, W.F. (1931) *Rec. Trav. Chim. Pays-Bas* **50**, 725
Keith, M.D., Vadimsky, R.G. and Padden, F.J. (1970) *J. Polym. Sci. PtA2* **8**, 1687
Kitamura, S., Yoneda, S. and Kuge, T. (1984) *Carbohydr. Polym.* **4**, 127
Klein, M. (1989) Thesis, Université Louis Pasteur, Strasbourg
Klein, M. and Guenet, J.M. (1989) *Macromolecules* **22**, 3716
Klein, M., Brûlet, A. and Guenet, J.M. (1990*a*) *Macromolecules* **23**, 540
Klein, M., Menelle, A., Mathis, A. and Guenet, J.M. (1990*b*) *Macromolecules* **23** 4591
Klein, M., Brûlet, A., Boué, F. and Guenet, J.M. (1991) *Polymer*, in press
Klug, A., Crick, F.H.C. and Wyckoff, H.W. (1958) *Acta Crystallogr.* **11**, 199
Koberstein, J.T., Picot, C. and Benoit, H. (1985) *Polymer* **26**, 641
Koltisko, B., Keller, A., Litt, M., Baer, E. and Hiltner, A. (1986) *Macromolecules* **19**, 1207
Koningsveld, R., Stockmayer, W.H. and Nies, E. (1990) *Makromol. Chem. Macro Macromol. Symp.* **39**, 1
Könnecke, K. and Rehage, G. (1983) *Collold Polym. Sci.* **259**, 1062
Könnecke, K. and Rehage, G. (1983) *Makromol. Chem.* **184**, 2679
Kratochvil, P., Petrus, V., Munk, P., Bohdanecký, M. and Solc, K. (1967) *J. Polym. Sci. part C* **16**, 1257
Krigbaum, W.R., Carpenter, D.K. and Newmann, S. (1958) *J. Phys. Chem.* **62**, 1586
Kusanagi, H., Tadokoro, H. and Chatani, Y. (1976) *Macromolecules* **9**, 531
Kusakov, M.M. and Mekenitskaya, L.I. (1973) *Vysokomol. Soed.* **B15**, 213
Kusuyama, H., Takase, M., Higashihata, Y., Tseng, H.T., Chatani, Y. and Tadokoro, H. (1982) *Polym. Commun.* **23**, 1256
Kusuyama, H., Miyamoto, N., Chatani, Y. and Tadokoro, H. (1983) *Polym. Commun.* **24**, 119
Landau, L. and Lifshitz, E. (1967*a*) in *Statistical Physics*, Mir edition, Moscow
Landau, L. and Lifshitz, E. (1967*b*) in *Theory of Elasticity*, Mir edition, Moscow
Lauritzen, J.I. and Hoffmann, J.D. (1960) *J. Res. Natl. Bur. Stand.* **64A**, 1
Lee, J., Kim, H. and Yu, H. (1988) *Macromolecules* **21**, 860

Léger, L., Hervet, H. and Rondelez, F. (1981) *Macromolecules* **14**, 1732
Leharne, S.A., Park, G.S. and Norman, R.H. (1979) *Br. Polym. J.* **11**, 7
Leick, A. (1904) *Ann. Physik* **14**, 139
Leloup, A. (1989) Thesis, Paris
Lemstra, P.J. and Challa, G. (1975) *J. Polym. Sci. Polym. Phys. Ed.* **13**, 1809
Lemstra, P.J., Keller, A. and Cudby, M. (1978) *J. Polym. Sci. Polym. Phys. Ed.* **16**, 1507
Letherby M.R. and Young D.A. (1981) *J. Chem. Soc. Faraday Trans. I* **77**, 1953
Liquori, A.M., Anzuino, G., Coiro, V.M., D'Alagni, M., de Santis, P. and Savino, M. (1965) *Nature* **206**, 358
Litt, M. (1961) *J. Colloïd Sci.* **16**, 297
Liu, H.Z. and Liu, K.J. (1968) *Macromolecules* **1**, 157
Lodge, A.S. (1961) *Polymer* **2**, 195
Loyd, D.J. (1926) in *Colloïd Chemistry*, vol. 1 (J. Alexander, ed.), p. 767, Chemical Catalog Company, New York
Luzzati, V. and Benoit, H. (1961) *Acta Crystallogr.* **14**, 297
Luzzati, V., Cesari, M., Spach, G., Masson, F. and Vincent, J.M. (1961) *J. Mol. Biol.* **3**, 566
Macsuga, D.D. (1972) *Biopolymers* **11**, 2521
Mal'tseva, I.I., Slonimskii, G.L. and Belavtseva, E.M. (1972) *Vysokomol. Soed.* **13B**, 707
Manning, G.S. (1972) *Biopolymers* **11**, 937
Maquet, J., Théveneau, H., Djabourov, M., Leblond, J. and Papon, J. (1986) *Polymer* **27**, 1103
McBain J.W. (1950) *Colloïd Science*, chapter 9, Heath, Boston
McKenna, G.B. and Guenet, J.M. (1988*a*) *J. Polym. Sci. Polym. Phys. Ed.* **26**, 267
McKenna, G.B. and Guenet, J.M. (1988*b*) *Polym. Commun.* **29**, 58
McKinnon, A.J. and Tobolsky, A.V. (1968) *J. Phys. Chem.* **72**, 1157
Meader, D., Atkins, E.D.T. and Happey, F. (1978) *Polymer* **19**, 1371
Mekenitskaya, L.I., Golova, L.K. and Amerik, Yu.B. (1980) *Polym. Sci. USSR* **22**, 987
Miles, M.J., Morris, V.J. and Ring, S.J. (1985) *Carbohydr. Res.* **135**, 257
Miller, W.G., Kou, L., Tohyama, K. and Voltaggio, V. (1978) *J. Polym. Sci. Polym. Symp.* **65**, 91
Mitchell, J.R. (1980) *J. Text. Stud.* **7**, 313
Mittelbach, P. and Porod, G. (1961) *Acta Phys. Austr.* **14**, 405
Miyamoto, T. and Inagaky, H. (1970) *Polym. J.* **1**, 46
Monteiro, E.E.C. and Mano, E.B. (1984) *J. Polym. Sci. Polym. Phys. Ed.* **22**, 533
Mooney, M. (1940) *J. Appl. Phys.* **11**, 582
Mooney, M. (1948) *J. Appl. Phys.* **18**, 434
Mrevlishvili, G.M. and Sharimanov, G. (1978) *Biofiz.* **23**, 242
Mrkvickova, L., Stejskal, J., Spěvaček, J., Horska, J. and Quadrat, O. (1983) *Polymer* **24**, 700
Munch, J.P., Candau, S., Hild, G. and Herz, J. (1977) *J. Phys. (Paris)* **38**, 971
Muroga, Y. (1988) *Macromolecules* **21**, 2751
Murthy, A.K. and Muthukumar, M. (1987) *Macromolecules* **20**, 564
Mutin, P.H. (1986) Thesis Université Louis Pasteur, Strasbourg
Mutin, P.H. and Guenet, J.M. (1986) *Polymer* **27**, 1098
Mutin, P.H. and Guenet, J.M. (1989) *Macromolecules* **22**, 843
Mutin, P.H., Guenet, J.M., Hirsch, E. and Candau, S.J. (1988) *Polymer* **29**, 31
Najeh, M., Munch, J.P. and Guenet, J.M. (1992) (submitted to *Macromolecules*)

Nakajima, A., Hayashi, S., Taka, T. and Utsumi, N. (1969) *Kolloid Z.Z. Polym.* **234**, 1097
Nakaoki, T. and Kobayashi, M. (1991) *J. Mol. Struct.* **242**, 315
Narh, K.A., Barham, P.J. and Keller, A. (1982) *Macromolecules* **15**, 464
Natta, G. (1955) *J. Polym. Sci.* **16**, 143
Natta, G., Corradini, P. and Bassi, I.W. (1960) *Nuov. Cim. Suppl. 1* **15**, 68
Newman, S., Krigbaum, W.R. and Carpenter, D.K. (1956) *J. Phys. Chem.* **60**, 648
te Nijenhuis, K.T. (1981*a*) *Colloid Polym. Sci.* **259**, 522
te Nijenhuis, K.T. (1981*b*) *Colloid Polym. Sci.* **259**, 1017
te Nijenhuis, K.T. and Dijkstra, H. (1975) *Rheol. Acta* **14**, 71
Ninamiya, K. and Ferry, J.D. (1967) *J. Polym. Sci. A2* **5**, 195
Northrop, J.H. (1927) *J. Gen. Physiol.* **10**, 893
Northrop, J.H. and Kunitz, M. (1927) *J. Gen. Physiol.* **10**, 905.
Northrop, J.H. and Kunitz, M. (1931) *J. Phys. Chem.* **35**, 162
Norton, I.T., Morris, E.R. and Rees, D.A. (1984) *Carbohydr. Res.* **134**, 89
Nunes, S.P. and Wolf, B.A. (1987) *Macromolecules* **20**, 1952
Nunes, S.P., Wolf, B.A. and Jeberien, H.E. (1987) *Macromolecules* **20**, 1948
Ogasawara, K., Nakajima, T., Yamaura, K. and Matsuzawa, S. (1975) *Prog. Coll. Polym. Sci.* **58**, 145
Ogasawara, K., Yuasa, K. and Matsuzawa, S. (1976) *Makromol. Chem.* **177**, 3403
Oster, G. and Riley, D.P. (1952) *Acta Crystallogr.* **5**, 272
Overbergh, N., Berghmans, H. and Smets, G. (1972), *J. Polym. Sci. Part C* **38**, 237
Papkov, S.P., Yefimova, S.G., Shablygin, M.V. and Mikhaiulov, N.V. (1966) *Vysokomol. Soyed.* **8**, 1035
Papkov, S.P., Shestnev, Yu.F., Iovleva, M.M. and Banduryan, S.I. (1971) *Vysokomol. Soed.* **13B**, 707
Parry, D.A.D. and Elliott, A. (1967) *J. Mol. Biol.* **25**, 1
Paul, D.R. (1967) *J. Appl. Polym. Sci.* **11**, 439
Pennings, A.J. (1977) *J. Polym. Sci. Polym. Symp.* **59**, 55
Pérez, E., Vanderhart, D.L. and McKenna, G.B. (1988) *Macromolecules* **21**, 2418
Peterlin, A. and Turner, D.T. (1963) *Nature* **197**, 488
Peterlin, A. and Turner, D.T. (1963) *J. Polym. Sci. Polym. Lett. Ed.* **3**, 517
Peterlin, A., Quan, C. and Turner, D.T. (1965) *J. Polym. Ski Polym. Lett. Ed.* **3**, 521
Pezron, E., Ricard, A., Lafuma, F. and Audebert, R. (1988) *Macromolecules* **21**, 1121
Pezron, I., Djabourov, M., Bosio, L. and Leblond, J. (1990) *J. Polym. Sci. Polym. Phys. Ed.* **28**, 1823
Pezron, I., Djabourov, M. and Leblond, J. (1991) *Polymer* **32**, 3201
Pfannemüller, B. (1987), *Int. J. Biol. Macromol.* **9**, 105
Phuong-Nguyen, H. and Delmas, G. (1985) *Macromolecules* **18**, 1235
Phuong-Nguyen, H., Bruderlein, H., Pepin, Y. and Delmas, G. (1982) *Can. J. Biochem.* **60**, 91
Piculell L. and Nilsson S. (1989) *J. Phys. Chem.* **93**, 5596
Piez, K.A. and Carrillo, A.L. (1964) *Biochemistry* **3**, 908
Pines, E. and Prins, W. (1973) *Macromolecules* **6**, 888
Plate, N.A. and Shibayev, V.P. (1971) *Vysokomol. Soyed.* **A13**, 410
Plate, N.A. and Shibayev, V.P. (1974) *Macromol. Rev.* **8**, 117
Plazek, D.J. and Altares, T. (1986) *J. Appl. Phys.* **60**, 2694
Plazek, D. J. and In-Chul-Chay (1991) *Polym. Prep.* **32**, 433

Point, J.J., Coutelier, C. and Vilers, D. (1985) *J. Polym. Sci. Polym. Phys. Ed.* **23**, 231
Porod, G. (1948) *Acta Phys. Austr.* **2**, 255
Porod, G. (1951) *Koll. Z.* **124**, 83
Pringle, O.A. and Schmidt, P.W. (1971) *J. Appl. Crystallogr.* **4**, 290
Pyrlik, M. and Rehage, G. (1975) *Rheol. Acta* **14**, 303
Pyrlik, M. and Rehage, G. (1976) *Colloïd Polym. Sci.* **254**, 329
Quintana, J.R., Stubbersfield, R.B., Price, C. and Katime, I.A. (1989) *Eur. Polym. J.* **25**, 973
Ramachandran, G.N. ed. (1967) *Treatise on Collagen*, Academic Press, New York
Ramachandran, G.N. and Chandrasekharan, R. (1968) *Biopolymers* **6**, 1649
Ramachandran, G.N. and Kartha, G. (1955) *Nature* **176**, 593
Rankine, A.O. (1906) *Phil. Mag.* **11**, 447
Rees, D.A. (1972) *Chem. Ind. London*, 630
Rees, D.A., Steele, I.W. and Williamson, F.B. (1969) *J. Polym. Sci. Pt. C* **28**, 261
Reisman, A. (1970) *Phase Equilibria*, Academic Press, New York
Rich, A. and Crick, F.H.C. (1955) *Nature* **176**, 915
Ring, S.J. (1985) *Int. J. Biol. Macromol.* **7**, 253
Ring, S.J., I'Anson, K.J. and Morris, V.J. (1985) *Macromolecules* **18**, 182
Ring, S.J., Colonna, P., I'Anson, K.J., Kalichevsky, M.T., Miles, M.J., Morris, V.J. and Orford, P.D. (1987) *Carbohydr. Res.* **162**, 277
Rivlin, R.S. (1948) *J. Appl. Phys.* **18**, 444
Rochas, C. (1982) Thesis, Grenoble
Rochas, C. and Lahaye, M. (1989) *Carbohydr. Polym.* **10**, 289
Rochas, C. and Landry, S. (1988) in *Gums and Stabilizers for Food Industry* (G.O. Phillips, D.J. Wedlock and P.A. Williams, eds), p. 445, IRL Press
Rochas, C. and Rinaudo, M. (1980) *Biopolymers* **19** (1675)
Rochas, C. and Rinaudo, M. (1982*a*) *Carbohydr. Res.* **105**, 227
Rochas, C. and Rinaudo, M. (1982*b*) *Biopolymers* **23**, 735
Rochas, C., Rinaudo, M. and Landry, S. (1984) *Carbohydr. Polym.* **12**, 255
Rochas, C., Taravel, F.R. and Turquois, T. (1990) *Int. J. Biol. Macromol.* **12**, 353
Rogovina, L.Z., Slonimskii, G.L., Gembitskii, L.S., Serova, E.A., Grogor'eva, V.A. and Gubenkova, E.N. (1973) *Vysokomol. Soed.* **15A**, 1256
Rosso, J.C., Guieu, R., Ponge, C. and Carbonnel, L. (1973) *Bull. Soc Chim.* **9–10**, 2780
Russo, P.S., Magestro, P. and Miller, W.G. (1987) *ACS Symp. Ser.* **350**, 152
Sadler, D.M. and Keller, A. (1976) *Polymer* **17**, 37
Sarkar, N. (1979) *J. Appl. Polym. Sci.* **24**, 1073
Sasaki, S., Hikata, M., Shiraki, C. and Uematsu, I. (1982) *Polym. J.* **14**, 205
Sasaki, S., Tokuma, K. and Uematsu, I. (1983) *Polym. Bull.* **10**, 539
Sasaki, N., Shiwa, S., Yagihara, S. and Hikishi, K. (1983) *Biopolymers* **22**, 2539
Sathyanarayana, B.K. and Rao, V.S.R. (1971) *Biopolymers* **10**, 1605
Schaaf, P., Wittmann J.C. and Lotz B. (1987) *Polymer*, **28**, 193
Schachovskaya, L.I. and Kraeva, L.V. (1976) *Vysokomol. Soed.* **B18**, 840
Schellman, J.A. (1987) *Biopolymers* **26**, 549
Schmidt, P.W. (1970) *J. Appl. Crystallogr.* **3**, 257
Schneider, B., Stokr, J., Spěvaček, J. and Baldrian, J. (1987) *Makromol. Chem.* **188**, 2705
Schomaker, E. and Challa, G. (1989) *Macromolecules* **22**, 3337
Shepherd, L., Chen, T.K. and Harwood, H.J. (1979) *Polym. Bull. Berlin* **1**, 445
Shibatani, K. (1970) *Polym. J.* **1**, 348

Simha, R. (1940) *J. Phys. Chem.* **44**, 25
Sledaček, B., Spěvaček, J., Mrkvickova, L., Stejskal, J., Horska, J., Baldrian, J. and Quadrat, O. (1984) *Macromolecules* **17**, 825
Smidsrød, O. (1972) *Farad. Disc. Chem. Soc.* **57**, 263
Smidsrød, O. and Grasdalen, H. (1982) *Carbohyrdr. Polym.* **2**, 270
Smith, P. and Lemstra, P.J. (1979) *Makromol. Chem.* **180**, 2983
Smith, P. and Lemstra, P.J. (1980*a*) *J. Mater. Sci.* **15**, 505
Smith, P. and Lemstra, P.J. (1980*b*) *Polymer* **21**, 1341
Smith, P. and Pennings, A.J. (1974) *Polymer* **15**, 413
Smith, P., Lemstra, P.J., Pijpers, J.P.L. and Kiel, A.M. (1981) *Colloïd and Polym. Sci.* **259**, 1070
Snoeren, T.H.M., Both, P. and Schmidt D.G. (1976) *Neth. Milk Dairy J.* **30**, 132
Sone, Y., Hirabayashi, K. and Sakurada, I. (1953) *Kobunshi Kagaku* **10**, 1
Spěvaček, J. (1978) *J. Polym. Sci., Polym. Phys. Ed.* **16**, 523
Spěvaček, J. and Schneider, B. (1974) *Makromol. Chem.* **175**, 2939
Spěvaček, J. and Schneider, B. (1975) *Makromol. Chem.* **176**, 729
Spěvaček, J. and Schneider, B. (1980) *Coll. Polym. Sci.* **258**, 621
Spěvaček, J. and Schneider, B. (1987) *Adv. Coll. Int. Sci.* **27**, 81
Spěvaček, J., Schneider, B., Bohdanecky, M. and Sikora, A. (1982) *J. Polym. Sci. Polym. Phys. Ed.* **20**, 1623
Stein, R.S. and Tobolsky, A.V. (1948) *Text. Res. J.* **18**, 302
Stoks, W., Berghmans, H., Moldenaers, P. and Mewis, J. (1988) *Br. Polym. J.* **20**, 361
Sundararajan, P.R. (1977) *J. Polym. Sci. Polym. Lett. Ed.* **15**, 699
Sundararajan, P.R. (1979) *Macromolecules* **12**, 575
Sundararajan, P.R. and Flory, P.J. (1974) *J. Am. Chem. Soc.* **96**, 5025
Sundararajan, P.R. and Tyrer, N. (1982) *Macromolecules* **15**, 1004
Sundararajan, P.R., Tyrer, N. and Bluhm, T.L. (1982) *Macromolecules* **15**, 286
Tabb, D.L. and Koenig, J.L. (1975) *Macromolecules* **8**, 929
Tadokoro, H., Chatani, Y., Kusanagi, H. and Yokoyama, M. (1970) *Macromolecules* **3**, 441
Takahashi, A. (1973) *Polym. J.* **4**, 379
Takahashi, A. and Hiramitsu, S. (1974) *Polym. J.* **44**, 105
Takahashi, A., Nakamura, T. and Kagawa, I. (1972) *Polym. J.* **3**, 207
Takayanagi, M., Harima, H. and Iwata, Y. (1963) *Mem. Fac. Eng. Kyushu Univ.* **23**, 1
Tako, M. and Nakamura, S. (1986) *Agric. Biol. Chem.* **50**, 2817
Tal'roze, R.V., Shibayev, V.P. and Plate, N.A. (1974) *J. Polym. Sci. Polym. Symp.* **44**, 35
Tan, H.M., Hiltner, A., Moet, H. and Baer, E. (1983) *Macromolecules* **16**, 28
Tanaka, F. (1989) *Macromolecules* **22**, 1988
Thirion, P. and Chasset, R. (1967) *Chim. Ind.-Génie Chimiq.* **97**, 617
Tohyama, K. and Miller, W.G. (1981) *Nature* **289**, 813
Tomka, I., Bohonek, J., Spühler, A. and Ribeaud, M. (1975) *J. Photogr. Sci.* **23**, 97
Turquois, T., Rochas. C. and Taravel, F.R. (1991) *Carbohydrate Polymers* **17**, 1992
Upstill, C., Atkins, E.D.T. and Attwool, P.T. (1986) *Int. J. Biol. Macromol.* **8**, 275
Vacher, P.J. (1940) *Chem. Ind.* **43**, 347
Vidotto, G. and Kovacs, A.J. (1967) *Kolloid Z. Polym.* **220**, 1
von Hippel, P.H. and Harrington, W.F. (1959) *Biochim. Biophys. Acta* **36**, 427
von Nägeli, (1858) *Pflanzenphysiologische Untersuchungen*, Zürich

Vorenkamp, E.J. and Challa, G. (1981) *Polymer* **22**, 1705
Vorenkamp, E.J., Bosscher, F. and Challa, G. (1979) *Polymer* **20**, 59
Vreeman, H.J., Snoeren, T.H.M. and Payens, T.A.J. (1980) *Biopolymers* **19**, 1357
Walter, A.T. (1954) *J. Polym. Sci.* **13**, 207
Ward A.G. and Courts, A. (1977) *The Science and Technology of Gelatin*, Academic Press, New York
Watanabe, J., Imai, K., Gehani, R. and Uematsu, I. (1981) *J. Polym. Sci. Polym. Phys. Ed.* **19**, 653
Watanabe, W.H., Ryan, C.F., Fleischer P.C. and Garrett, B.S. (1961) *J. Phys. Chem.* **65**, 896
Watase M. and Nishinari, K. (1987) *Makromol. Chem.* **188**, 1177
Watase, M., Nishinari, K., Clark, A.H. and Ross-Murphy, S.B. (1989) *Macromolecules* **22**, 1196
Weill, G. and Des Cloizeaux J. (1979) *J. Phys. (Paris)* **40**, 99
Wellinghoff, S.J., Shaw, J. and Baer, E. (1979) *Macromolecules* **12**, 32
Werbowyj, R.S. and Gray, D.G. (1976) *Mol. Cryst. Liq. Cryst. (lett.)* **34**, 97
Werbowyj, R.S. and Gray, D.G. (1980) *Macromolecules* **13**, 69
Wilkes, C.E., Folt, V. and Krimm, S. (1973) *Macromolecules* **6**, 235
Williams, J.W., Saunders, W.M. and Cicirelli, J.S. (1954) *J. Phys. Chem.* **58**, 774
Wittmann, J.C. and St John Manley, R. (1977) *J. Polym. Sci. Polym. Phys. Ed.* **15**, 1089
Wu, H.C. and Sarko, A. (1978*a*) *Carbohydr. Res.* **61**, 7
Wu, H.C. and Sarko, A. (1978*b*) *Carbohydr. Res.* **61**, 27
Wu, W., Shibayama, M., Roy, S., Kurokawa, H., Coyne, L.D., Nomura, S. and Stein, R.S. (1990) *Macromolecules* **23**, 2245
Yamakawa, H. and Fujii, M. (1974) *Macromolecules* **7**, 128
Yang, Y.C. and Geil, P.H. (1983) *J. Macromol. Sci. B* **(22)3**, 463
Yasuda, H., Olf, H., Crist, B., Camaze, C.E. and Peterlin, A. (1979) *Water Structure at the Water–Polymer Interfaces* (H. Jellinek, ed.), p. 35, Plenum Press, New York
Yoshisaki, T. and Yamakawa, H. (1980) *Macromolecules* **13**, 1518
Zaides, A.L. (1954) *Kolloid Zhur.* **16**, 265
Zhang, J. and Rochas, C. (1990) *Carbohydr. Polym.* **13**, 257

Index

Note: Figures and Tables are indicated by *italic page numbers*